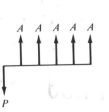

To Find $A$
Given $\quad P$ $\quad (A/P,i\%,n)$

$$A = P\left[\frac{i(1 + i)^n}{(1 + i)^n - 1}\right]$$

**CAPITAL RECOVERY**

To Find $P$
Given $\quad A$ $\quad (P/A,i\%,n)$

$$P = A\left[\frac{(1 + i)^n - 1}{i(1 + i)^n}\right]$$

**SERIES PRESENT WORTH**

$\left(\dfrac{BIGGEST - \begin{array}{c}WHAT\ IT\ IS\\ SUPPOSED\ TO\ EQUAL\end{array}}{BIGGEST - SMALLEST}\right)\left(\dfrac{1}{5}\dfrac{-\dfrac{1}{2}}{L}-\dfrac{1}{5}\right)$

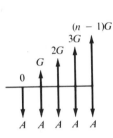

To Find $A$
Given $\quad G$ $\quad (A/G,i\%,n)$ $\qquad A = G\left[\dfrac{1}{i} - \dfrac{n}{(1 + i)^n - 1}\right]$

**GRADIENT UNIFORM SERIES**
(Arithmetic Gradient To Uniform Series)

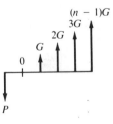

To Find $P$
Given $\quad G$ $\quad (P/G,i\%,n)$ $\qquad P = \dfrac{G}{i}\left[\dfrac{(1 + i)^n - 1}{i} - n\right]\left[\dfrac{1}{(1 + i)^n}\right]$

**GRADIENT PRESENT WORTH**
(Arithmetic Gradient To Present Worth)

# ENGINEERING ECONOMIC ANALYSIS

# Engineering Economic Analysis

## Donald G. Newnan
San Jose State University

Revised Edition

Engineering Press          San Jose, California 95103

**Library of Congress Cataloging in Publication Data**

Newnan, Donald G
  Engineering economic analysis.

  Bibliography: p.
  Includes index.
  1. Engineering economy.      I. Title.
TA177.4.N48        1977        658.1′55        77-4927
ISBN 0-910554-23-4

Cover illustration courtesy of Monex International Ltd.

Printed in the United States of America

\* \* \* \*

ENGINEERING  PRESS      P.O. Box 5      San Jose, California 95103

# Preface

This book is designed to teach the fundamental concepts of engineering economy to engineers. By limiting the intended audience to engineers it is possible to provide an expanded presentation of engineering economic analysis and do it more concisely than if the book were written for a wider audience. Other features of the book include the following.

1.  Over 130 example problems, with detailed solutions, are included in the written text to illustrate and explain the principles being described.
2.  Each economic analysis method (present worth, rate of return, and so on) is presented separately so all important aspects, including the treatment of multiple alternatives, may be studied together.
3.  The impact of inflation (and deflation) on economic analysis is discussed.
4.  The digital computer aspects of economic analysis are presented in the last chapter.
5.  A set of compound interest tables is provided for values ranging from ¼% to 60%. The eight single payment, uniform series, and gradient series factors are assembled on the same page to facilitate their use.

Because this is a textbook, the techniques presented and their order of presentation, are those which have proved most satisfactory in the classroom. We begin by looking at the process of decision-making in the first few chapters. Engineering decision-making, where problems are resolved into their economic consequences, is a more constrained view of the general decision-making process. The conversion of a flow of cash receipts and disbursements into a different, but equivalent, cash flow is then illustrated. The next four chapters demonstrate present worth, uniform annual cost, rate of return, and a number of other methods for comparing alternatives. Depreciation methods are described in Chapter 10, followed by the effect of income taxes on decision-making in Chapter 11

Equipment replacement, a rather special kind of engineering economic analysis, is described in Chapter 12, followed by the ever-present problem of inflation and deflation. Chapter 14 discusses the methods and difficulties in estimating the future consequences of alternatives. The final four chapters present the diverse topics of choosing an interest rate for economic analysis, government economic analyses, rationing money among competing projects when there are more good projects than money, and finally, ways in which the digital computer may be used in engineering economic analysis.

The goal has been to present ideas clearly and concisely using whatever mathematics are necessary to represent the relationships. The standardized functional notation, which is most familiar to instructors and the easiest notation for students to understand, has been adopted here. Solutions to all problems in the book and other materials have been prepared for instructors. These are contained in the Solutions Manual and the Instructor's Manual.

A book is a portion of an author's total experience in the particular field. It must, therefore, represent a sort of sieve, retaining the best ideas and concepts from an author's own education, his colleagues in the field of engineering economy, and his students. I am most grateful to the many people who have indirectly contributed to the content of this book. I am also grateful to a number of people, and Prof. Dick Bernhard in particular, for their suggestions for this revised edition. As in the preliminary edition, the production coordination was done by Bernard Scheier and Bob Holmes. I would appreciate being told about errors and misprints that are noted.

*Donald G. Newnan*

# Contents

# ENGINEERING
# ECONOMIC ANALYSIS

# 1
# Introduction

This is a book about decision-making. Decision-making, however, is a broad topic, for it is one of the major elements in human existence. Here, we will isolate the problems that are more amenable to the analytical approach of engineering economic analysis and then develop the tools to solve them. If we can accomplish this, then we will have a means of solving problems that might otherwise seem too complex to resolve. We will seek to show that if one understands the problem of decision-making and the tools for obtaining realistic comparisons between alternatives, then we should expect better decisions to result.

Although a lot of effort will be spent on solving problems that confront firms in the marketplace, it is equally true that the same techniques may be applied to one's own problems. This will be demonstrated in many of the problems at the end of each chapter. Since problem solving (which is possibly a less glamorous name for decision-making) is what we are after, let us start by looking at some problems.

## MAN SURROUNDED BY A SEA
## OF PROBLEMS

A careful look at the world around us clearly demonstrates that we are surrounded by a sea of problems. Just in a single day we find that problems start at the first moment of awakening to the sound of an alarm clock.

Should Joe shut off the alarm clock? Oh, yes, by all means. The darn thing is far too noisy to let it ring on. Wow! Awake for three seconds and there's the first decision of the day. Should Joe get up and go to work today? That's really two more questions, and hence, the need for two more decisions. For early morning decision-making it might be better to split the questions and take them one at a time. Shall Joe get up? It's obvious that unless Joe plans to just lie there and die, the answer at some point in time has got to be yes.

About 30 or 40 decisions later, Joe finds himself dressed and at the breakfast table. A question is posed: Would Joe like toast with the rest of his breakfast? That kind of decision-making is straightforward. And even if Joe later decides he made the wrong decision, the consequences are not of much importance. But there are harder decisions to come.

A look at the morning newspaper poses more serious questions. There is a lengthy article concerning world disarmament. That doesn't look easy to solve, so Joe had better turn to another page. The financial pages offer possibly 2000 different stocks and bonds in which he can invest his money. Some of these are bound to be excellent. In fact, it wouldn't take more than a couple of hundred dollars and two or three years to become a millionaire—if only Joe could make the right decisions. But, the path to that kind of riches is not easy.

It has been a decision-a-minute situation until now, and it probably will remain that way for the rest of the day. However, Joe would probably point out that he actually did not make a conscious decision whether or not to shut off the alarm clock when he was awakened by it—he always shuts off the alarm clock.

One could ask, however, "Did he or did he not make a decision to shut off the alarm clock?" That would seem to depend on whether or not he had an alternative. Did he? Yes, of course he did—he could have let the alarm clock keep ringing. If it were a spring-wound clock, it would run down in time, or if an electric clock, it might just ring all day. It is clear, then, that Joe *did* have an alternative but he didn't even consider it. As a result of not recognizing that there was an alternative (and hence, not considering it) Joe thought he was taking the only course of action. By ignoring all other alternatives, Joe slipped into the well-worn path that he was accustomed to taking each workday morning.

There is a second, and equally important, item to be seen from Joe's experience. Although we may not actually recognize it, we are surrounded by a great variety of "problems" or events which call for decisions. There does not seem to be any exact way of classifying problems, simply because they are so diverse. One approach would be to arrange problems by their difficulty. A complex problem might be, how should the U.S. Government budget its money? or should a firm build an assembly plant in a foreign country? An example of an intermediate problem might be, should a manual or semi-automatic machine be purchased for the factory? or shall I buy or lease my next car? A simple problem could be, should I stop smoking?

or shall I have toast for breakfast? or shall we replace a burned-out motor or close the plant? Problems appear to group themselves into three broad classifications.

## Complex Problems

These problems are indeed complex. They represent a mixture of economic, political, and humanistic elements. The preparation of the annual budget of the United States is economic to the extent that it attempts to allocate the available money to all the various federal agencies so that the best possible use is made of the money. But the allocation of money clearly has political consequences; consequences for the President, the Congress, and for individual members of Congress. The closing of a large military installation in a particular congressional district, for example, may create unemployment in the area and, consequently, dissatisfaction of the people with their representative in Congress. No congressman is anxious to create such an unfavorable climate in his district. Thus, the congressman can be expected to oppose government actions that are economically sound, but that create political problems. Similarly, it is difficult to know what might be an economic level of defense spending. Economics, in fact, may have little or nothing to do with this problem. Rather, defense spending appears to be based primarily on certain critical decisions by the President on a complex world strategy.

Other problems at the upper end of the classification scale also seem to involve complex relationships between people, with economics playing a subordinate role.

## Intermediate Problems

At this level of complexity we have a group of problems that are primarily economic. The selection between a manual or a semi-automatic machine is the kind of problem where the economics of the situation will be the primary basis for decision-making. There will, of course, be other aspects as well. A semi-automatic machine implies that less labor would be needed than if the manual machine were selected. But whether or not an extra machine tender is hired probably is more an economic consideration than a social one in this kind of problem.

## Simple Problems

On the lower end of the problem classification are some less difficult problems. It is not clear that they are truly simple. After all, consider how many people smoke cigarettes although they realize that smoking may be harmful to health. The problem, while seemingly simple, could have all sorts of social and psychological overtones.

Deciding on the components of a breakfast may not be easy; and yet, the problem does not seem of great importance. In fact, when compared to complex problems, it may not be much of a problem at all.

### The Role of Engineering Economic Analysis

What kinds of problems can be solved by engineering economic analysis? Our classification of problems suggests that the problems we classify as complex are a mixture of people problems with economics being one of the lesser elements. We will not expect engineering economic analysis to be too helpful in solving these kinds of complex problems.

At the other end of the scale are simple problems. They seem sufficiently simple that they can be solved rather quickly in one's head. There does not seem much need for analytical techniques to aid in their solution. Since many problems are of little consequence (after all, if you don't have toast for breakfast today, you can have it tomorrow), one could hardly justify making any calculations, even though the calculations might be meaningful.

It is the intermediate problems that appear best suited for solution by engineering economic analysis. In this classification the economics of the problem are a major component in decision-making. There are, of course, a great many other aspects as well, but the economic aspects seem to dominate the problem, and therefore, to be dominant in determining the best solution. Also, these intermediate level problems are of sufficient importance that we can afford to sit down and spend some time in trying to solve them. And that is important. We certainly would not want to spend ten dollars' worth of time and effort to solve a fifty-cent problem. The problem simply would not justify that amount of effort.

We may generalize by saying that the problems most suitable for solution by engineering economic analysis have these qualities.

1. The problem is sufficiently important that we are justified in giving it some serious thought and effort.
2. The problem can't be worked in one's head—that is, a careful analysis requires that we organize the problem and all the various consequences, and this is just too much to be done all at once.
3. The problem has economic aspects that are sufficiently important that they will be a significant component of the analysis leading to a decision.

When problems have these three criteria, engineering economic analysis is an appropriate technique for seeking a solution. Since there are a vast number of problems that one will encounter in the business world (and in one's personal life) that meet these criteria, engineering economic analysis can be a valuable tool in a great many situations.

# 2
# The Decision-Making Process

Decision-making may take place by default; that is, without consciously recognizing that an opportunity for decision-making exists. This fact leads us to an important first element in a definition of decision-making. To have a decision-making situation, there must be at least two alternatives available. If only one course of action is available, there can be no decision-making, for there is nothing to decide. We would have no alternative but to proceed with the single available course of action. (One might argue, however, that it is a rather unusual situation when there are no alternative courses of action. More frequently, alternatives simply are not recognized.)

At this point we might conclude that the decision-making process consists of choosing from among alternative courses of action. But this is an inadequate definition. Consider the following.

> At a horse race, a bettor was uncertain which of the five horses to bet on in the next race. He closed his eyes and pointed his finger at the list of horses printed in the racing program. Upon opening his eyes he saw that he was pointing to horse number four. He hurried off to place his bet on the horse.

Does the race horse selection represent the process of decision-making? Yes, it clearly was a process of choosing among alternatives (assuming the bettor had already ruled out the do-nothing alternative of placing no bet). But the decision-making process described seems inadequate and irrational. We want to deal with rational decision-making.

Rational decision-making is a complex process containing a number of essential elements. Rather than attempt a definition in a single sentence, it may be better to consider the essential elements. Although somewhat arbitrary, we will define the rational decision-making process in terms of eight elements: (a) recognition of a problem; (b) definition of the goal or objective; (c) assembly of relevant data; (d) identification of feasible alternatives; (e) selection of the criterion for judging which is the best alternative; (f) construction of the interrelationships between the objective, alternatives, data, and the outcome; (g) prediction of the outcomes for each alternative; and (h) choice of the best alternative to achieve the objective. The following sections will describe these elements in greater detail.

## Recognition of a Problem

The starting point in any conscious attempt at rational decision-making must be recognition that a problem exists. Only when a problem has been recognized can the work toward its solution begin in a logical manner. In the early 1970's, for example, it was discovered that a number of species of ocean fish contained substantial concentrations of mercury. The decision-making process began with this recognition of a problem, and the rush was on to determine what should be done. Research into the problem revealed that fish taken from the ocean decades before, and preserved in laboratories, also contained similar concentrations of mercury. Thus, the problem had existed for a long time; yet, it was not until recently that the problem was recognized.

In typical situations, recognition is obvious and immediate. An auto accident, an overdrawn check, a burned-out motor, an exhausted supply of parts or whatever, produces the recognition of a problem. Once we are aware of the problem, we can take action to solve it as best we can.

## Define the Goal or Objective

In a sense, every problem is a situation that prevents us from achieving previously determined goals. If a personal goal is to lead a pleasant and meaningful life, then any situation that would prevent it is viewed as a problem. Similarly, in a business situation, if a company objective is to operate profitably, then problems are those occurrences which prevent the company from achieving its previously defined profit objective.

But an objective need not be a grand, overall goal of a business or an individual. It may be quite narrow and specific. "I want to pay off the loan on my car by May," or "The plant must produce 300 golf carts in the next two weeks," are more limited objectives. Thus, defining the objective is the act of exactly describing the task or goal.

## Assembly of Relevant Data

To make a good decision, one must first assemble good information. It has been said that 100 years ago an individual could assemble all the published knowledge of the world in his library at home. Today it is doubtful that all published knowledge could even be assembled. Sheer volume alone dictates that, even if we were able to gather it, we probably could not organize it in any very meaningful way. Pertinent published data thus have become both more voluminous, and more difficult to assemble.

In addition to published information, there is a vast quantity of information that is not written down anywhere, but is stored as part of knowledge and experience of individuals. And finally, there is information that remains ungathered. A question like: "How many people in Lafayette, Indiana would be interested in buying a pair of left-handed scissors?" cannot be answered by examining published data or by asking any one person. Market research or other data gathering would be required to obtain the desired information.

From all of this information, which of it is relevant in a specific decision-making process? It may be a complex task to decide which data are important and which data are not. The availability of data further complicates this task. Some data are available immediately at little or no cost in published form; other data are available by consulting with specific knowledgeable people; still other data require surveys or research to assemble the information. Information that can only be gathered by the two latter means may prove to be both expensive and time consuming to collect.

In developing and selecting relevant data, the analyst must often decide whether the value of certain information justifies the cost to obtain it. (This may in itself constitute another problem in rational decision-making.) In decision-making, we generally find the assembly of relevant data to be one of the more difficult parts of the process.

## Identification of Feasible Alternatives

For decision-making to take place, there must be alternative courses of action available. We can usually devise a variety of ways of achieving any objective after some thought. But there is an ever present danger that in devising alternatives, we may overlook the best alternative of all. If this happens, we are left with the situation where the best of the identified alternatives will be selected, but the result will not be the best possible solution.* There is no way to ensure that the best alternative is among the alternatives considered. Probably, one should be certain that all conventional alternatives are enumerated, and that a serious effort is made to

---

*A group of techniques called "value analysis" are sometimes used to examine past decisions. Where the previously made decision is not the best solution, value analysis (which is a re-examination of the decision process) may help identify a better solution, and hence, improve decision-making.

suggest innovative solutions. Sometimes a group of people considering alternatives in an innovative atmosphere ("brainstorming") can be helpful.

Any listing of alternatives will produce both practical and impractical alternatives. It would be of little use to seriously consider an alternative that cannot be adopted. An alternative may be infeasible for a variety of reasons, such as, it violates fundamental laws of science, or it requires resources or materials that cannot be obtained, or it cannot be available in time specified in the problem objective. After elimination, only feasible alternatives remain, and these become an input for further analysis.

## Selection of the Criterion for Judging Which is the Best Alternative

The central task of decision-making is the choice from among alternatives. How is the choice made? Logically, one wants to choose the best alternative. This can only be done, however, if we can define what we mean by "best." There must be a criterion for judging which alternative is best. Now we recognize that best is a relative adjective. It is on one end of the

> poorest    poorer    poor    good    better    best

spectrum. Since we are dealing in relative terms, rather than absolute values, the selection will be the alternative that is relatively the most desirable. Consider a person found guilty of speeding by a judge and given the alternatives of a $30 fine or three days in jail. On an absolute criterion, neither alternative is desirable. On a relative basis, one would choose the better of the undesirable alternatives. In this case we would be following the old adage to "make the best of a poor situation."

There must be an almost unlimited number of ways in which one may judge the results of decision-making. Several possible criteria are listed.

1. Create the least disturbance to the ecology.
2. Improve the distribution of wealth among people.
3. Use money in ways that are economically efficient.
4. Minimize the expenditure of money.
5. Ensure that the benefits to those who gain from the decision are greater than the losses of those who are harmed by the decision.*
6. Minimize the time to accomplish the goal or objective.
7. Minimize unemployment.

The selection of the criterion for choosing the best alternative may not be easy. If one were to apply the seven criteria above to some situation

---

*Kaldor Criterion.

in which there were a number of alternatives, it seems likely that the different criteria would result in different decisions. It may be impossible, for example, to minimize unemployment without at the same time increasing the expenditure of money. The disagreement between management and labor in collective bargaining concerning wages and conditions of employment reflect a disagreement over the criterion for selecting the best alternative. Management's idea of the best alternative, based on its criterion, is seldom the best alternative, using organized labor's criterion.

**Construction of the Interrelationships**

At some point in the decision-making process the various elements must be brought together. The objective, the relevant data, the feasible alternatives, and the selection criterion must be merged. The relationships may be obscure and complex, as in trying to measure the impact of a domestic decision on world peace. They may be impossible to define on paper in any meaningful way. On the other hand, if one were considering borrowing money to pay for an automobile, for example, there is a readily defined mathematical relationship between the following variables: amount of the loan, loan interest rate, duration of the loan, and monthly payment.

The construction of the interrelationships between the decision-making elements is frequently called model building or construction of the model. To an engineer, modeling may be of two forms: a scaled physical representation of the real thing or system; or a mathematical equation, or set of equations, that describe the desired interrelationships. In a laboratory there may be a physical model, but in decision-making the model is mathematical. In modeling it is usual to represent only that part of the real system that is important to the problem at hand. Thus, the mathematical model of the student capacity of a classroom might be

$$\text{Capacity} = \frac{lw}{k}$$

where $l$ = length of classroom in feet

$w$ = width of classroom in feet

$k$ = classroom arrangement factor

| $k$ | *Arrangement* |
|---|---|
| 6 | Auditorium seating |
| 17 | Classroom |
| 32 | Design/drafting room |
| 100 | Electronics laboratory |

The equation for student capacity of a classroom is a very simple model; yet it may be adequate for the problem being solved. Other situations might have much more elaborate mathematical models.

## Prediction of the Outcome for Each Alternative

A model is used to predict the outcome for each of the feasible alternatives. As was suggested earlier, each alternative produces a variety of outcomes. Selecting a motorcycle, rather than a bicycle, for example, may make the fuel supplier happy, the neighbors unhappy, the environment more polluted, and one's savings account smaller. But to avoid unnecessary complications we assume that decision-making is based on a single criterion for measuring the relative attractiveness of the various alternatives. The other outcomes or consequences are ignored and this single criterion* is used to judge the alternatives. Using the model, the magnitude of the selected criterion is computed and recorded for each alternative.

## Choice of the Best Alternative

When the prior elements of the rational decision-making process have been completed, the final step is choosing the best alternative. If the other elements of decision-making have been carefully done, the choice of the best alternative is simply accomplished by selecting the alternative which best meets the chosen criterion.

## DECISION PROCESS SYSTEM

We know that decision-making cannot begin until the existence of a problem is recognized. But from that point on there is no fixed path to choosing the best alternative. Problems seldom can be solved by the sequential approach of Figure 2-1. This is because it is usually difficult, or impossible, to complete one element in the process without considering the effect on other elements in decision-making. The gathering of relevant data may suggest feasible alternatives. But it could just as easily be that in identifying feasible alternatives one will need additional data not yet assembled. Thus, decision-making cannot be seen as an eight-step process that proceeds from step 1 to step 8.

A somewhat better diagram of the decision process is illustrated in Figure 2-2. This diagram groups the elements in a more flexible, and therefore more realistic, manner. There is no attempt to dictate which comes first—the objective or goal, the feasible alternatives, or the relevant data. In fact, the implication is that once one has recognized the problem then

---

*If appropriate, one could devise a single composite criterion that is the weighted average of several different choice criteria.

several of the decision process elements may be considered concurrently. This makes sense. We said that the eight elements of Figures 2-1 and 2-2 are rather arbitrary and artificial. It therefore is not surprising that we have difficulty in drawing a diagram to properly represent the interrelationship between the elements.

Even Figure 2-2 seems to suggest that once the relevant data, for example, are determined, that element of the decision process has been concluded. We objected to that concept in the linear relationship in Figure 2-1 and so we are equally critical of Figure 2-2 in this respect. The missing aspect of both Figures 2-1 and 2-2 is *feedback*. No matter where one is in the decision-making process, there will frequently be a need to go back and redo or extend the work on some other element in the process. In other words, one may pass through a particular element several times while in

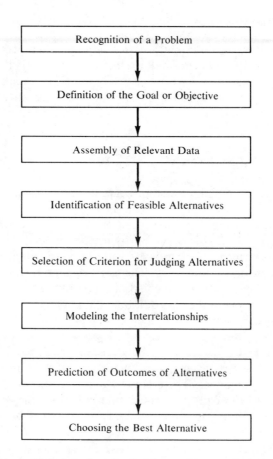

**Figure 2-1.**  An unlikely flow chart for decision-making.

the decision process system. This feedback, where subsequent elements in-
fluence previously determined elements, is difficult to show in a diagram,
for in fact there would seem to be a prospective path from any element to
most other elements. To redraw Figure 2-2 with these additional pathways
would be to obscure the decision process mechanism rather than to explain
it.

## WHEN IS A DECISION MADE?

Having seen the structure of the decision-making process it seems appropriate to ask, "When is a decision made and who makes it?" If one person
performs all the steps in decision-making then he is the decision maker.
*When* he makes the decision is less clear. The selection of the feasible alternatives may be the key item, with the rest of the analysis a methodical

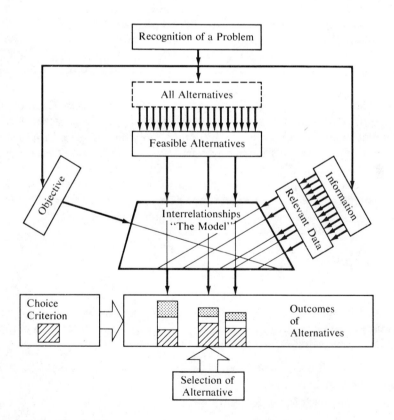

**Figure 2-2.** The decision process.

process leading to the inevitable decision. We can see that the decision may be drastically affected, or even predetermined, by the way in which the decision-making process is carried out. This is illustrated by the following example.

Dave, a young engineer, was assigned to make an analysis of what additional equipment to add to the machine shop. The criterion for selection was that the equipment selected should be the most economical, considering both initial costs and future operating costs. A little investigation by Dave revealed three practical alternatives.

1.  A new specialized lathe.
2.  A new general-purpose lathe.
3.  A rebuilt lathe available from a used equipment dealer.

A preliminary analysis indicated that the rebuilt lathe would be the most economical. Dave did not like the idea of buying a rebuilt lathe so he decided to discard that alternative. He prepared a two-alternative analysis which showed the general purpose lathe was more economical than the specialized lathe.

Dave presented his completed analysis to his manager. The manager assumed that the two alternatives presented were the best of all feasible alternatives and he approved Dave's recommendation.

At this point we should ask: Who was the decision maker, Dave or his manager? Although the manager signed his name at the bottom of the economic analysis worksheets to authorize purchasing the general-purpose lathe, it appears he was merely authorizing what already had been made inevitable, and thus he was not the decision maker. Rather, Dave had made the key decision when he decided to discard the most economical alternative from further consideration. The result was a decision to buy the better of the two less economically desirable alternatives.

## SUMMARY

For rational decision-making to take place there must be an effort to select the best alternative from among the feasible alternatives by a logical method of analysis. While difficult to isolate into discrete items, the analysis can be thought of as including eight elements.

1.  Recognition of a problem. The realization that a problem exists is the first step in problem solving.
2.  Definition of the goal or objective to be accomplished. What is the task?

3. Assembly of relevant data. What are the facts? Do we need to gather additional data? Is the additional information worth at least what it costs us to obtain it?
4. Identification of feasible alternatives. What are the practical alternative ways of accomplishing our objective or task?
5. Selection of the criterion for judging the best alternative. There are many possible criteria from which to choose. They may be political, economic, ecological, humanitarian, or whatever.
6. Construction of the various interrelationships. This phase is frequently called mathematical modeling.
7. Prediction of outcomes for each alternative.
8. Choice of the best alternative to achieve the objective.

The decision process system is not a matter of proceeding from the first element to the last one for there is no mandatory sequence that must be followed. In fact, as one proceeds it is often necessary to go back and reexamine earlier elements in a feedback process. Finally, we note that the actual decision maker is more likely to be the person who performs the analysis than the person who selects the resulting alternative to be adopted.

## PROBLEMS

**2-1** Think back to your first hour after awakening this morning. List 15 decision-making opportunities that existed during that one hour. After you have done that, mark the decision-making opportunities that you actually recognized this morning and upon which you made a conscious decision.

**2-2** A college student determines that he will have only $25 per month available for his housing for the coming semester. He is determined to continue in the university, so he has decided to list all feasible alternatives for his housing. To help him, list five feasible alternatives.

**2-3** An electric motor on a conveyor burned out. The foreman told the plant manager the motor had to be replaced. The foreman indicated that "there are no alternatives," and asked for authorization to order the replacement. In this situation is there any decision-making taking place? By whom?

**2-4** Bill Jones' parents insisted that Bill buy himself a new sportshirt. Bill's father gave rather specific instructions, saying the shirt must be in "good taste," that is, neither too wildly colored nor too extreme in tailoring. Bill went shopping in the local department store and found there were really three types of sportshirts available: (a) rather somber shirts that Bill's father would want him to buy; (b) really good-looking shirts that appealed to Bill; and (c) weird shirts that were even too much for Bill. Bill was uncertain what to do. He wanted a good-looking shirt, but wondered how to convince his father to let him keep it. The clerk suggested that Bill take home two shirts for his father to see and return the one he did not like. Bill thought that the suggestion was good, for it would avoid further arguments. He selected a good-looking blue shirt he liked, and also a weird lavender shirt and took them both home. Bill's father took one look and insisted that Bill keep the blue shirt and return the lavender one. Bill did as his father instructed. What was the key decision in this decision process, and who made it?

**2-5** Bob Johnson, shortly after graduating from college, decided he wished to purchase a new home. After looking at tracts of new homes he decided a custom built home was preferable. He hired an architect to prepare the drawings. In due time the architect completed the drawings and submitted them to Bob. Bob liked the plans; he was less pleased that he had to pay the architect a fee of $2000 to design the house. Bob asked a building contractor to provide a bid to construct the home on a lot Bob already owned. While the contractor was working to assemble the bid, Bob came across a book of standard house plans. In the book was a home that Bob and his wife liked very much. In fact, they liked the floor plan better than the one designed for them by the architect. Bob paid $25 and obtained a complete set of plans for this other house. Bob then asked the contractor to provide a bid to construct this "stock plan" home. In this way Bob felt he could compare the costs and make a decision. The building contractor submitted the following bids:

Custom designed home      $38,000+2000<br>
Stock plan home              38,500  +25+2000

Both Bob and his wife by this time had decided they liked the stock plan home better—and that they were willing to pay the $500 extra for it. Bob's wife, however, told Bob that they would have to go ahead with the custom designed home, for, as she put it, "We can't afford to throw away a set of plans that cost $2000." Bob agrees, but he dislikes the thought of building a home that is less desirable than the stock plan home. He asks your advice. Which house would you advise him to build? Explain.

**2-6** A salesman travels from city to city in the conduct of his business. Every other year he buys a new car for about $3500. The auto dealer allows about $1500 as a trade-in allowance with the result that the salesman spends $2000 every other year for a car. The salesman keeps accurate records that show that all other expenses on his car amount to 9.2¢ per mile for each mile he drives. The Internal Revenue Service allows two different methods by which the salesman may deduct his car expenses.

1. He may itemize his operating expenses and deduct them, and, in addition, recover the decline in value of the automobile by an annual deduction of $1000 for depreciation.
2. He may multiply his total business mileage by the Standard Mileage Rate of 15¢ per mile.

If the salesman travels 14,000 miles per year, which method of computation gives him the larger deduction? At what annual mileage do the two methods give the same deduction?

2-7 Jeff Martin, a college student, is getting ready for three final examinations at the end of the school year. Between now and the start of exams, he has 15 hours of study time available. He would like to get as high a grade average as possible in his Math, Physics, and Engineering Economy classes. He feels he must study at least two hours for each course, and if necessary, will settle for the low grade that the limited study would yield. How much time should Jeff devote to each class if he estimates his grade in each subject as follows.

| Mathematics | | Physics | | Engineering Economy | |
|---|---|---|---|---|---|
| Hrs. of study | Grade | Hrs. of study | Grade | Hrs. of study | Grade |
| 2 | 25 | 2 | 35 | 2 | 50 |
| 3 | 35 | 3 | 41 | 3 | 61 |
| 4 | 44 | 4 | 49 | 4 | 71 |
| 5 | 52 | 5 | 59 | 5 | 79 |
| 6 | 59 | 6 | 68 | 6 | 86 |
| 7 | 65 | 7 | 77 | 7 | 92 |
| 8 | 70 | 8 | 85 | 8 | 96 |

2-8 A grower estimates that if he picks his apple crop now, he will obtain 1000 boxes of apples which he can sell at $1 per box. However, he estimates that his crop will increase by an additional 120 boxes of apples for each week he delays picking, but that the price will drop at a rate of 5¢ per box per week, and in addition, it is likely that he will experience spoilage of approximately 20 boxes per week for each week he delays picking. When should he pick his crop to obtain the largest total cash return, and how much will be receive for his crop at that time?

2-9 In the Fall, Jay Thompson decided to live in a university dormitory. He signed a dorm contract under which he was obligated to pay the room rent for the full college year. One clause stated that if he moved out during the year he could sell his dorm contract to another student who would move into the dormitory as his replacement. The dorm cost was $400 for the two semesters, which Jay already has paid.

A month after he moved into the dorm he decided he would prefer to live in an apartment. That week, after some searching for a replacement to fulfill his dorm contract, Jay had two offers. One student offered to move in immediately and to pay Jay $20 per month for the eight remaining months of the school year. A second student offered to move in the second semester and pay $125 to Jay.

Jay now has $800 left (after paying the $400 dorm bill and food for a month) which must provide for all his room and board expenses for the balance of the year. He estimates his food cost per month is $80 if he lives in the dorm and $70 if he lives in an apartment with three other students. His share of the apartment rent and utilities will be $65 per month. Assume each semester is $4\frac{1}{2}$ months long. Disregard the small differences in the timing of the disbursements or receipts.

(a) What are the three alternatives available to Jay?

(b) Evaluate the cost for each of the alternatives.

(c) What do you recommend that Jay do?

# 3
# Engineering Decision-Making

Not every problem can be solved by the decision-making process. One need only pick up a daily newspaper to read of situations where decision makers do not seem even to know what the desired objective or task is. The problems of real life do not necessarily lend themselves to as orderly a presentation as has been suggested in the decision-making process chapter. There are other problems, however, that *are* more readily solvable.

From this set of problems, to which we may apply the decision-making process, we narrow our objectives even further. In this book we are interested in engineering decision-making and engineering economic analysis. We will examine substantial problems to be solved by engineers where the economic aspects dominate and economic efficiency is the criterion for choosing among the alternatives.

Engineering decision-making is based on the same eight elements presented in the previous chapter on the decision-making process, plus one final element—the post audit of results. We will examine the individual elements as they apply to engineering decision-making. Following this we will use the engineering decision-making techniques to solve design problems and other short-range economic problems. In the last part of the chapter we will convert problems into a series of money receipts or disbursements, or what is called a cash flow.

The nine elements of engineering decision-making are listed.

1. Recognition of a problem.
2. Definition of the goal or objective.
3. Assembly of relevant data.
4. Identification of feasible alternatives.
5. Selection of the criterion for judging the alternatives.

6. Modeling of the interrelationships.
7. Prediction of outcomes for each alternative.
8. Choice of the best alternative to achieve the objective.
9. Post audit of results.

In general everything that has been said about the decision-making process applies in engineering decision-making. There are, however, some details that are specific to engineering decision-making. These are discussed in the sections that follow.

**Recognition of the Problem**

Some problems arise from circumstances outside the firm and beyond its control. A newly enacted law, for example, may have a serious impact on a firm. There are also problems that occur within the firm. Faulty manufacturing practices, for example, would be an internally created problem. But the fact that a problem exists is not enough. It must be recognized by people who can do something about it. Frequently a problem is well known to a group of workers in a particular area of the plant, but not to those who might initiate the decision-making process. Employee suggestion boxes and similar programs are frequently used to encourage internal communication about problems.

**Assembly of Data**

Obtaining the relevant data for decision-making is always a problem. In engineering decision-making, an important source of data is the firm's own accounting system. These data must be examined carefully. Financial and cost accounting is designed to show the flow of costs in a company's operations. Where the costs are directly related to specific operations there is no difficulty. But there are other costs that are not related to specific operations. These overhead or indirect costs are usually allocated to a company's operations and products by some arbitrary allocation method. The results are generally satisfactory for cost accounting purposes, but may be incorrect for use in economic analysis. In economic analysis we must determine the true differences between the alternatives. To do this some adjustment of cost accounting data may be required.

*EXAMPLE 3-1*

The cost accounting records of a large company show the following average monthly costs for the three man printing department.

Direct labor and salaries
  (including employee benefits)          $3000
Materials and supplies consumed          3000
Allocated overhead costs*
  1500 sq ft of floor area at $2/sq ft    3000
                                         $9000

The printing department charges the other departments for the use of its services to recover its $9000 monthly cost. For example, the charge to run 1000 copies of an announcement is:

Direct labor              $ 5.50
Materials and supplies      2.24
Overhead costs              5.50
Cost to other departments  $13.24

The shipping department checked with an outside printer and found they could have the same 1000 copies printed for $9.95. Although the shipping department only has about 30,000 copies printed a month, they have decided to stop using the printing department and have their printing done at less cost by a regular commercial printer. The printing department objects to this. As a result, the general manager has asked you to study the situation and recommend what should be done.

Most of the printing department's work concerns costs, prices, and accounting information. The printing department is considered necessary to reduce the possibility of disclosing this information to people outside the company. A review of the cost accounting charges reveals nothing unusual. The charges made to the printing department cover direct labor, materials and supplies, and an allocated overhead cost. The indirect costs must be allocated to the various departments in some manner. This firm, like many others, uses floor space as the basis. The printing department in turn must distribute its costs to the work that is done.

| | Printing department | | Outside printer | |
| --- | --- | --- | --- | --- |
| | *1000 copies* | *30,000 copies* | *1000 copies* | *30,000 copies* |
| Direct labor | $ 5.50 | $165.00 | | |
| Materials and supplies | 2.24 | 67.20 | $9.95 | $298.50 |
| Overhead costs | 5.50 | 165.00 | | |
| | $13.24 | $397.20 | $9.95 | $298.50 |

The shipping department would reduce its cost from $397.20 to $298.50

*The allocated overhead costs include heat, light, power, other building costs and a share of other indirect costs. They are allocated to each department in proportion to the floor space assigned.

by using the outside printer. Now how much would the printing department costs decline? We will examine each of the cost components.

1.  Direct labor. If the printing department had been working overtime, then the overtime could be reduced or eliminated. But, assuming no overtime, how much would the saving be? It seems unlikely that a printer could be fired or even put on less than a 40 hour work week. Thus, there might be a $165 saving but it is more likely that there will be no reduction in direct labor.

2.  Materials and supplies. There would be a $67.20 saving in materials and supplies.

3.  Overhead costs. There will be no reduction in the $3000 allocated overhead cost for there will be no reduction in department floor space. Actually, of course, there may be a slight reduction in power costs if the printing department does less work.

The net result is that the firm will save $67.20 in materials and supplies and may or may not save $165.00 in direct labor if the printing department no longer does the shipping department work. The maximum saving would be $67.20 + $165.00 or $232.20. Thus if the shipping department is permitted to obtain its printing from the outside printer, the firm must pay $298.50 a month. The savings from not doing the shipping department work in the printing department would not exceed $232.20, and probably would be only $67.20. The results would be a net increase in cost to the firm. For this reason the shipping department should be discouraged from sending its printing to the outside commercial printer.

Gathering cost data presents other difficulties. One way to look at the various financial consequences (costs and benefits) of various alternatives is as follows.

*Market Consequences.* These consequences have an established price in the marketplace. We can quickly determine raw material prices, machinery costs, labor costs, and so forth in this manner.

*Extra-Market Consequences.* There are other items which are not directly priced in the marketplace. But by indirect means a price may be assigned to these items. (Economists call these prices *shadow prices.*) Examples might be the cost of an employee injury or the value to employees of going from a five day to a four day, 40-hour week.

*Consequences Not Included in the Monetary Analysis.* The numerical economic analysis probably never fully describes the differences between the alternatives. The tendency is to leave out the consequences where they do not have a significant impact on the analysis or the conversion into money is difficult. How does one evaluate the potential loss of worker's jobs due to automation or the value of landscaping around a factory? These and a variety of other consequences may be left out of the numerical calculations. They should be considered in conjunction with the numerical results in reaching a decision on the problem.

## Identification of Feasible Alternatives

One must keep in mind that unless the best alternative is considered, the result will always be suboptimal. Two types of alternatives are sometimes ignored. First, in many situations a do-nothing alternative is feasible. This may be the "let's keep doing what we are now doing" alternative or it may be the "let's not spend any money on that problem" alternative. Second, there are often feasible (but unglamorous) alternatives such as "patch it up and keep it running for another year before replacing it."

## Criteria for Judging Alternatives

All economic analysis problems inevitably fall into one of three categories:

1. Fixed input. The amount of money or other input resources are fixed. The objective is to effectively utilize them. *Examples:* (a) A project engineer has a budget of $350,000 to overhaul a portion of a petroleum refinery. (b) You have $100 to buy clothes for the start of school. For economic efficiency the appropriate criterion is to maximize the benefits or other outputs.

2. Fixed output. There is a fixed task or other output to be accomplished. *Examples:* (a) A civil engineering firm has been given the job to survey a tract of land and prepare a "Record of Survey" map. (b) You wish to purchase a new car with no optional equipment. The economically efficient criterion for a situation of fixed output is to minimize the costs or other inputs.

3. Neither input nor output fixed. The third category is the general situation where neither the amount of money or other inputs, nor the amount of benefits or other outputs are fixed. *Examples:* (a) A consulting engineering firm has more work available than it can handle. It is considering paying the staff for working evenings to increase the amount of design work it can perform. (b) One might wish to invest in the stock market, but neither the total cost of the investment nor the benefits are fixed. (c) An automobile battery is needed. They are available at different prices, and although each

will provide the energy to start the vehicle, their useful lives are different. What should be the criterion in this category? Obviously, we want to make as much money as possible. This will occur when we maximize the difference between the return from the investment (benefits) and the cost of the investment. Since the difference between the benefits and the cost is simply profit, a businessman would define this criterion as maximizing profit.

For the three categories the proper economic criteria are:

| *Category* | *Economic criterion* |
|---|---|
| Fixed input | Maximize the benefits or other outputs |
| Fixed output | Minimize the costs or other inputs |
| Neither input nor output fixed | Maximize (benefits–costs), or stated another way, maximize profit |

## Obtaining Comparable Outcomes

To obtain a meaningful basis for choosing the best alternative, the outcomes for each alternative must be arranged in a *comparable* way. Since we wish to choose the best alternative, we must arrange the mathematical calculations to provide a comparison among the alternatives. The initial step in that direction is the decision to resolve the consequences of each alternative into money in the form of costs and benefits. This resolution is done with all market and extra-market consequences. The intangible consequences are unable to be included in the numerical calculations. In the initial problems to be examined the costs and benefits occur over a short time period and can be considered as occurring at the same time. In other situations the various costs and benefits take place in a longer time period. The result may be costs at one point in time followed by periodic benefits. These we will resolve into a cash flow table to show the timing of the various costs and benefits. A number of methods (present worth, annual cost, rate of return, and so forth) will be used to resolve the cash flow table of each alternative into comparable values.

## Choosing the Best Alternative

Earlier we indicated that choosing the best alternative may be simply a matter of determining which alternative best meets the selection criterion. This is an important aspect, but it is not the only consideration. Since the intangible consequences of the various alternatives are left out of the numerical calculations, they should be introduced into the decision-making process at this point. We said that engineering economic analysis techniques are generally used where the economic consequences dominate. That statement implies that the intangible consequences will be of lesser importance than the numerical economic consequences. The alternative to be chosen is the one that best meets the choice criterion after looking at both the numerical consequences and the consequences not included in the monetary analysis.

## Post Audit of Results

In any operating system it is important to see that the results are in reasonable agreement with the projections. If a new machine tool was purchased because of both labor savings and improvements in quality, it is only logical to see if those savings are being realized. If they are, the economic analysis projections would seem to be accurate. If the savings are not being obtained, we need to see what has been overlooked. The post audit review may help insure that the projected operating advantages are obtained. On the other hand, the economic analysis projections may have been unduly optimistic. We want to know this too so that these mistakes may be avoided in the future. Finally, of course, an effective way to insure realistic economic analysis calculations is for all people concerned to know that there will be an audit of the results.

## ELEMENTARY ENGINEERING
## DECISION-MAKING

Some of the easiest forms of engineering decision-making deal with problems of alternate designs, methods, or materials. Since results of the decision occur in a very short period of time, one can quickly add up the costs and benefits for each alternative. Then using the suitable economic criterion, the best alternative can be identified. Example 3-2 concerns the economic selection of alternate materials.

*EXAMPLE 3-2*

A concrete aggregate mix is required to contain at least 31% sand by volume for proper batching. One source of material which has 25% sand and 75% coarse aggregate sells for $1 per cubic yard. Another source, which has 40% sand and 60% coarse aggregate sells for $1.24 per cubic yard. Determine the least cost per cubic yard of blended aggregates which will meet the specifications and the percent from each source.

The least cost of blended aggregates will result from maximum use of the lower cost material. The higher cost material will be used to increase the proportion of sand up to the minimum level (31%) specified.

Let $x$ = portion of blended aggregates from $1 per
 cubic yard source
$1 - x$ = portion of blended aggregates from $1.24 per
 cubic yard source

Sand balance: $x(0.25) + (1 - x)(0.40) = 0.31$

$$0.25x + 0.40 - 0.40x = 0.31$$

$$x = \frac{0.31 - 0.40}{0.25 - 0.40} = \frac{-0.09}{-0.15} = 0.60$$

Thus the blended aggregates will contain

60% of $1 per cubic yard material
40% of $1.24 per cubic yard material

The least cost per cubic yard of blended aggregates

$$= 0.60(\$1.00) + 0.40(\$1.24) = 0.60 + 0.50$$
$$= \$1.10 \text{ per cubic yard}$$

Example 3-3 describes a situation of selecting between alternate methods.

## EXAMPLE 3-3

A machine part is manufactured at a unit cost of 40¢ for material and 15¢ for direct labor. An investment of $500,000 in tooling is required. The order calls for 3 million pieces. Half way through the order a new method of manufacture can be put into effect which will reduce the unit costs to 34¢ for material and 10¢ for direct labor but it will require $100,000 for additional tooling. If all tooling costs are to be amortized during the production of the order, and other costs are 250% of direct labor cost, would it be profitable to make the change?

*Alternative A: continue with present method*

| | | |
|---|---|---|
| Material cost | 1,500,000 pieces × 0.40 = | $ 600,000 |
| Direct labor cost | 1,500,000 pieces × 0.15 = | 225,000 |
| Other costs | 2.50 × direct labor cost = | 562,500 |
| Cost for remaining 1,500,000 pieces | | $1,387,500 |

*Alternative B: change the manufacturing method*

| | | |
|---|---|---|
| Additional tooling cost | | = $ 100,000 |
| Material cost | 1,500,000 pieces × 0.34 = | 510,000 |
| Direct labor cost | 1,500,000 pieces × 0.10 = | 150,000 |
| Other costs | 2.50 × direct labor cost = | 375,000 |
| Cost for remaining 1,500,000 pieces | | $1,135,000 |

Before making a final decision one should closely examine the "other costs" to see that they do in fact vary as the direct labor cost varies.

Assuming they do, the decision would be to change the manufacturing method.

Example 3-4 illustrates selection between alternate designs.

## EXAMPLE 3-4

In the design of a cold storage warehouse the specifications call for a maximum heat transfer through the warehouse walls of 30,000 joules/hr/sq meter of wall when there is a 30°C temperature difference between the inside surface and the outside surface of the insulation. The two insulation materials being considered are as follows:

| Insulation material | Cost/cubic meter | Conductivity $J$-$m/m^2$-$°C$-$hr$ |
|---|---|---|
| Rock wool | $12.50 | 140 |
| Foamed insulation | 14.00 | 110 |

The basic equation for heat conduction through a wall is:

$$Q = \frac{K(\Delta T)}{L}$$

where $Q$ = heat transfer in $J/hr/m^2$ of wall
$K$ = conductivity in $J$-$m/m^2$-$°C$-$hr$
$\Delta T$ = difference in temperature between the two surfaces in $°C$
$L$ = thickness of insulating material in meters

Which insulation material should be selected?

There are two steps required to solve the problem. First, the required thickness of each of the alternate materials must be calculated. Then, since the problem is one of providing a fixed output (heat transfer through the wall limited to a fixed maximum amount), the criterion is to minimize the input (cost).

Required insulation thickness:

Rock wool          $30{,}000 = \dfrac{140(30)}{L}$     $L = 0.14$ m

Foamed insulation     $30{,}000 = \dfrac{110(30)}{L}$     $L = 0.11$ m

Cost of insulation per square meter of wall:

Unit cost = Cost/$m^3$ × Insulation thickness in meters
Rock wool:          Unit cost = $12.50 × 0.14 m = $1.75/$m^2$

Foamed insulation: Unit cost $= \$14.00 \times 0.11$ m $= \$1.54/m^2$

The foamed insulation is the lesser cost alternative.

## COMPUTING CASH FLOWS

In the examples presented so far we have selected the least cost alternative to meet a specification or requirement (Examples 3-2 and 3-4) or the savings have been obtained in a short period of time (Example 3-3). There are other situations where the alternatives have different consequences (costs and benefits) that continue for an extended period of time. In these circumstances we will not add up the various consequences. Instead we will attempt to portray each alternative as cash receipts or disbursements at different points in time. In this way each alternative is resolved into what we will call a cash flow. This is illustrated by Examples 3-5 and 3-6.

*EXAMPLE 3-5*

The manager has decided to purchase a new $30,000 mixing machine. The machine may be paid for by one of two ways.

1. Pay the full price now minus a 3% discount.
2. Pay $5000 now; at the end of one year pay $8000; at the end of four subsequent years pay $6000 per year.

List the alternatives in the form of a table of cash flow.

In this problem the two alternatives represent different ways to pay for the mixing machine. While the first plan represents a lump sum of $29,100 now, the second one calls for payments continuing until the end of the fifth year. In the next chapter we will have much to say on how to treat a series of payments extending over a significant period of time. For now, however, the problem is to convert an alternative into cash receipts or disbursements and show the timing of each receipt or disbursement. The result is called a cash flow or cash flow table.

For this problem the cash flows for each of the two alternatives are very simple. The cash flow table, with disbursements given negative signs, is as follows:

| End of Year | Pay in full now | Pay over 5 years |
|---|---|---|
| 0 (now) | −$29,100 | −$5000 |
| 1 | 0 | −8000 |
| 2 | 0 | −6000 |
| 3 | 0 | −6000 |
| 4 | 0 | −6000 |
| 5 | 0 | −6000 |

## EXAMPLE 3-6

A man borrowed $1000 from a bank at 8% interest. He agreed to repay the loan in two end-of-year payments. At the end of the first year he will repay half of the $1000 principal amount plus the interest that is due. At the end of the second year he will repay the remaining half of the principal amount plus the interest for the second year. Compute the borrower's cash flow.

In engineering economic analysis we normally refer to the beginning of the first year as time 0. At this point the man receives $1000 from the bank. We will use a positive sign to represent a receipt of money and a negative sign for a disbursement. Thus, at time 0 the cash flow is +$1000.

At the end of the first year the man pays 8% interest for the use of $1000 for one year. The interest is 0.08 × $1000 = $80. In addition, he repays half the $1000 loan or $500. Therefore, the end-of-year one cash flow is −$580.

At the end of the second year the payment is 8% for the use of the balance of the principal ($500) for the one-year period or 0.08 × 500 = $40. The $500 principal is also repaid for a total end-or-year two cash flow of −$540. The cash flow is:

| End of year | Cash flow |
|---|---|
| 0 (now) | +$1000 |
| 1 | −580 |
| 2 | −540 |

Chapter 4 will deal with the techniques for resolving cash flows into comparable forms so that decisions can be made on the relative desirability of the alternatives.

## SUMMARY

Engineering economic analysis refers to the solution of substantial engineering problems where economic aspects dominate and economic efficiency is the criterion for choosing among the alternatives. It is a special case of the decision-making process. Some of the special aspects of engineering economic analysis are as follows.

1.  Cost accounting systems, while an important source of cost data, contain allocations of indirect costs that may be inappropriate for use in economic analysis.
2.  The various consequences (costs and benefits) of an alternative may be of three types.

    (a) Market consequences: where there are established market prices available.

    (b) Extra-market consequences: there are no direct market prices, but prices can be assigned by indirect means.

    (c) Consequences not included in the monetary analysis: the consequences cannot be valued in any practical way.

3. The economic criteria for judging alternatives can be reduced to three cases.

    (a) For fixed input: maximize benefits or other outputs.

    (b) For fixed output: minimize costs or other inputs.

    (c) When neither input nor output are fixed: maximize the difference between output and input or more simply stated, maximize profit. The third case states the general rule from which both the first and second cases may be derived.

4. To choose among the alternatives, the market consequences and extra-market consequences are organized into a cash flow. We will see that differing cash flows can be resolved into comparable values of outcomes. These outcomes are compared against the selection criterion. From this comparison plus the consequences not included in the monetary analysis, the best alternative is selected.

5. An essential part of engineering economic analysis is the post audit of results. This step helps to insure that projected benefits are obtained and to discourage unrealistic estimates in analyses.

In elementary engineering decision-making, the problem is often one of selecting appropriate designs, methods, or materials. In some cases all the consequences occur in such a short time period that it is reasonable to add together the various consequences.

When the consequences of alternatives occur over a longer time period (say over one year) an intermediate step in the analysis is to resolve the alternatives into a table of cash flows. Chapter 4 will deal with the manipulation of cash flows.

## PROBLEMS

**3-1** Bill's dad read that at the end of each year an automobile is worth 25% less than it was at the beginning of the year. After a car is three years old the rate of decline reduces to 15%. Maintenance and operating costs, on the other hand, increase as the age of the car increases. Because of the manufacturer's warranty, the first year maintenance is very low.

| Age of car | Maintenance expense |
|------------|---------------------|
| Year 1     | $ 50                |
| 2          | 100                 |
| 3          | 150                 |
| 4          | 300                 |
| 5          | 340                 |
| 6          | 375                 |
| 7          | 450                 |

Bill decided this is a good problem on which to apply his knowledge of economic analysis. Bill's dad wants to minimize his annual cost of automobile ownership. If the make of car Bill's dad prefers costs $4200 new, should he buy a new car or a used car? If a used car, how old should the car be when purchased. How long should the car be kept before selling it? (*Answer:* Buy a three-year old car and keep it three years.)

**3-2** A city is in need of increasing its rubbish disposal facilities. There is a choice of two rubbish disposal areas, as follows.

*Area A:* A gravel pit with a capacity of 16 million cubic yards. Due to the possibility of high ground water, however, the Regional Water Pollution Control Board has restricted the lower 2 million cubic yards of fill to inert material only; for example, earth, concrete, asphalt, paving, brick, and so forth. The inert material, principally clean earth, must be purchased and hauled to this area for the bottom fill.

*Area B:* Capacity is 14 million cubic yards. The entire capacity may be used for general rubbish disposal. This area will require an average increase in round-trip haul of five miles for 60% of the city, a decreased haul of two miles for 20% of the city. For the remaining 20% of the city the haul is the same distance as for area *A*.

Assume the following conditions:

1.  Cost of inert material placed in area *A* will be $1.20 per cubic yard.
2.  Average speed of trucks from last pickup to disposal site is 15 miles per hour.
3.  The rubbish truck and a two-man crew will cost $15 per hour.
4.  Truck capacity of $4\frac{1}{2}$ tons per load or 20 cubic yards.
5.  Sufficient cover material is available at all areas; however, inert material for the bottom fill in area *A* must be hauled in.

Which of the sites do you recommend? (*Answer:* Area *B*)

**3-3** The three economic criteria for choosing the best alternative are: minimize input; maximize output; or maximize the difference between output and input. For each of the following situations, which is the appropriate economic criterion?

(a) A manufacturer of plastic drafting triangles can sell all the triangles he can produce at a fixed price. His unit costs increase as he increases production due to overtime pay, and so forth. The manufacturer's criterion should be:

_____ .

(b) An architectural and engineering firm has been awarded the contract to design a wharf for a petroleum company for a fixed sum of money. The engineering firm's criterion should be: _____.

(c) A book publisher is about to set the list price (retail price) on a textbook. If they choose a low list price, they plan on less advertising than if they select a higher list price. The amount of advertising will affect the number of copies sold. The publisher's criterion should be: _____.

(d) At an auction of antiques, a bidder for a particular statue would be trying to _____.

3-4  For each of the following situations, which is the appropriate economic criterion?

(a) The engineering school held a raffle of an automobile with tickets selling for 50¢ each or three for $1. When the students were selling tickets they noted that many people were undecided whether to buy one or three tickets. This indicates the buyers criterion was _____.

(b) A student organization bought a soft drink machine for use in a student area. There was considerable discussion as to whether they should set the machine to charge 10¢, 15¢, or 20¢ per drink. The organization recognized that the number of soft drinks sold would depend on the price charged. Eventually the decision was made to charge 15¢. Their criterion was _____ _____.

(c) In many cities grocery stores find that their sales are much greater on days when they have advertised their special bargains. The advertised special prices do not appear to increase the total physical volume of groceries sold by a store. This leads us to conclude that many shopper's criterion is _____ _____.

(d) A recently graduated engineer has decided to return to school in the evenings to obtain a Master's degree. He feels it should be accomplished in a manner that will allow him the maximum amount of time for his regular day job plus time for recreation. In working for the degree he will _____ _____.

3-5  A small machine shop with 30 horsepower of connected load purchases electricity under the following monthly rates (assume any demand charge is included in this schedule):

*First:* 50 kw-hr per HP of connected load at 2.8¢ per kw-hr
*Next:* 50 kw-hr per HP of connected load at 1.8¢ per kw-hr
*Next:* 150 kw-hr per HP of connected load at 1.0¢ per kw-hr
All over, 250 kw-hr per HP of connected load at 0.75¢ per kw-hr

The shop uses 2800 kw-hr per month.

(a) Calculate the monthly bill for this shop.

(b) Suppose the proprietor of the shop has the chance to secure additional business that will require him to operate his existing equipment more hours per day. This will use an extra 1200 kw-hr per month. What is the lowest figure that he might reasonably consider to be the "cost" of this additional energy? What is this per kw-hr?

(c) He contemplates installing certain new machines that will reduce the labor time required on certain operations. These will increase the connected load by 10 HP, but, as they will operate only on certain special jobs, will add only 100 kw-hr per month. In a study to determine the economy of installing these new machines, what should be considered as the "cost" of this energy? What is this per kw-hr?

**3-6** On his first engineering job, Jim Hayes was given the responsibility of determining the production rate for a new product. He has assembled data as indicated on the two graphs:

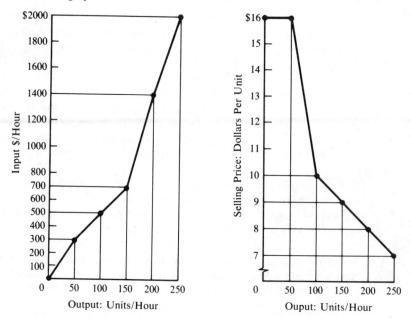

(a) Select an appropriate economic criterion and estimate the production rate based upon it.

(b) Jim's boss told Jim: "I want you to maximize output with minimum input." Jim wonders if it is possible to achieve his boss's criterion. He asks your advice. What would you tell him?

**3-7** An oil company is considering adding an additional grade of fuel at its service stations. To do this an additional 3000 gallon tank must be buried at each station. Discussions with tank fabricators indicates that the least expensive tank would be cylindrical with minimum surface area. What size tank should be ordered? (*Answer:* 8 ft diameter by 8 ft long.)

# 4
# Cash
# Flow
# Equivalence

In the last chapter we saw the full range of the engineering decision-making process. As part of that process there was the prediction of outcomes for each alternative. In many situations the economic consequences of an alternative were immediate, that is, took place either right away or in a very short period of time. The decision on the design of a concrete aggregate mix (Example 3-2) or the change of manufacturing method (Example 3-3) both represent problems where the economic consequences occur over a short period of time. Thus we can quickly total the various positive and negative aspects and reach a decision. But can we do the same if the economic consequences occur over a considerable period of time?

The installation of an expensive piece of machinery in a plant obviously has economic consequences that occur over a protracted period of time. If the machinery were bought on credit, then the simple process of paying for it is one that may take several years. What about the usefulness of the machinery? Certainly it must have been purchased because it would be a beneficial addition to the plant. These favorable consequences may last as long as the equipment performs its useful function. Anyone who has visited an industrial plant probably has seen equipment 10, 20, 30 years old or older still in use. There may then be economic consequences that continue over a substantial time period.

In the previous chapter there were a number of situations where the outcome from selecting a particular alternative continued over a considerable length of time. In those situations we created a cash flow table to show the various receipts and disbursements of money and their timing. In this chapter we will examine the value of money at different points in time. We will be able to compare the value of money at different dates. And as we will see, the ability to compare the value of money at different times is essential to engineering economic analysis. We must be able to compare, for example, a low cost motor with a higher cost motor. If there were no other consequences, we would obviously prefer the low cost one. But if the higher cost motor were more efficient and thereby reduced the annual electric power cost, we are faced with the question of whether to spend more money now on the motor to reduce power costs in the future. Through the equivalence relationship we will be able to relate a present sum of money to future sums. In this chapter we will develop the basic tools for engineering economic analysis.

## TIME VALUE OF MONEY

We often find that the money consequences of any alternative occur over a substantial period of time—say a year or more. When money consequences occur in a short period of time we simply add up the various sums of money and obtain a net result. But can we treat money this same way when the time span is greater? Which would you prefer, $100 cash today or the assurance of receiving $100 a year from now? You might decide you would prefer the $100 now because that is one way to be certain of receiving it. But suppose you were convinced that you would receive the $100 one year hence. Now what would be your answer? A little thought should convince you that it *still* would be more desirable to receive the $100 now. If you had the money now, rather than a year hence, you would have the use of it for an extra year. And if you had no current use for $100 you could let someone else use it. The use of money is a valuable asset—so valuable that people are willing to pay to have money available for their use. Money can be rented in roughly the same way one rents an apartment, only with money the charge for its use is called interest instead of rent.

The existence of interest is demonstrated by the continuing offer by banks and savings institutions to pay for the use of people's money, to pay interest. If the current interest rate is 5% per year, and you put $100 into the bank for one year, how much will you receive back at the end of the year? Clearly you will receive back your original sum of $100 together with $5 interest for a total of $105. This example demonstrates the time preference for money. We would rather have $100 today than the assured promise of $100 a year hence. In fact, we probably would need to be assured of at least $105 a year hence before we would consider taking the deferred payment alternative. This is because there is a time value of money in the form of the willingness of people to pay interest for the use of money.

### Repaying a Debt

To better understand the mechanics of interest let us consider a situation where we borrow $5000 and agree to repay it in five years together with

6% annual interest. There are a great many ways in which the debt might be repaid but for simplicity we have selected four specific ways for our example. Table 4-1 tabulates the four plans.

In plan 1 $1000 will be paid at the end of each year plus the interest

**Table 4-1.** Four plans for repayment of $5000 in five years with interest at 6%

| (1) Year | (2) Amount owed at beginning of year | (3) Interest owed for year 6% × (2) | (4) Total money owed at end of year (2) + (3) | (5) Principal payment | (6) Total end-of-year payment |
|---|---|---|---|---|---|
| *Plan 1:* At end of each year pay $1000 principal plus interest due | | | | | |
| 1 | $5000 | $300 | $5300 | $1000 | $1300 |
| 2 | 4000 | 240 | 4240 | 1000 | 1240 |
| 3 | 3000 | 180 | 3180 | 1000 | 1180 |
| 4 | 2000 | 120 | 2120 | 1000 | 1120 |
| 5 | 1000 | 60 | 1060 | 1000 | 1060 |
| | | $900 | | $5000 | $5900 |
| *Plan 2:* Pay interest due at end of each year and principal at end of 5 years | | | | | |
| 1 | $5000 | $300 | $5300 | $ 0 | $300 |
| 2 | 5000 | 300 | 5300 | 0 | 300 |
| 3 | 5000 | 300 | 5300 | 0 | 300 |
| 4 | 5000 | 300 | 5300 | 0 | 300 |
| 5 | 5000 | 300 | 5300 | 5000 | 5300 |
| | | $1500 | | $5000 | $6500 |
| *Plan 3:* Pay in 5 equal end-of-year payments | | | | | |
| 1 | $5000 | $300 | $5300 | $ 887 | $1187 |
| 2 | 4113 | 247 | 4360 | 940 | 1187 |
| 3 | 3173 | 190 | 3363 | 997 | 1187 |
| 4 | 2176 | 131 | 2307 | 1056 | 1187 |
| 5 | 1120 | 67 | 1187 | 1120 | 1187 |
| | | $935 | | $5000 | $5935 |
| *Plan 4:* Pay principal and interest in one payment at end of 5 years | | | | | |
| 1 | $5000 | $ 300 | $5300 | $ 0 | $ 0 |
| 2 | 5300 | 318 | 5618 | 0 | 0 |
| 3 | 5618 | 337 | 5955 | 0 | 0 |
| 4 | 5955 | 357 | 6312 | 0 | 0 |
| 5 | 6312 | 379 | 6691 | 5000 | 6691 |
| | | $1691 | | $5000 | $6691 |

due at the end of the year for the use of money to that point Thus, at the end of the first year we will have had the use of $5000. The interest owed is 6% × $5000 = $300. The end-of-year payment is therefore $1000 principal plus $300 interest for a total payment of $1300. At the end of the second year another $1000 principal plus interest will be repaid on the money owed during the year. This time the amount owed has declined from $5000 to $4000 because of the $1000 principal payment at the end of the first year. The interest payment is 6% × $4000 = $240, making the end-of-year payment a total of $1240. As indicated in Table 4-1 the series of payments continue each year until the loan is fully repaid at the end of the fifth year.

Plan 2 is another way to repay $5000 in five years with interest at 6%. This time the end-of-year payment is limited to the interest due with no principal payment. Instead the $5000 owed is repaid in a lump sum at the end of the fifth year. In this plan the end-of-year payment in each of the first four years is 6% × $5000 = $300. The fifth year the payment is $300 interest plus the $5000 principal for a total of $5300.

Plan 3 calls for five equal end-of-year payments of $1187 each. At this point we have not shown how the figure of $1187 was computed. However, it is clear that there would be some equal end-of-year amount that would repay the loan. By following the computations in Table 4-1 we see that this series of five payments of $1187 repays a $5000 debt in five years with interest at 6%.

Plan 4 is still another method of repaying the $5000 debt. In this plan no payment is made until the end of the fifth year when the loan is completely repaid. Note what happens at the end of the first year. The interest due for the first year 6% × $5000 = $300 is not paid. Instead it is added to the debt. So the second year the debt has increased to $5300 and the end of the second year interest is 6% × $5300 = $318. It is again unpaid and so it too is added to the debt, increasing it further to $5618. At the end of the fifth year the total sum due has grown to $6691 and is paid at that time. Note that when the $300 interest was not paid at the end of the first year it was added to the debt and in the second year there was interest charged on this unpaid interest. That is, the $300 of unpaid interest resulted in 6% × $300 = $18 of additional interest charge in the second year. That $18 together with 6% × $5000 = $300 interest on the $5000 original debt brought the total interest charge at the end of the second year to $318. Charging interest on unpaid interest is called compound interest. In this book we will deal extensively with compound interest calculations.

In Table 4-1 we have illustrated four different ways of accomplishing the same task, that is, to repay a debt of $5000 in five years with interest at 6%. Having described the alternatives we will now seek to use them to present the important concept of equivalence.

## EQUIVALENCE

When we are indifferent whether we have a quantity of money now or the assurance of some other sum of money in the future, or series of future sums of money, we say that the present sum of money is *equivalent* to the future sum or series of future sums.

If an industrial firm believed 6% was an appropriate interest rate it would have no particular preference whether it received $5000 now or was repaid by plan 1 of Table 4-1. Thus $5000 today is equivalent to the series of five end-of-year payments. In the same fashion the industrial firm would accept repayment plan 2 as equivalent to $5000 now. Logic tells us that if plan 1 is equivalent to $5000 now and plan 2 is also equivalent to $5000 now it must follow that plan 1 is equivalent to plan 2. In fact, all four repayment plans must be equivalent to each other and to $5000 now.

Equivalence is an essential factor in engineering economic analysis. In Chapter 3 we saw how an alternative could be represented by a cash flow table. How might two alternatives with different cash flows be compared? For example consider these two possibilities.

| Year | Alternative A | Alternative B |
|------|---------------|---------------|
| 1 | −$1300 | −$300 |
| 2 | −1240 | −300 |
| 3 | −1180 | −300 |
| 4 | −1120 | −300 |
| 5 | −1060 | −5300 |
| | −$5900 | −$6500 |

If you were given your choice between the two alternatives which one would you choose? Obviously the two alternatives have cash flows that are different. Alternative *A* requires that there be larger payments in the first four years, but the total payments are smaller than the sum of the alternative *B* payments. To make a decision the cash flows must be altered so that they can be compared. The technique of equivalence is the answer. We can determine by mathematical manipulation an equivalent value at some point in time for *A* and a comparable equivalent value for *B*. Then we can judge the relative attractiveness of the two alternatives, not from their cash flows, but from comparable equivalent values. In this example we know from Table 4-1 that *A* and *B* are actually plan 1 and plan 2 and that they each repay a present sum of $5000 with interest at 6%. Since they both are equivalent to $5000 now, the alternatives are equally attractive. This could not have been deduced from the given cash flows. It was necessary to learn this by determining the present equivalent values for each alternative.

**Difference in Repayment Plans**

The four plans computed in Table 4-1 are different. Table 4-2 repeats the end-of-year payment schedule from the previous figure. In addition, each plan is graphed to show the debt still owed at any point in time. Since $5000 was borrowed at the beginning of the first year all the graphs begin at that point. We see however that the four plans result in quite different situations on the amount of money owed at any other point in time. In plans 1 and 3 the money owed declines as time passes. In plan 2 the debt remains constant while in plan 4 the debt increases until the end of the fifth year. These graphs show an important difference among the repayment plans—the areas under the curves differ greatly. Since the axes are money owed and time in years, the area is their product: (money owed) × (time in years).

In the discussion of the time value of money, we saw that the use of money over a time period was valuable and that people are willing to pay interest to have the use of money for periods of time. When people borrow money they are acquiring the use of money as represented by the area under the curve on the money owed vs time in years curve. It follows, therefore, that at a given interest rate the amount of interest to be paid will be proportional to the area under the curve. Since in each case the $5000 loan is repaid, the interest for each plan is the total minus the $5000 principal.

| Plan | Interest |
|------|----------|
| 1 | $ 900 |
| 2 | 1500 |
| 3 | 935 |
| 4 | 1691 |

Using Table 4-2 and the data from Table 4-1, we can compute the area under each of the four curves. The area bounded by the abscissa, ordinate and the curve may be computed. It is the sum of the ordinate (money owed) times the abscissa (1 year) for each of the five years.

Area = (Money owed in year 1)(1 year) + (Money owed in year 2)
(1 year) + · · · + (Money owed in year 5)(1 year)

Area under curve
(money owed)(time) = dollar-years

|  | Plan 1 | Plan 2 | Plan 3 | Plan 4 |
|---|--------|--------|--------|--------|
| (Money owed in year 1)(1 year) | $ 5,000 | $ 5,000 | $ 5,000 | $ 5,000 |
| (Money owed in year 2)(1 year) | 4,000 | 5,000 | 4,113 | 5,300 |
| (Money owed in year 3)(1 year) | 3,000 | 5,000 | 3,173 | 5,618 |
| (Money owed in year 4)(1 year) | 2,000 | 5,000 | 2,176 | 5,955 |
| (Money owed in year 5)(1 year) | 1,000 | 5,000 | 1,120 | 6,312 |
|  | $15,000 | $25,000 | $15,582 | $28,185 |

**Table 4-2.** Four plans for repayment of $5000 in five years with interest at 6%

*Plan 1:* At end of each year pay $1000 principal plus interest due

| Year | End-of-year payment |
|------|---------------------|
| 1 | $1300 |
| 2 | 1240 |
| 3 | 1180 |
| 4 | 1120 |
| 5 | 1060 |
|   | $5900 |

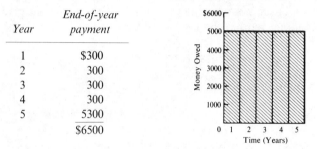

*Plan 2:* Pay interest due at end of each year and principal at end of 5 years

| Year | End-of-year payment |
|------|---------------------|
| 1 | $300 |
| 2 | 300 |
| 3 | 300 |
| 4 | 300 |
| 5 | 5300 |
|   | $6500 |

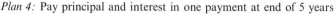

*Plan 3:* Pay in 5 equal end-of-year payments

| Year | End-of-year payment |
|------|---------------------|
| 1 | $1187 |
| 2 | 1187 |
| 3 | 1187 |
| 4 | 1187 |
| 5 | 1187 |
|   | $5935 |

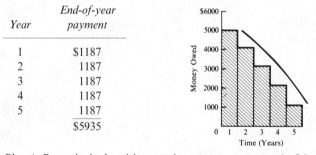

*Plan 4:* Pay principal and interest in one payment at end of 5 years

| Year | End-of-year payment |
|------|---------------------|
| 1 | $ 0 |
| 2 | 0 |
| 3 | 0 |
| 4 | 0 |
| 5 | 6691 |
|   | $6691 |

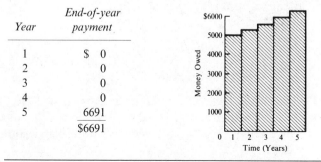

With the area under each curve computed in dollar-years, the ratio of total interest paid to area under the curve may be obtained.

| Plan | Area under curve dollar-years | Total interest paid dollars | Ratio: $\dfrac{\text{Total interest paid}}{\text{Area under curve}}$ |
|------|------|------|------|
| 1 | $15,000 | $ 900 | 0.06 |
| 2 | 25,000 | 1500 | 0.06 |
| 3 | 15,582 | 935 | 0.06 |
| 4 | 28,185 | 1691 | 0.06 |

We see that the ratio of total interest paid to area under curve is constant and equal to 6%. Stated another way, the total interest paid is equal to the interest rate times the area under the curve.

From the calculations we now can understand why the repayment plans can be equivalent and yet require the repayment of different total sums of money. The answer is, of course, that the four repayment plans provide the borrower with different quantities of dollar-years. Since dollar-years multiplied by the interest rate equals the interest charge, the four plans result in different total interest charges.

### Equivalence is Dependent on Interest Rate

In the example of Table 4-1 all calculations have been made for a 6% interest rate. At this interest rate it has been shown that all four plans are equivalent to a present sum of $5000. But what would happen if we were to change the problem by changing the interest rate?

If the interest rate were increased to 7%, we know that the required interest payment for each plan would increase, and the calculated repayment schedules would no longer repay a $5000 debt with interest at 7%. Instead, each plan would repay a sum less than the principal of $5000, (because more money would have to be used to repay the higher interest rate). By some calculations that will be explained later, the equivalent present sum that each plan will repay at 7% interest is:

| Plan | Repay a present sum of |
|------|------|
| 1 | $4867 |
| 2 | 4796 |
| 3 | 4862 |
| 4 | 4775 |

As predicted, we see that at the higher 7% interest the repayment plans of Table 4-1 each repay a sum less than $5000. But they do not repay the same present sum. Plan 1 would repay $4867 with 7% interest, while plan 2 could repay $4796 with interest at 7%. Thus, with interest at 7% plan 1 and plan 2 are no longer equivalent, for they will not repay the same present sum. The two series of payments (plan 1 and plan 2) were equivalent at 6%, but not at 7%. This leads to the conclusion that *equivalence is dependent on the interest rate.* Changing the interest rate destroys the equivalence between two series of payments.

Could one create revised repayment schemes that would be equivalent to $5000 now with interest at 7%? Yes, of course, we could. To revise plan 1 of Table 4-1 we would need to increase the total end-of-year payment in order to pay 7% interest on the outstanding debt.

| Year | Amount owed at beginning of year | 7% interest for year | Total end-of-year payment equals $1000 plus interest for year |
|------|------|------|------|
| 1 | $5000 | $350 | $1350 |
| 2 | 4000 | 280 | 1280 |
| 3 | 3000 | 210 | 1210 |
| 4 | 2000 | 140 | 1140 |
| 5 | 1000 | 70 | 1070 |

Plan 2 of Table 4-1 could be revised for 7% interest by increasing the first four payments to 7% × $5000 = $350 and the final payment to $5350. Now two plans that repay $5000 in five years with interest at 7% are presented.

| Year | Revised plan 1 end-of-year payment | Revised plan 2 end-of-year payment |
|------|------|------|
| 1 | $1350 | $ 350 |
| 2 | 1280 | 350 |
| 3 | 1210 | 350 |
| 4 | 1140 | 350 |
| 5 | 1070 | 5350 |

We have determined that the revised plan 1 is equivalent to a present sum of $5000 and the revised plan 2 has been devised to be equivalent to $5000 now, so it follows that at 7% interest, the revised plan 1 is equivalent to the revised plan 2.

**Application of Equivalence Calculations**

To understand the usefulness of equivalence calculations consider the following.

|  | Alternative A: | Alternative B: |
| | Lower initial cost | Higher initial cost |
| Year | Higher operating cost | Lower operating cost |
| 0 (now) | −$600 | −$850 |
| 1 | −115 | −80 |
| 2 | −115 | −80 |
| 3 | −115 | −80 |
| . | . | . |
| . | . | . |
| . | . | . |
| 10 | −115 | −80 |

Is the least cost alternative the one that has the lower initial cost and higher operating costs or the one with higher initial cost and lower continuing costs? Because of the time value of money one cannot add up sums of money at different points in time directly. This means that a comparison between alternatives cannot be made in actual dollars at different points in time, but must be made in some equivalent comparable sums of money.

It is not sufficient to compare the initial $600 against $850. Instead we must compute a value that represents the entire stream of payments. In other words, we want the sum that is equivalent to the alternative *A* cash flow. Similarly, we need to compute the equivalent present sum for alternative *B*. By computing equivalent sums at the same point in time ("now"), we will have values which may be validly compared. The methods for accomplishing this will be presented later in this chapter.

Thus far we have discussed computing equivalent present sums for a cash flow. But the technique of equivalence is not limited to a present computation. Instead, we could compute the equivalent sum for a cash flow at any point in time. We could compare alternatives in equivalent year 10 dollars rather than now (year 0) dollars. And further, the equivalence need not be a single sum, but instead could be a series of payments or receipts. In plan 3 of Table 4-1 we had a situation where the series of equal payments was equivalent to $5000 now. But the equivalency works both ways. If we asked the question: What is the equivalent equal annual payment continuing for five years, given a present sum of $5000 and interest at 6%? The answer, of course, is $1187.

## COMPOUND INTEREST

To facilitate equivalence computations, a series of interest formulas will be derived. To simplify the presentation, the following notation will be used.

$i$ = interest rate per interest period. In the equations the interest rate is stated as a decimal (that is, 5% interest is 0.05).

$n$ = number of interest periods.

$P$ = a present sum of money.

$F$ = a future sum of money. The future sum $F$ is an amount, $n$ interest periods from the present, that is equivalent to $P$ with interest rate $i$.

$A$ = an end-of-period cash receipt or disbursement in a uniform series continuing for $n$ periods, the entire series equivalent to $P$ or $F$ at interest rate $i$.

## Single Payment Formulas

Suppose a present sum of money $P$ is invested for one year* at interest rate $i$. At the end of the year we should receive back our initial investment $P$ together with interest equal to $iP$ or a total amount $P + iP$. Factoring $P$ the sum at the end of one year is $P(1 + i)$. Let us assume that instead of removing our investment at the end of one year, we agree to let it remain for another year. How much would our investment be worth at the end of the second year? The end of the first year sum $P(1 + i)$ will in the second year draw interest of $iP(1 + i)$. The principal sum $P(1 + i)$ plus the interest $iP(1 + i)$ means that at the end of the second year the total investment will become $P(1 + i) + iP(1 + i)$. This may be rearranged by factoring $P(1 + i)$ to $P(1 + i)(1 + i)$ or $P(1 + i)^2$. If the process is continued for a third year, the end of third year total amount will be $P(1 + i)^3$ and at the end of $n$ years will be $P(1 + i)^n$. The progression looks like this.

|  | Amount at beginning of interest period | + | Interest for period | = | Amount at end of interest period |
|---|---|---|---|---|---|
| First year | $P$ | | $+ iP$ | $=$ | $P(1 + i)$ |
| Second year | $P(1 + i)$ | | $+ iP(1 + i)$ | $=$ | $P(1 + i)^2$ |
| Third year | $P(1 + i)^2$ | | $+ iP(1 + i)^2$ | $=$ | $P(1 + i)^3$ |
| $n$th year | $P(1 + i)^{n-1}$ | | $+ iP(1 + i)^{n-1}$ | $=$ | $P(1 + i)^n$ |

In other words, a present sum $P$ increases in $n$ periods to $P(1 + i)^n$. We therefore have a relationship between a present sum $P$ and its equivalent future sum $F$.

Future sum = (present sum)$(1 + i)^n$

$$F = P(1 + i)^n \tag{4-1}$$

This is the single payment compound amount formula and in functional notation is

$$F = P(F/P, i\%, n)$$

---

*A more general statement is to specify "one interest period" rather than "one year." It is easier to visualize one year so the derivation will assume that one year is the interest period.

Functional notation is designed so the compound interest factors may be written in an equation in an algebraically correct form. In the equation above, for example, the functional notation is interpreted as

$$F = \cancel{P}\left(\frac{F}{\cancel{P}}\right)$$

which is dimensionally correct. Without proceeding further we can see that if we were to derive a compound interest factor to find a present sum $P$, given a future sum $F$, the factor would be $(P/F,i\%,n)$ so the resulting equation would be

$$P = F(P/F,i\%,n)$$

which is dimensionally correct.

## EXAMPLE 4-1

If \$500 were deposited in a bank savings account, how much would be in the account three years hence if the bank paid 4% interest compounded annually?

We need to identify the various elements of the equation. The present sum $P$ is \$500. The interest rate per interest period is 4%, and in three years there are three interest periods. The future sum $F$ is to be computed.

$$P = \$500 \qquad i = 0.04 \qquad n = 3 \qquad F = \text{unknown}$$
$$F = P(1 + i)^n = 500(1 + 0.04)^3 = \$562.50$$

### Alternate Solution

A diagram of the problem:

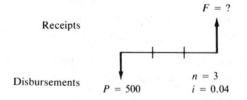

The equation $F = P(1 + i)^n$ is not solved by calculator or slide rule. Instead, the single payment compound factor $(1 + i)^n$ is determined from computed tables. The factor is written in convenient notation as

$$(1 + i)^n = (F/P,i\%,n).*$$

In this example problem,

$$(F/P,4\%,3)$$

*This continues to be called the "Single Payment Compound Amount" factor.

Compound interest tables appear in the Appendix at the end of the book. Each table is computed for a particular value of $i$. Knowing $n = 3$, locate the proper row in the 4% table. Read in the first column which is headed "Single payment, compound amount factor," the value 1.125. Thus,

$$500(F/P, 4\%, 3) = 500(1.125) = \$562.50$$

If we take $F = P(1 + i)^n$ and solve for $P$, then

$$P = F\frac{1}{(1 + i)^n} = F(1 + i)^{-n}$$

This is the single payment present worth formula. The equation

$$P = F(1 + i)^{-n} \tag{4-2}$$

in our notation becomes

$$P = F(P/F, i\%, n).$$

*EXAMPLE 4-2*

If you wished to have $800 in a savings account at the end of four years, and 5% interest was paid annually, how much should you put into the savings account now?

$$F = \$800 \qquad i = 0.05 \qquad n = 4 \qquad P = \text{unknown}$$
$$P = F(1 + i)^{-n} = 800(1 + 0.05)^{-4} = 800(0.8227) = \$658.16$$

To have $800 in the savings account at the end of four years, we must deposit $658.16 now.

*Alternate Solution*

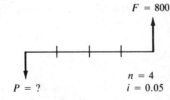

$$F = 800$$
$$n = 4$$
$$i = 0.05$$
$$P = ?$$

$$P = F(P/F, i\%, n) = \$800(P/F, 5\%, 4)$$

From the compound interest tables

$$(P/F, 5\%, 4) = 0.8227$$
$$P = \$800(0.8227) = \$658.16$$

Here the problem has an exact answer. In many situations, however, the answer is rounded off, recognizing that it is only as accurate as the input information upon which it is based.

### EXAMPLE 4-3

Suppose the bank changed their interest policy in Example 4-1 to "4% interest, compounded quarterly." For this situation a $500 deposit now would result in how much money in the account at the end of three years? First, we must be certain to understand the meaning of "4% interest, compounded quarterly." There are two elements.

1. *4% interest:* Unless otherwise described it is customary to assume the stated interest is for a one year period. If the stated interest is for other than a one year period, then it must be explained.
2. *Compounded quarterly:* This indicates there are four interest periods per year; that is, an interest period is three months long.

We know that the 4% interest is an annual rate, *because if it were anything different it must be stated.* Since we are dealing with four interest periods per year, it follows that the interest rate per interest period is 1%. For the total three year duration, there are 12 interest periods.

$$P = \$500 \qquad i = 0.01 \qquad n = 4 \times 3 = 12 \qquad F = \text{unknown}$$
$$F = P(1 + i)^n = P(F/P,i\%,n)$$
$$F = \$500(1 + 0.01)^{12} = \$500(F/P,1\%,12)$$
$$F = \$500(1.127) = \$563.50$$

### Single Payment Formulas—Continuous Compounding

If we define $r$ as the annual interest rate (more accurately called the nominal annual interest rate) and $m$ as the number of interest periods per year, then the interest rate per interest period $i = r/m$ and the number of interest periods in $n$ years is $mn$. The single payment compound amount formula may be rewritten as

$$F = P\left(1 + \frac{r}{m}\right)^{mn}$$

If we increase $m$, the number of interest periods per year, without limit, $m$ becomes very large and approaches infinity and $r/m$ becomes very small and approaches zero.

This is the condition of continuous compounding, that is, where the duration of the interest period decreases from some finite duration $\Delta t$ to an infinitely small duration $dt$ and the number of interest periods per year becomes infinite. In this situation of continuous compounding:

$$F = P \lim_{m \to \infty} \left(1 + \frac{r}{m}\right)^{mn} \tag{4-3}$$

An important limit in calculus is

$$\lim_{x \to 0} (1 + x)^{1/x} = 2.71828 = e \tag{4-4}$$

If we set $x = r/m$ then $mn$ may be written as $(1/x)(rn)$. As $m$ becomes infinite, $x$ becomes 0. Eq. 4-3 becomes

$$F = P\left[\lim_{x \to 0} (1 + x)^{1/x}\right]^{rn}$$

Eq. 4-4 tells us the quantity within the brackets equals $e$, so

$$F = Pe^{rn} \tag{4-5}$$

This is the continuous compounding single payment compound amount formula. The continuous compounding single payment present worth formula is

$$P = Fe^{-rn} \tag{4-6}$$

A table for values of $e^{rn}$ and $e^{-rn}$ will be found in the Appendix.

### EXAMPLE 4-4

If you were to deposit $2000 in a bank that pays 4% interest, compounded continuously, how much would be in the account at the end of two years? The single payment compound amount equation for continuous compounding is

$$F = Pe^{rn} \quad \text{where } r = \text{nominal interest rate} = 0.04$$
$$n = \text{number of years} = 2$$
$$F = 2000e^{(0.04 \times 2)} = 2000(1.0833) = \$2166.60$$

### EXAMPLE 4-5

A bank offers to sell savings certificates that will pay the purchaser $5000 at the end of ten years, but will pay nothing to the purchaser in the meantime. If interest is computed at 6%, compounded continuously, at what price is the bank selling the certificates?

$$P = Fe^{-rn} \quad \text{where } F = \$5000, r = 0.06, n = 10 \text{ years}$$
$$P = 5000e^{-(0.06 \times 10)} = 5000(0.5488) = \$2744.$$

Therefore, the bank is selling the $5000 certificates for $2744.

In addition to the two factors for continuous compounding presented here, there are others. One could derive factors for a uniform series of receipts or disbursements with interest compounded continuously. On the other hand, one could assume continuously flowing receipts or disbursements with either interest compounded periodically or continuously.

Calculations involving funds flowing continuously or continuous compounding are more likely to be used in conjunction with calculations done on a digital computer.

## Nominal and Effective Interest

*EXAMPLE 4-6*

Consider the situation if a person deposited $100 in a bank that pays 5% interest, compounded semi-annually. How much would be in the savings account at the end of one year?

Five percent interest, compounded semi-annually, means that the bank pays $2\frac{1}{2}\%$ every six months. Thus, the initial amount $P = \$100$ would be credited with $0.025(100) = \$2.50$ interest at the end of six months, or

$$P \rightarrow P + Pi = 100 + 100(0.025) = 100 + 2.50 = \$102.50.$$

The $102.50 is left in the savings account, and at the end of the second six-month period, the interest earned is $0.025(102.50) = \$2.56$ for a total in the account at the end of one year of $102.50 + 2.56 = \$105.06$, or

$$(P + Pi) \rightarrow (P + Pi) + i(P + Pi) = P(1 + i)^2 = 100(1 + 0.025)^2$$
$$= \$105.06$$

With the data from Example 4-6, we can now define nominal and effective interest and how they are computed.

*Nominal interest is the annual interest rate without considering the effect of any compounding.*

In the example, the bank pays $2\frac{1}{2}\%$ interest every six months. The nominal interest rate, therefore, is $2(2\frac{1}{2}\%) = 5\%$.

*Effective interest is the annual interest rate, taking into account the effect of compounding during the year.*

In Example 4-6 we saw that $100 left in the savings account for one year increased to $105.06. The interest paid was $5.06, making the effective interest rate

$$\frac{5.06}{100} = 5.06\%$$

The example problem demonstrates the method for computing the effective interest rate. The initial sum $P$ became $P(1 + i)^m$ where

$i =$ interest rate/interest period,

$m =$ number of compoundings per year.

If a \$1 deposit were made, the total at the end of one year would be

$$1(1 + i)^m$$

and if we then deducted the \$1 principal sum, the expression would be

$$1(1 + i)^m - 1$$

This would be the interest received in one year for a deposit of \$1. This provides a method for computing effective interest.

Effective interest rate $= (1 + i)^m - 1$           (4-7)

where $i$ = interest rate/interest period

$m$ = number of compoundings per year

It can be seen that the single payment compound amount factor (for either periodic or continuous compounding, as appropriate) may be used in computing the effective interest rate.

## EXAMPLE 4-7

If a savings bank pays $1\frac{1}{2}\%$ interest every three months, what are the nominal and the effective interest rates?

The nominal interest rate is the annual interest rate ignoring the effect of compounding, or $mi = 4(1.5\%) = 6\%$.

The effective interest rate $= (1 + i)^m - 1$. In this case

$$i = 0.015 \qquad m = 4$$

Effective interest rate $= (F/P, 1.5\%, 4) - 1$

$$= 1.061 - 1 = 0.061 = 6.1\%.$$

## EXAMPLE 4-8

If the savings bank in Example 4-7 changed their interest policy to 6% interest, compounded continuously, what are the nominal and the effective interest rates?

The nominal interest rate remains at 6%.

The effective interest rate is computed using the single payment compound amount factor for continuous compounding.

Effective interest rate $= e^{rn} - 1$    where $n$ equals one year

$$= e^{0.06(1)} - 1 = 0.062 = 6.2\%$$

Table 4-3 tabulates the effective interest rate for a range of compounding frequencies and nominal interest rates. It should be noted that when a nominal interest rate is compounded annually, the nominal interest rate equals the effective interest rate. Also, it will be noted that increasing the

**Table 4-3.**   Nominal and effective interest

| Nominal interest rate | Effective interest rate when nominal rate is compounded | | | | |
|---|---|---|---|---|---|
| i | Yearly | Semi-annually | Monthly | Daily | Continuously |
| 1% | 1.0000% | 1.0025% | 1.0046% | 1.0050% | 1.0050% |
| 2 | 2.0000 | 2.0100 | 2.0184 | 2.0201 | 2.0201 |
| 3 | 3.0000 | 3.0225 | 3.0416 | 3.0453 | 3.0455 |
| 4 | 4.0000 | 4.0400 | 4.0742 | 4.0809 | 4.0811 |
| 5 | 5.0000 | 5.0625 | 5.1162 | 5.1268 | 5.1271 |
| 6 | 6.0000 | 6.0900 | 6.1678 | 6.1831 | 6.1837 |
| 8 | 8.0000 | 8.1600 | 8.3000 | 8.3278 | 8.3287 |
| 10 | 10.0000 | 10.2500 | 10.4713 | 10.5156 | 10.5171 |
| 15 | 15.0000 | 15.5625 | 16.0755 | 16.1798 | 16.1834 |
| 25 | 25.0000 | 26.5625 | 28.0732 | 28.3916 | 28.4025 |

frequency of compounding (for example, from monthly to continuously) has only a small impact on the effective interest rate.

## Uniform Series Formulas

Many times we will find situations where there are a uniform series of receipts or disbursements. Automobile loans, house payments, and many other loans are based on a uniform payment series. It will often be convenient to use tables based on a uniform series of receipts or disbursements. The series $A$ is defined:

$A$ = an end-of-period* cash receipt or disbursement in a uniform series continuing for $n$ periods, the entire series equivalent to $P$ or $F$ at interest rate $i$.

The horizontal line in Figure 4-1 is a representation of time with four interest periods illustrated. Four uniform receipts $A$ have been placed at the end of each interest period and there are as many $A$'s as there are interest periods $n$. Both of these conditions are specified in the definition of $A$.

In the previous section on single payment formulas, we saw that a sum $P$ at one point in time would increase to a sum $F$ in $n$ periods according to the equation

$$F = P(1 + i)^n$$

*In textbooks on economic analysis it is customary to define $A$ as an end-of-period event rather than beginning-of-period or possibly middle-of-period. The derivations that follow are based on this end-of-period assumption. One could, of course, derive other equations based on beginning-of-period or mid-period assumptions.

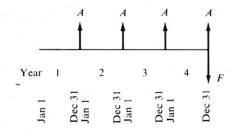

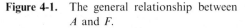

**Figure 4-1.** The general relationship between
*A* and *F*.

We will use this relationship in our uniform series derivation.

Looking at Figure 4-1, we see that if an amount *A* is invested at the end of each year for *n* years, the total amount *F* at the end of *n* years will obviously be the sum of the compound amounts of the individual investments.

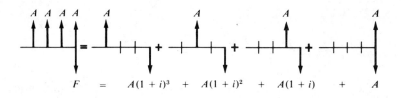

In the general case for *n* years,

$$F = A(1 + i)^{n-1} + \cdots + A(1 + i)^3 + A(1 + i)^2 + A(1 + i) + A \tag{4-8}$$

Multiplying by $(1 + i)$,

$$(1 + i)F = A(1 + i)^n + \cdots + A(1 + i)^4 + A(1 + i)^3 + A(1 + i)^2 + A(1 + i) \tag{4-9}$$

Factoring out *A*, and subtracting Eq. 4-8 gives

$$(1 + i)F = A[(1 + i)^n + \cdots + (1 + i)^4 + (1 + i)^3 + (1 + i)^2 + (1 + i)] \tag{4-10}$$
$$F = A[(1 + i)^{n-1} + \cdots + (1 + i)^3 + (1 + i)^2 + (1 + i) + 1] \tag{4-11}$$
$$iF = A[(1 + i)^n - 1]$$

Solving for *F*,

$$F = A\left[\frac{(1 + i)^n - 1}{i}\right] \tag{4-12}$$

Thus we have an equation for $F$ when $A$ is known. The term within the brackets

$$\left[\frac{(1 + i)^n - 1}{i}\right]$$

is called the uniform series compound amount factor and is referred to by the notation

$$(F/A,i\%,n)$$

### EXAMPLE 4-9

A man deposits $500 in a credit union at the end of each year for five years. The credit union pays 5% interest, compounded annually. At the end of five years, immediately following his fifth deposit, how much will he have in his account?

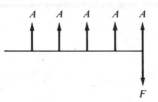

The diagram of the five deposits and the desired computation of the future sum $F$ duplicates the situation for the uniform series compound amount formula

$$F = A\left[\frac{(1 + i)^n - 1}{i}\right] = A(F/A,i\%,n)$$

where $A$ = $500, $n$ = 5, $i$ = 5%, $F$ = unknown.
$$F = \$500(F/A,5\%,5) = \$500(5.526) = \$2763.$$

If Eq. 4-12 is solved for $A$, we have

$$A = F\left[\frac{i}{(1 + i)^n - 1}\right] \qquad (4\text{-}13)$$

where

$$\left[\frac{i}{(1 + i)^n - 1}\right]$$

is called the uniform series sinking fund* factor and is written as $(A/F,i\%,n)$.

---

*A *sinking fund* is a separate fund into which one makes a uniform series of money deposits ($A$) with the goal of accumulating some desired future sum ($F$) at a given future point in time.

## EXAMPLE 4-10

Jim Hayes read that in the western United States a ten-acre parcel of land could be purchased for $1000 cash. Jim decided to save a uniform amount the end of each month so that at the end of a year he would have the required $1000. The local credit union pays 6% interest, compounded monthly. How much would Jim have to deposit each month? In this example,

$$F = \$1000 \qquad n = 12 \qquad i = \tfrac{1}{2}\% \qquad A = \text{unknown}$$
$$A = 1000(A/F, \tfrac{1}{2}\%, 12) = 1000(0.0811) = \$81.10.$$

If we use the sinking fund formula, Eq. 4-13, and substitute for $F$ the single payment compound formula (Eq. 4-1) we obtain

$$A = F\left[\frac{i}{(1 + i)^n - 1}\right] = P(1 + i)^n\left[\frac{i}{(1 + i)^n - 1}\right]$$

$$A = P\left[\frac{i(1 + i)^n}{(1 + i)^n - 1}\right] \qquad\qquad (4\text{-}14)$$

We have an equation for determining the value of the series of end-of-period payments or disbursements $A$ when the present sum $P$ is known.

The portion within the brackets

$$\left[\frac{i(1 + i)^n}{(1 + i)^n - 1}\right]$$

is called the uniform series capital recovery factor and has the notation $(A/P, i\%, n)$.

## EXAMPLE 4-11

On January 1 a man deposits $5000 in a credit union that pays 6% interest, compounded annually. He wishes to make five equal end-of-year withdrawals, beginning December 31st of the first year. How much should he withdraw each year?

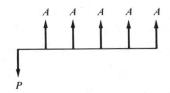

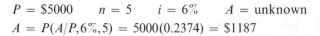

$$P = \$5000 \qquad n = 5 \qquad i = 6\% \qquad A = \text{unknown}$$
$$A = P(A/P, 6\%, 5) = 5000(0.2374) = \$1187$$

The annual withdrawal is $1187.

In the example, with interest at 6% a present sum of $5000 is equivalent to five equal end-of-period disbursements of $1187. This is another way of stating plan 3 of Table 4-1. The method for determining the annual payment that would repay $5000 in five years with 6% interest has now been explained. The calculation is simply

$$A = 5000(A/P,6\%,5) = 5000(0.2374) = \$1187$$

If the capital recovery formula, Eq. 4-14, is solved for the present sum $P$, we obtain the uniform series present worth formula.

$$P = A\left[\frac{(1 + i)^n - 1}{i(1 + i)^n}\right] \qquad\qquad (4\text{-}15)$$

Within the brackets we have

$$\left[\frac{(1 + i)^n - 1}{i(1 + i)^n}\right]$$

which is the uniform series present worth factor $(P/A,i\%,n)$.

### EXAMPLE 4-12

An investor holds a time payment purchase contract on some machine tools. The contract calls for the payment of $140 at the end of each month for a five-year period. The first payment is due in one month. He offers to sell you the contract for $6800 cash today. If you otherwise can make 1% per month on your money, would you accept or reject the investor's offer?

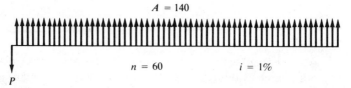

$$A = 140$$

$$n = 60 \qquad\qquad i = 1\%$$

$$P$$

In this problem we are being offered a contract that will pay $140 per month for 60 months. We must determine if the contract is worth $6800 if we consider 1% per month to be a suitable interest rate. Using the uniform series present worth formula, we will compute the present worth of the contract.

$$P = A(P/A,i\%,n) = 140(P/A,1\%,60) = 140(44.955)$$
$$P = \$6293.70$$

It is clear that if we pay the $6800 asking price for the contract, we will receive something less than the 1% per month interest we desire. We will, therefore, reject the investor's offer.

## EXAMPLE 4-13

Suppose we decided to pay the $6800 for the time purchase contract in Example 4-12. What monthly rate of return would we obtain on our investment?

In this situation we know $P$, $A$, and $n$, but we do not know $i$. The problem may be solved using either the uniform series present worth formula $P = A(P/A,i\%,n)$ or the uniform series capital recovery formula $A = P(A/P,i\%,n)$. Either way we have one equation with one unknown.

$$P = \$6800 \qquad A = \$140 \qquad n = 60 \qquad i = \text{unknown}$$
$$P = A(P/A,i\%,n)$$
$$\$6800 = \$140(P/A,i\%,60)$$
$$(P/A,i\%,60) = \frac{6800}{140} = 48.57$$

We know the value of the uniform series present worth factor, but we do not know the interest rate $i$. As a result, we need to look through several compound interest tables and compute the rate of return $i$ by interpolation.

From the tables in the Appendix we find

| *Interest rate* | $(P/A,i\%,60)$ |
|:---:|:---:|
| $\frac{1}{2}\%$ | 51.726 |
| $\frac{3}{4}\%$ | 48.173 |
| $1\%$ | 44.955 |

The rate of return is between $\frac{1}{2}\%$ and $\frac{3}{4}\%$, and may be computed by a linear interpolation:

$$\text{Rate of return } i = 0.50\% + 0.25\%\left[\frac{51.726 - 48.57}{51.726 - 48.173}\right]$$
$$= 0.50\% + 0.25\%\left[\frac{3.156}{3.553}\right] = 0.50\% + 0.22\%$$
$$= 0.72\% \text{ per month}$$

## RELATIONSHIPS BETWEEN COMPOUND INTEREST FACTORS

From the derivations we saw there are several simple relationships between the compound interest factors. They are summarized here.

**Single Payment**

$$\text{Compound amount factor} = \frac{1}{\text{Present worth factor}}$$

$$(F/P,i\%,n) = \frac{1}{(P/F,i\%,n)}$$

**Uniform Series**

$$\text{Capital recovery factor} = \frac{1}{\text{Present worth factor}}$$

$$(A/P,i\%,n) = \frac{1}{(P/A,i\%,n)}$$

$$\text{Compound amount factor} = \frac{1}{\text{Sinking fund factor}}$$

$$(F/A,i\%,n) = \frac{1}{(A/F,i\%,n)}$$

The uniform series present worth factor is simply the sum of the $n$ terms of the single payment present worth factor

$$(P/A,i\%,n) = \sum_{J=1}^{n} (P/F,i\%,J)$$

For example:

$$(P/A,5\%,4) = (P/F,5\%,1) + (P/F,5\%,2) + (P/F,5\%,3) + (P/F,5\%,4)$$
$$3.546 = 0.952 + 0.907 + 0.864 + 0.823$$

The uniform series compound amount factor equals 1 plus the sum of $(n-1)$ terms of the single payment compound amount factor

$$(F/A,i\%,n) = 1 + \sum_{J=1}^{n-1} (F/P,i\%,J)$$

For example:

$$(F/A,5\%,4) = 1 + (F/P,5\%,1) + (F/P,5\%,2) + (F/P,5\%,3)$$
$$4.310 = 1 + 1.050 + 1.102 + 1.158$$

The uniform series capital recovery factor equals the uniform series sinking fund factor plus $i$.*

$$(A/P,i\%,n) = (A/F,i\%,n) + i$$

For example:

$$(A/P,5\%,4) = (A/F,5\%,4) + 0.05$$
$$0.282 = 0.232 + 0.05$$

## UNIFORM GRADIENT

We frequently encounter the situation where the cash flow series is not of constant amount $A$. Instead, there is an increasing series as shown:

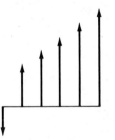

Cash flows of this form may be resolved into two components:

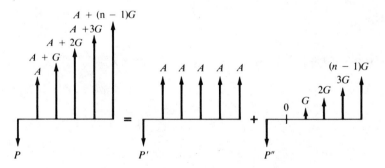

*This may be proved as follows.

$$(A/P,i\%,n) = (A/F,i\%,n) + i$$

$$\left[\frac{i(1 + i)^n}{(1 + i)^n - 1}\right] = \left[\frac{i}{(1 + i)^n - 1}\right] + i$$

Multiply by $(1 + i)^n - 1$

$$i(1 + i)^n = i + i(1 + i)^n - i = i(1 + i)^n$$

Note that by resolving the problem in this manner, it makes the first cash flow in the gradient series equal to zero. We already have an equation for $P'$, and we need to derive an equation for $P''$. In this way, we will be able to write

$$P = P' + P'' = A(P/A,i\%,n) + G(P/G,i\%,n)$$

## Derivation of Uniform Gradient Factors

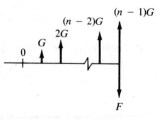

The gradient series may be thought of as a series of individual cash flows:

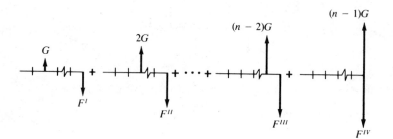

The value of $F$ for the sum of the cash flows $= F^{\mathrm{I}} + F^{\mathrm{II}} + \cdots + F^{\mathrm{III}} + F^{\mathrm{IV}}$

$$F = G(1+i)^{n-2} + 2G(1+i)^{n-3} + \cdots + (n-2)(G)(1+i)^1 + (n-1)G \tag{4-16}$$

Multiply Eq. 4-16 by $(1 + i)$ and factor out G

$$(1+i)F = G[(1+i)^{n-1} + 2(1+i)^{n-2} + \cdots + (n-2)(1+i)^2 + (n-1)(1+i)^1] \tag{4-17}$$

Rewrite Eq. 4-16 to show other terms in the series

$$F = G[(1+i)^{n-2} + \cdots + (n-3)(1+i)^2 + (n-2)(1+i)^1 + n-1] \tag{4-18}$$

Subtract Eq. 4-18 from Eq. 4-17

$$F + iF - F = G[(1+i)^{n-1} + (1+i)^{n-2} + \cdots + (1+i)^2 + (1+i)^1 + 1] - nG \tag{4-19}$$

In the derivation of Eq. 4-12 the terms within the brackets of Eq. 4-19 were shown to equal the series compound amount factor.

$$[(1 + i)^{n-1} + (1 + i)^{n-2} + \cdots + (1 + i)^2 + (1 + i)^1 + 1] = \frac{(1 + i)^n - 1}{i}$$

Thus, Eq. 4-19 becomes

$$iF = G\left[\frac{(1 + i)^n - 1}{i}\right] - nG$$

Rearranging and solving for $F$

$$F = \frac{G}{i}\left[\frac{(1 + i)^n - 1}{i} - n\right] \tag{4-20}$$

Multiplying Eq. 4-20 by the single payment present worth factor

$$P = \frac{G}{i}\left[\frac{(1 + i)^n - 1}{i} - n\right]\left[\frac{1}{(1 + i)^n}\right]$$

gives an equation for $(P/G,i\%,n)$, the gradient present worth factor. Multiplying Eq. 4-20 by the sinking fund factor

$$A = \frac{G}{i}\left[\frac{(1 + i)^n - 1}{i} - n\right]\left[\frac{i}{(1 + i)^n - 1}\right] = G\left[\frac{1}{i} - \frac{n}{(1 + i)^n - 1}\right]$$

yields $(A/G,i\%,n)$, the gradient uniform series factor.

### EXAMPLE 4-14

It has been estimated that the maintenance cost of an automobile is as follows:

| Year | Maintenance cost |
|------|------------------|
| 1    | $120             |
| 2    | 150              |
| 3    | 180              |
| 4    | 210              |
| 5    | 240              |

A man purchased a new automobile. He wishes to set aside enough money in a bank account to pay the maintenance on the car for the first five years. Assume the maintenance costs occur at the end of each year and that the bank pays 5% interest. How much should he deposit in the bank now?

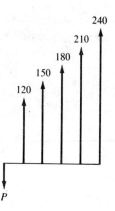

The cash flow may be broken into its two components:

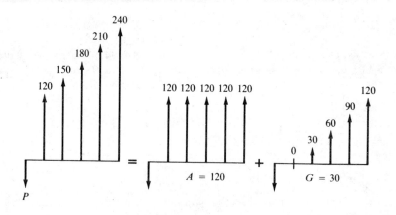

Both of the components represent cash flows for which compound interest factors have been derived. The first is uniform series present worth, and the second is gradient series present worth.

$$P = A(P/A,5\%,5) + G(P/G,5\%,5)$$
$$P = 120(4.329) + 30(8.237)$$
$$P = 519 + 247 = \$766$$

*EXAMPLE 4-15*

On a certain piece of machinery, it is estimated that the maintenance expense will be as follows:

| Year | Maintenance |
|------|-------------|
| 1 | $100 |
| 2 | 200 |
| 3 | 300 |
| 4 | 400 |

What is the equivalent uniform annual maintenance cost for the machinery if 6% interest is used?

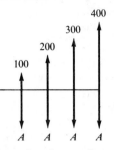

This is *not* the proper form for the gradient uniform series equation. The cash flow must be resolved into two components as in Example 4-14.

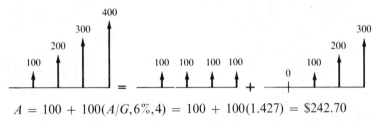

$$A = 100 + 100(A/G,6\%,4) = 100 + 100(1.427) = \$242.70$$

The equivalent uniform annual maintenance cost is $242.70.

### EXAMPLE 4-16

A textile mill in India installed a number of new looms. It is expected that initial maintenance and repairs will be high, but that they will then decline for several years. The projected cost is

| Year | Maintenance and repair cost |
|------|-----------------------------|
| 1 | 24,000 rupees |
| 2 | 18,000 |
| 3 | 12,000 |
| 4 | 6,000 |

What is the projected equivalent annual maintenance and repair cost if interest is 10%?

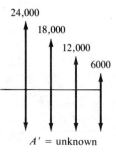

$A'$ = unknown

The projected cash flow is not in the form of the uniform gradient factors. Both factors were derived for an increasing gradient over time. The factors cannot be used directly for a declining gradient. Instead, we will subtract an increasing gradient from an assumed uniform series of payments.

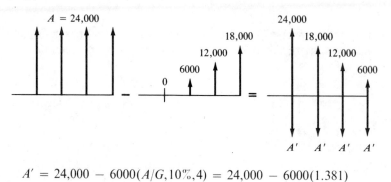

$A' = 24,000 - 6000(A/G, 10\%, 4) = 24,000 - 6000(1.381)$
    $= 15,714$ rupees

The projected equivalent uniform maintenance and repair cost is 15,714 rupees per year.

## SUMMARY

In this chapter, the concepts of time value of money and equivalence are described in detail. Also, the various compound interest formulas are derived. It is essential that these concepts and the applicability of the interest formulas be understood as the remainder of the book is based upon them.

*Time Value of Money.* The continuing offer of banks to pay interest for the temporary use of other people's money is ample proof that there is a time value of money. Thus, we would always choose to receive $100 today rather than the promise of $100 to be paid at a future date.

*Equivalence.* What sum would a person be willing to accept a year hence instead of $100 today. If a 5% interest rate is considered to be appropriate, a year hence he would require $105. If $100 today or $105 a year hence were considered equally desirable, we would say the two sums of money are *equivalent*. But if on further consideration we decided that a 7% interest rate is applicable, then $105 a year hence would no longer be equivalent to $100 today. This illustrates that equivalence is dependent on the interest rate.

*Compound Interest.* The notation used is:

$i$ = interest rate per interest period

$n$ = number of interest periods

$P$ = a present sum of money

$F$ = a future sum of money. The future sum $F$ is an amount, $n$ interest periods from the present, that is equivalent to $P$ with interest rate $i$

$A$ = An end-of-period cash receipt or disbursement in a uniform series continuing for $n$ periods, the entire series equivalent to $P$ or $F$ at interest rate $i$

$G$ = a uniform arithmetic gradient representing a period-by-period increase in payments or disbursements.

Ten compound interest formulas were derived as follows.

*Single Payment Formulas*

| | | |
|---|---|---|
| Compound amount | $F = P(1 + i)^n$ | $= P(F/P,i\%,n)$ |
| Present worth | $P = F(1 + i)^{-n}$ | $= F(P/F,i\%,n)$ |

*Uniform Series Formulas*

Compound amount

$$F = A\left[\frac{(1 + i)^n - 1}{i}\right] = A(F/A,i\%,n)$$

Sinking fund

$$A = F\left[\frac{i}{(1 + i)^n - 1}\right] = F(A/F,i\%,n)$$

Capital recovery

$$A = P\left[\frac{i(1 + i)^n}{(1 + i)^n - 1}\right] = P(A/P,i\%,n)$$

Present worth

$$P = A\left[\frac{(1 + i)^n - 1}{i(1 + i)^n}\right] = A(P/A,i\%,n)$$

*Uniform Gradient Formulas*

Gradient present worth $P = \dfrac{G}{i}\left[\dfrac{(1 + i)^n - 1}{i} - n\right]\left[\dfrac{1}{(1 + i)^n}\right]$

$$= G(P/G,i\%,n)$$

Gradient uniform series $A = G\left[\dfrac{1}{i} - \dfrac{n}{(1 + i)^n - 1}\right] = G(A/G, i\%, n)$

*Single Payment—Continuous Compounding*

Compound amount         $F = Pe^{rn}$

Present worth           $P = Fe^{-rn}$

*Nominal interest* is the annual interest rate computed without considering the effect of any compounding. *Effective interest* is the annual interest rate, taking into account the effect of compounding during the year.

## PROBLEMS

**4-1**  A man went to his bank and borrowed $750. He agreed to repay the sum at the end of three years together with interest at 8% per year, compounded annually. How much will he owe the bank at the end of three years?  (*Answer:* $945)

**4-2**  What sum of money now is equivalent to $8250 two years hence, if interest is 8% per annum, compounded semiannually?  (*Answer:* $7052)

**4-3**  The local bank offers to pay 5% interest, compounded annually, on savings deposits. In a nearby town the bank pays 5% interest, compounded quarterly. A man who has $3000 to put in a savings account wonders if the increased interest paid in the nearby town justifies driving his car there to make the deposit. Assuming he will leave all money in the account for two years, how much additional interest would he obtain from the out-of-town bank over the local bank?

**4-4**  A sum of money invested at 4% interest, compounded semi-annually, will double in amount in approximately how many years?  (*Answer:* $17\frac{1}{2}$ years)

**4-5**  A local bank will lend a customer $1000 on a two-year car loan as follows:

| | |
|---|---|
| Money to pay for car | $1000 |
| Two years interest at 7%: 2 × 0.07 × 1000 | 140 |
| | $1140 |

24 monthly payments $= \dfrac{1140}{24} = \$47.50$

The first payment must be made in 30 days. What is the nominal interest rate (per annum) the bank is receiving?

**4-6**  A local lending institution advertises their "51–50" Club. A person may borrow $2000 and repay $51 for the next 50 months beginning 30 days after he receives the money. Compute the nominal annual interest rate for this loan. What is the effective interest rate?

**4-7**  A loan company has been advertising on television a plan where one may borrow $1000 and make a payment of $10.87 per month. This payment is for interest only and includes no payment on the principal. What is the nominal interest rate which they charge?

**4-8**  What effective interest rate per annum corresponds to a nominal rate of 12% compounded monthly?  (*Answer:* 12.7%)

**4-9** Mr. Sansome withdrew $1000 from a savings account and invested it in common stock. At the end of five years he sold the stock and received a check for $1307. If Mr. Sansome had left his $1000 in the savings account, he would have received an interest rate of 5%, compounded quarterly. Mr. Sansome would like to compute a comparable interest rate on his common stock investment. Based on quarterly compounding, what nominal interest rate did Mr. Sansome receive on his investment in stock? What effective interest rate did he receive?

**4-10** A woman opened an account in a local store. In the charge account agreement the store indicated it charges $1\frac{1}{2}\%$ each month on the unpaid balance. What nominal interest rate is being charged? What is the effective interest rate?

**4-11** A man buys a car for $3000 with no money down. He pays for the car in 30 equal monthly payments with interest at 12% per annum, compounded monthly. What is his monthly car loan payment? (*Answer:* $116.10)

**4-12** What amount will be required to purchase, on a man's 40th birthday, an annuity to provide him with 30 equal semi-annual payments of $1000 each, the first to be received on his 50th birthday, if interest is 4% compounded semi-annually?

**4-13** On the birth of his first boy, Dick Jones decided to establish a savings account to partly pay for his son's education. He plans to deposit $20 per month in the account beginning when the boy is 13 months old. The savings and loan association has a current interest policy of 6% per annum, compounded monthly, ~~paid quarterly~~. Assuming no change in the interest rate, how much will be in the savings account when Dick's son becomes 16 years old?   5817

**4-14** An engineer borrowed $3000 from the bank, payable in six equal end-of-year payments at 8%. The bank agreed to reduce the interest on the loan if interest rates declined in the United States before the loan was fully repaid. At the end of three years, at the time of the third payment, the bank agreed to reduce the interest rate from 8% to 7% on the remaining debt. What was the amount of the equal annual end-of-year payments for each of the first three years? What was the amount of the equal annual end-of-year payments for each of the last three years?

**4-15** On a new car it is estimated that the maintenance cost will be $40 the first year. Each subsequent year it is expected to be $10 more than the previous one. How much would you need to set aside when you bought a new car to pay all future maintenance costs if you planned to keep it seven years? Assume interest is 5% per annum. (*Answer:* $393.76)

**4-16** A man decides to deposit $50 in the bank today and make ten additional deposits every six months beginning six months from now, the first of which will be $50 and increasing $10 per deposit after that. A few minutes after he makes the last deposit, he decides to withdraw all the money deposited. If the bank pays 6% compounded semi-annually, how much money will he receive?

**4-17** A young engineer wishes to become a millionaire by the time he is 60 years old. He believes that by careful investment he can obtain a 15% rate of return. He plans to add a uniform sum of money to his investment program each year beginning on his 20th birthday and continuing through his 59th birthday. How much money must the engineer set aside in this project each year?

**4-18** The councilmen of a small town have decided that the earth levy that protects the town from a nearby river should be rebuilt and strengthened. The town engineer estimates that the cost of the work at the end of the first year will be $85,000. He estimates that in subsequent years the annual repair costs will decline by $10,000, making the second year cost $75,000; the third year $65,000, and so forth. The councilmen want to know what the equivalent present cost is for the first five years of repair work if interest is 4%. (*Answer:* $292,870)

**4-19** The Apex Company sold a water softener to a customer. The price of the unit was $350. The customer asked for a deferred payment plan and a contract was written. Under the contract the buyer could delay paying for the water softener provided that he purchased the coarse salt for re-charging the softener from Apex. At the end of two years, the buyer was to pay for the unit in a lump sum, including 6% interest, compounded quarterly. The contract provided that if the customer ceased buying salt from Apex at any time prior to two years, the full payment due at the end of two years would automatically become due. Six months later the customer decided to buy salt elsewhere and stopped buying from Apex. Apex thereupon asked for the full payment that was to have been due 18 months hence. The customer was unhappy, so Apex offered as an alternate to accept the $350 with interest at 20% per annum compounded semi-annually for the six months that the customer had had the water softener. Which alternative should the customer accept? Explain.

**4-20** A $150 bicycle was purchased on December 1st with a $15 down payment. The balance was to be paid at the rate of $10 at the end of each month with the first payment due on December 31st. The last payment may be some amount less than $10. If interest on the unpaid balance is computed at $1\frac{1}{2}\%$ per month, when will the bicycle be paid for, and what is the total amount that will have been paid?

**4-21** A company buys a machine for $12,000, which it agrees to pay for in five equal annual payments, beginning one year after the date of purchase, at an interest rate of 4% per annum. Immediately after the second payment, the terms of the agreement are changed to allow the balance due to be paid off in a single payment the next year. What is the final single payment?

**4-22** An engineering student bought a car at a local used car lot. Including tax and insurance, the total price was $1500. He was to pay for the car in 12 equal monthly payments beginning with the first payment immediately (in other words, the first payment was the down payment). Interest on the loan is at 12% compounded monthly. After he made six payments (the down payment plus five additional payments), he decides to sell the car. A buyer agrees to pay a cash amount to pay off the loan in full at the time the next payment is due, and also to pay the engineering student $500. If there are no penalty charges for this early payment of the loan, how much will the car cost the new buyer?

**4-23** A bank recently announced an "instant cash" plan for holders of its bank credit cards. A cardholder may receive cash from the bank up to a present limit (about $500). There is a special charge of 4% made at the time the "instant cash" is sent the cardholder. The debt may be repaid in monthly installments. Each month the bank charges $1\frac{1}{2}\%$ on the unpaid balance. The monthly payment, including interest, may be as little as $10. Thus, for $150 of "instant cash" an initial charge of $6 is made and added to the balance due. Assume the cardholder makes a monthly payment of $10 (this includes both principal and interest). How many months are required to repay the debt? If your answer includes a fraction of a month then round up to the next month.

**4-24** The treasurer of a firm noted that many invoices were received by his firm with the following terms of payment:

"2%-10 days, net 30 days"

Thus, if he were to pay the bill within 10 days of its date, he could deduct 2%. On the other hand, if he did not promptly pay the bill, the full amount would be due 30 days from the date of the invoice. This 2% deduction for prompt payment is equivalent to what effective interest rate? (Hint: What is the time difference between the two alternatives?)

**4-25** In 1555, King Henry borrowed money from his bankers on the condition that he pay 5% of the loan at each fair (there were four fairs per year) until he had made 40 payments. At that time the loan was considered repaid. What effective annual interest did King Henry pay?

**4-26** A man wants to provide a college education for his young son. He can afford to invest $600/yr. for the next four years beginning on the boy's 4th birthday. He wishes to give his son $1500 on his 18th, 19th, 20th, and 21st birthdays, for a total of $6000. Assuming 5% interest, what uniform annual investment will he have to make on the boy's 8th through 17th birthdays? (*Answer:* $87.96)

**4-27** A man has $5000 on deposit in a bank that pays 5% interest compounded annually. He wonders how much more advantageous it would be to transfer his funds to another bank whose dividend policy is 5% interest, compounded continuously. Compute how much he would have in his savings account at the end of three years under each of these situations.

**4-28** A friend was left $50,000 by his uncle. He has decided to put it into a savings account for the next year or so. He finds there are varying interest rates at savings institutions: $4\frac{3}{8}\%$ compounded annually; $4\frac{1}{4}\%$ compounded quarterly; and $4\frac{1}{8}\%$ compounded continuously. He wishes to select the savings institution that will give him the highest return on his money. What interest rate should he select?

**4-29** One of the local banks indicates that it computes the interest it pays on savings accounts by the continuous compounding method. Suppose you deposited $100 in the bank and they pay 4% per annum, compounded continuously. After five years, how much money will there be in the account?

**4-30** A company expects to install smog control equipment on the exhaust of a gasoline engine at a pumping station immediately. The local smog control district has agreed to pay to the firm a lump sum of money to provide for the first cost of the equipment and its maintenance during its ten-year useful life. At the end of ten years the equipment, which initially cost $10,000, is valueless. The company and smog control district have agreed that the following are reasonable estimates of the end-of-year maintenance costs:

| Year 1 | $500 | Year 6 | $200 |
|--------|------|--------|------|
| 2 | 100 | 7 | 225 |
| 3 | 125 | 8 | 250 |
| 4 | 150 | 9 | 275 |
| 5 | 175 | 10 | 300 |

Assuming interest at 6% per year, how much should the smog control district pay to the company now to provide for the first cost of the equipment and its maintenance for ten years? (*Answer:* $11,693)

**4-31** One of the largest automobile dealers in the city advertises a one year old car for sale as follows.

Cash price $3575, or a down payment of $375 with 45 monthly payments of $93.41.

Susan DeVaux bought the car and made a down payment of $800. The dealer charged her the same interest rate used in his advertised offer. How much will Susan pay each month for 45 months? What effective interest rate is being charged? (*Answers:* $81.03, 16.1%)

# 5
# Present
# Worth
# Analysis

In the previous chapter two important tasks were accomplished. First, the concept of equivalence was presented. We would be powerless to compare series of cash flows unless we could resolve them into some equivalent arrangement. Second, equivalence, with alteration of cash flows from one series to an equivalent sum or series of cash flows, created the need for compound interest factors. Ten compound interest factors were derived— eight for periodic compounding and two for continuous compounding. This background sets the stage for the chapters that follow.

We have at this point shown how to manipulate cash flows in a variety of ways. By so doing we have been able to solve a great variety of compound interest problems. But engineering economic analysis is more than simply solving interest problems. The decision process (Figure 2-2) requires that the outcomes of the feasible alternatives be arranged so that they may be judged in terms of the selection criterion. Depending on the situation, the economic criterion will be one of the following:

| Situation | Criterion |
|---|---|
| For fixed input | Maximize output |
| For fixed output | Minimize input |
| Neither input nor output fixed | Maximize (output-input) |

We must now examine ways to resolve engineering problems so that the criteria for economic efficiency can be applied.

Equivalence provides the logic by which we may adjust the cash flow for a given alternative into some equivalent sum or series. To apply the selection criterion to the outcomes of the feasible alternatives, we must first resolve them into comparable units. The question is, how should they be compared? In this chapter the analysis will resolve the alternatives into equivalent present consequences, referred to simply as "present worth analysis." The next chapter will convert the alternatives into an equivalent uniform annual cash flow, and the next solves for the interest rate at which the favorable consequences (benefits) are equivalent to the unfavorable consequences (costs).

## ECONOMIC CRITERIA

One of the easiest ways to compare mutually exclusive* alternatives is to resolve their consequences to the present time. The three criteria for economic efficiency are presented in Table 5-1.

**Table 5-1.**   Present worth analysis

|  | *Situation* | *Criterion* |
|---|---|---|
| Fixed input | Amount of money or other input resources are fixed | Maximize present worth of benefits or other outputs |
| Fixed output | There is a fixed task, benefit, or other output to be accomplished | Minimize present worth of costs or other inputs |
| Neither input nor output fixed | Neither amount of money or other inputs nor amount of benefits or other outputs are fixed | Maximize (present worth of benefits minus present worth of costs) or more simply, maximize net present worth |

## APPLYING PRESENT WORTH TECHNIQUES

As a general rule, any economic analysis problem may be solved by the methods presented here or the methods presented in the two following chapters. This is true because present worth, annual cash flow, and rate of return are exact methods that will always yield the same solution in selecting the best alternative from among a set of mutually exclusive alternatives. Some problems, however, may be more easily solved by one method than another. For this reason we will emphasize here the kinds of problems that can be most readily solved by present worth analysis.

Present worth analysis is most frequently used to determine the present value of future money receipts and disbursements. It would help us, for example, to determine a present worth of income producing property, like

---

*Mutually exclusive is where selecting one alternative precludes selecting any of the other alternatives. An example of mutually exclusive alternatives would be deciding between constructing a gas station or a drive-in restaurant on a particular piece of vacant land.

an oil well or an apartment house. If the future income and costs are known, then using a suitable interest rate, the present worth of the property may be calculated. This should provide a good estimate of the price at which the property could be bought or sold. Another application might be determining the valuation of stocks or bonds based on the anticipated future benefits from owning them.

In present worth analysis careful consideration must be given to the time period covered by the analysis. Usually the task to be accomplished has a time period associated with it. In that case the consequences of each alternative must be considered for this period of time which is usually called the analysis period, or sometimes the planning horizon.

There are three different analysis period situations encountered in economic analysis problems.

1.   The useful life of each alternative equals the analysis period.

2.   The alternatives have useful lives different from the analysis period.

3.   There is an infinite analysis period, ($n = \infty$).

Since different lives and an infinite analysis period present some complications, we will begin with four examples where the useful life of each alternative equals the analysis period.

## EXAMPLE 5-1

A firm is considering which of two devices to install to reduce costs in a particular situation. Both devices cost $1000 and have useful lives of five years and no salvage value. Device *A* can be expected to result in $300 savings annually. Device *B* will provide cost savings of $400 the first year, but will decline $50 annually, making the second year savings $350, the third year savings $300, and so forth. With interest at 7%, which device should the firm purchase?

The analysis period can conveniently be selected as the useful life of the devices or five years. Since both devices cost $1000, we have a situation where in choosing either *A* or *B*, there is a fixed input (cost) of $1000. The appropriate decision criterion is to choose the alternative that maximizes the present worth of benefits.

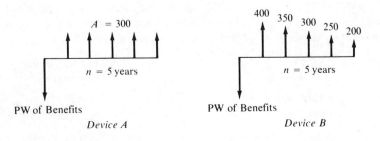

*Device A*          *Device B*

PW of benefits $A = 300(P/A,7\%,5) = 300(4.100) = \$1230$

PW of benefits $B = 400(P/A,7\%,5) - 50(P/G,7\%,5)$

$$= 400(4.100) - 50(7.647) = \$1257.65$$

Device $B$ has the larger present worth of benefits and is, therefore, the preferred alternative. It is worth noting that if we ignore the time value of money, both alternatives provide $1500 worth of benefits over the 5 year period. Device $B$ provides greater benefits in the first two years, and smaller benefits in the last two years. This more rapid flow of benefits from $B$, although the total magnitude equals that of $A$, results in a greater present worth of benefits.

*EXAMPLE 5-2*

Wayne County will build an aqueduct to bring water in from the upper part of the state. It can be built at a reduced size now for $300 million and be enlarged 25 years hence for an additional $300 million. An alternative is to construct the full sized aqueduct now for $400 million.

Both alternatives would provide the needed capacity for the 50 year analysis period. Maintenance costs are small and may be ignored. At 6% interest, which alternative should be selected?

This problem illustrates stage construction. The aqueduct may be built in a single stage or in a smaller first stage followed many years later by a second stage to provide the additional capacity when needed.

For the two stage construction:

PW of cost = $300 million + 300 million $(P/F,6\%,25)$

= $300 million + 69.9 million = $369.9 million

For the single stage construction:

PW of cost = $400 million

The two stage construction has a smaller present worth of cost and is the preferred construction plan.

*EXAMPLE 5-3*

The purchasing agent is considering the purchase of some new equipment for the mail room. Two different manufacturers have provided quotations. An analysis of the quotations indicates the following.

| Manufacturer | Cost | Useful life | End of useful life salvage value |
|---|---|---|---|
| Speedy | $1500 | 5 yrs | $200 |
| Allied | 1600 | 5 | 325 |

The equipment of both manufacturers is expected to perform at the desired level of (fixed) output. For a five year analysis period, which manufacturer's equipment should be selected? Assume 7% interest and equal maintenance costs. For fixed output, the criterion is to minimize the present worth cost.

*Speedy*

$$\text{PW of cost} = 1500 - 200(P/F,7\%,5) = 1500 - 200(0.713)$$
$$= 1500 - 143 = \$1357$$

*Allied*

$$\text{PW of cost} = 1600 - 325(P/F,7\%,5) = 1600 - 325(0.713)$$
$$= 1600 - 232 = \$1368$$

Since it is only the *differences between alternatives* that are relevant, maintenance costs may be left out of the economic analysis. Although the PW of cost for each of the alternatives is nearly identical, we would, nevertheless, choose the one with minimum present worth of cost unless there were other tangible or intangible differences that would change the decision. Buy the Speedy equipment.

## EXAMPLE 5-4

A firm is trying to decide which of two alternate weighing scales it should install to check a package filling operation in the plant. The scales would allow better control of the filling operation and result in less overfilling. If both scales have lives equal to the six year analysis period, which one should be selected? Assume an 8% interest rate.

| Alternative | Cost | Uniform annual benefit | End of useful life salvage value |
|---|---|---|---|
| Atlas scales | $2000 | $450 | $100 |
| Tom Thumb scales | 3000 | 600 | 700 |

*Atlas scales:*

$$\text{PW of benefits} - \text{PW of cost} = 450(P/A,8\%,6) + 100(P/F,8\%,6) - 2000$$
$$= 450(4.623) + 100(0.6302) - 2000$$
$$= 2080 + 63 - 2000 = \$143$$

*Tom Thumb scales:*

$$\text{PW of benefits} - \text{PW of cost} = 600(P/A,8\%,6) + 700(P/F,8\%,6) - 3000$$
$$= 600(4.623) + 700(0.6302) - 3000$$
$$= 2774 + 441 - 3000 = \$215$$

The salvage value of the scales, it should be noted, is simply treated as another benefit of the alternative. Since the criterion is to maximize the present worth of benefits minus the present worth of cost, the Tom Thumb scales is the preferred alternative.

## Net Present Worth

In Example 5-4 we compared two alternatives and selected the one where present worth of benefits minus present worth of cost was a maximum. This criterion is called the *Net Present Worth* criterion and written simply as NPW.

Net present worth = present worth of benefits − present worth of cost
$$NPW = PW \text{ of benefits} - PW \text{ of cost}$$

## Useful Lives Different from the
## Analysis Period

In present worth analysis there always must be an identified analysis period. It follows then that each alternative must be considered for the entire period. In the previous examples the useful life of each alternative was equal to the analysis period. Often we can arrange it this way. But there will be many more situations where the alternatives have useful lives different from the analysis period. This section will examine the problem and describe how to overcome this difficulty.

Suppose that in Example 5-3 the Allied equipment was expected to have a ten year useful life or twice that of the Speedy equipment. Assuming the Allied salvage value would still be $325 ten years hence, which equipment should now be purchased? We will recompute the present worth of cost of the Allied equipment.

*Allied.*

$$PW \text{ of cost} = 1600 - 325(P/F, 7\%, 10) = 1600 - 325(0.5083)$$
$$= 1600 - 165 = \$1435$$

The present worth of cost has increased. This is, of course, due to the more distant recovery of the salvage value. More importantly, we now find ourselves attempting to compare Speedy equipment, with its five-year life, against the Allied equipment with a ten-year life. This variation in the useful life of the equipment means we no longer have a situation of *fixed output*. Speedy equipment in the mailroom for five years is certainly not the same as ten years of service with Allied equipment. For present worth calculations it is important that we select an analysis period and judge the consequences of each of the alternatives during the selected analysis period.

The analysis period for an economy study should be determined from the situation. In some industries with rapidly changing technologies a rather short analysis period or planning horizon might be in order. Industries with more stable technologies (like steel making) might use a longer period (say 10–20 years) while government agencies frequently use analysis periods extending to 50 years or more.

Not only is the firm and its economic environment important in selecting an analysis period, but also the specific situation being analyzed is important. If the Allied equipment of Example 5-3 has a useful life of ten years, and the Speedy equipment will last five years, one obvious solution is to select a ten-year analysis period. The choice of the least common multiple analysis period in this case means that we would compare the ten-year life of Allied equipment against an initial purchase of Speedy equipment plus its replacement with new Speedy equipment in five years. The result is to judge the alternatives on the basis of a ten-year requirement in the mailroom. On this basis the economic analysis is as follows.

*Speedy.*

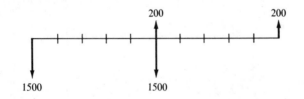

PW of cost $= 1500 + (1500 - 200)(P/F,7\%,5) - 200(P/F,7\%,10)$
$= 1500 + 1300(0.7130) - 200(0.5083)$
$= 1500 + 927 - 102 = \$2325$

*Allied.*

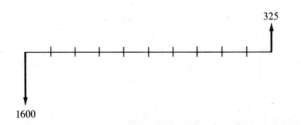

PW of cost $= 1600 - 325(P/F,7\%,10) = 1600 - 325(0.5083) = \$1435$

For the fixed output of ten years of service in the mailroom, the Allied equipment, with its smaller present worth of cost, is preferred.

We have seen that setting the analysis period equal to the least common multiple of the lives of the two alternatives seems reasonable in the revised Example 5-3. What would one do, however, if in another situation the alternatives had useful lives of 7 and 13 years, respectively? Here the least common multiple of lives is 91 years. An analysis period of 91 years hardly seems realistic. Instead, a suitable analysis period should be based on how long the equipment is likely to be needed. This may require that terminal values be estimated for the alternatives at some point prior to the end of their useful lives. Figure 5-1 graphically represents this concept. As Figure

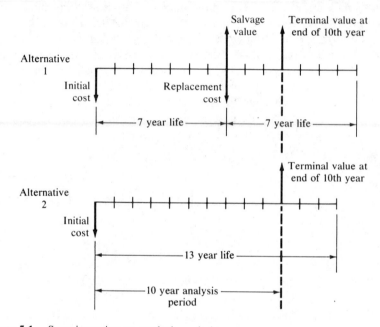

**Figure 5-1.**   Superimposing an analysis period
on 7 and 13 year alternatives.

5-1 indicates, it is not necessary for the analysis period to equal the useful life of an alternative or some multiple of the useful life. To properly reflect the situation at the end of the analysis period an estimate is required of the market value of the equipment at that time. The calculations might be easier if everything came out even, but it is not essential.

## Infinite Analysis Period—Capitalized Cost

Another difficulty in present worth analysis arises when we encounter an infinite analysis period ($n = \infty$). In governmental analyses there are at times circumstances where a service or condition is to be maintained for an infinite period. The need for roads, dams, pipelines, or whatever are sometimes considered permanent. In these situations a present worth of cost analysis would have an infinite analysis period. We call this particular analysis *capitalized cost*.

Capitalized cost is the present sum of money that would need to be set aside now at some interest rate to yield the funds required to provide the service or whatever indefinitely. To accomplish this, means that the money set aside for future expenditures must not decline. The interest received on the money set aside can be spent, but not the principal. When one stops to think about an infinite analysis period (as opposed to something relatively short like 100 years) we see that an undiminished principal sum is essential, otherwise one will of necessity run out of money prior to infinity.

In Chapter 4 we saw that:

Principal sum + interest for the period = amount at end of period

$$P \quad + \quad iP \quad = P + iP$$

If we spend $iP$, then in the next interest period the principal sum $P$ will again increase to $P + iP$. Thus we can again spend $iP$. This concept may be illustrated by a numerical example. Suppose you deposited $200 in a bank that paid 4% interest annually. How much money could be withdrawn each year without reducing the balance in the account below the initial $200?

At the end of the first year the $200 would earn 4%($200) = $8 interest. If this interest were withdrawn, the $200 would remain in the account. At the end of the second year the $200 balance would again earn 4%($200) = $8. This $8 could also be withdrawn and the account would still have $200. This procedure could be continued indefinitely and the bank account would always contain $200. The year-by-year situation would be as follows:

$200
initial $\longrightarrow$ 200 + 8 = 208
$P$
  withdrawal $iP$ =  8
       200 $\longrightarrow$ 200 + 8 = 208   Year two
            withdrawal $iP$ =  8
                200 $\longrightarrow$ 200 + 8 = 208   Year three
                     withdrawal $iP$ =  8
                           200

and so on

Thus for an initial present sum $P$ there can be an end-of-period withdrawal of $A$ equal to $iP$ each period, and these withdrawals may continue forever without diminishing the initial sum $P$. This gives us the basic relationship:

for $n = \infty$   $A = iP$

This relationship is the key to capitalized cost calculations. We previously defined capitalized cost as the present sum of money that would need to be set aside at some interest rate to yield the funds to provide the desired task or service forever. Capitalized cost is therefore the $P$ in the equation $A = iP$. If we can resolve the desired task or service into an equivalent $A$, the capitalized cost may be computed. The following examples illustrate the computations.

*EXAMPLE 5-5*

How much should one set aside to pay for $50 per year maintenance on a gravesite if interest is assumed to be 4%? For perpetual maintenance the principal sum must remain undiminished after making the annual disbursement.

$$\text{Capitalized cost } P = \frac{\text{Annual disbursement } A}{\text{Interest rate } i}$$

$$P = \frac{50}{0.04} = \$1250$$

*EXAMPLE 5-6*

A city plans a pipeline to transport water from a distant watershed area to the city. The pipeline will cost $8 million and have an expected life of 70 years. The city anticipates it will need to keep the water line in service indefinitely. Compute the capitalized cost assuming 7% interest.

We have the capitalized cost equation

$$P = \frac{A}{i}$$

that is simple to apply when there are end-of-period disbursements $A$. Here we have renewals of the pipeline every 70 years. To compute the capitalized cost, it is necessary to first compute an end-of-period disbursement $A$ that is equivalent to $8 million every 70 years.

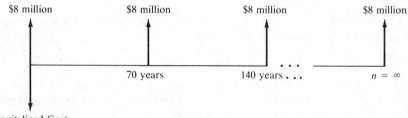

The $8 million disbursement at the end of 70 years may be resolved into an equivalent $A$.

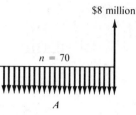

$$A = F(A/F,i\%,n) = \$8 \text{ million} (A/F,7\%,70)$$
$$= \$8 \text{ million} (0.0006) = \$4800$$

Each 70 year period is identical to this one and the infinite series is shown in Figure 5-2.

$$\text{Capitalized cost } P = \$8 \text{ million} + \frac{A}{i} = \$8 \text{ million} + \frac{4800}{0.07}$$

$$= \$8,069,000$$

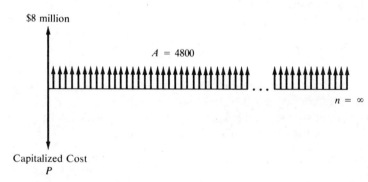

**Figure 5-2.** Infinite series computed using the sinking fund factor.

*Alternate Solution*

Instead of solving for an equivalent end-of-period payment $A$ based on a *future* $8 million disbursement, we could find $A$, given a *present* $8 million disbursement.

$$A = P(A/P,i\%,n) = \$8 \text{ million}(A/P,7\%,70)$$
$$= \$8 \text{ million}(0.0706) = \$564,800$$

On this basis the infinite series is shown in Figure 5-3. Carefully note the difference between this and Figure 5-2. Now:

$$\text{Capitalized cost } P = \frac{A}{i} = \frac{564,800}{0.07} = \$8,069,000$$

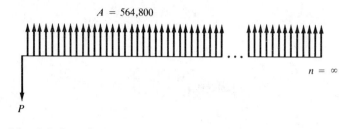

**Figure 5-3.**   Infinite series computed using the
capital recovery factor.


**Multiple Alternatives**

So far in the chapter the discussion has been based on examples with
only two alternatives. But multiple alternative problems may be solved
by exactly the same methods employed for problems with two alternatives.
The only reason for avoiding multiple alternatives has been to simplify the
examples. Examples 5-7 and 5-8 have multiple alternatives.


*EXAMPLE 5-7*

A contractor has been awarded the contract to construct a six mile long
tunnel in the mountains. During the five year construction period the
contractor will need water from a nearby stream. He will construct a
pipeline to convey the water to the main construction yard. An analysis
of costs for various pipe sizes is as follows:

|  | Pipe size | | | |
|---|---|---|---|---|
|  | *2"* | *3"* | *4"* | *6"* |
| Installed cost of pipeline and pump | $22,000 | 23,000 | 25,000 | 30,000 |
| Cost per hour for pumping | $1.20 | 0.65 | 0.50 | 0.40 |

The pipe and pump will have a salvage value at the end of five years
equal to the cost to remove them. The pump will operate 2000 hours per
year. The lowest interest rate at which the contractor is willing to invest
money is 7%. The minimum required interest rate for invested money is
called the minimum attractive rate of return (MARR). We can compute
the present worth of cost for each alternative.

For each pipe size the present worth of cost is equal to the installed
cost of the pipeline and pump plus the present worth of 5 years of pumping costs.

| | Pipe size | | | |
|---|---|---|---|---|
| | 2" | 3" | 4" | 6" |
| Installed cost of pipeline and pump | $22,000 | $23,000 | $25,000 | $30,000 |
| $1.20 \times 2000$ hrs $\times (P/A,7\%,5yr)$ | 9,840 | | | |
| $0.65 \times 2000$ hrs $\times 4.10$ | | 5,330 | | |
| $0.50 \times 2000$ hrs $\times 4.10$ | | | 4,100 | |
| $0.40 \times 2000$ hrs $\times 4.10$ | | | | 3,280 |
| Present worth of cost | $31,840 | $28,330 | $29,100 | $33,280 |

Select the alternative with the least present worth of cost—select the 3" pipe size.

## EXAMPLE 5-8

An investor paid $8000 to a consulting firm to analyze what he might do with a small parcel of land on the edge of town that can be bought for $30,000. In their report the consultants suggested four alternatives:

| Alternative | Total investment including land* | Uniform net annual benefit | Terminal value at end of 20 yrs |
|---|---|---|---|
| A: do nothing | $ 0 | $ 0 | $ 0 |
| B: vegetable market | 50,000 | 5,100 | 30,000 |
| C: gas station | 95,000 | 10,500 | 30,000 |
| D: small motel | 150,000 | 15,000 | 40,000 |

Assuming 10% is the minimum attractive rate of return, what should the investor do?

Alternative *A* represents the "do-nothing" alternative. Generally, one of the feasible alternatives in any situation is to remain in the present status and do nothing. In this problem the investor could decide that the most attractive alternative is not to purchase the property and develop it. This is clearly a "do-nothing" decision. We note, however, that even if he does nothing the total venture would not be a very satisfactory one. This is due to the fact that the investor spent $8000 for professional advice on the possible uses of the property. But because the $8000 is a past cost, it is a *sunk cost*.† It should not deter the investor from making the best decision now irrespective of the costs that brought him to this situation and point of time.

---

*Includes the land and structures but does not include the $8000 fee to the consulting firm.

†Sunk cost is the name given to past costs. The only relevant costs in an economic analysis are present and future costs. Past events and past costs are gone and cannot be allowed to affect future planning. The only place where past costs may be relevant is in computing depreciation charges and income taxes.

This problem is one of neither fixed input nor fixed output, so our criterion will be to maximize the present worth of benefits minus the present worth of cost. Or more simply stated, maximize net present worth.

*Alternative A: do nothing*

   NPW = 0

*Alternative B: vegetable market*

   NPW $= -50,000 + 5100(P/A,10\%,20) + 30,000(P/F,10\%,20)$
   $= -50,000 + 5100(8.514) + 30,000(0.1486)$
   $= -50,000 + 43,420 + 4460$
   $= -2120$

*Alternative C: gas station*

   NPW $= -95,000 + 10,500(P/A,10\%,20) + 30,000(P/F,10\%,20)$
   $= -95,000 + 89,400 + 4460$
   $= -1140$

*Alternative D: small motel*

   NPW $= -150,000 + 15,000(P/A,10\%,20) + 40,000(P/F,10\%,20)$
   $= -150,000 + 127,710 + 5940$
   $= -16,350$

The criterion is to maximize net present worth. In this situation one alternative has NPW equal to zero and three alternatives have negative values for NPW. We will select the best of the four alternatives, namely the do-nothing alternative $A$ with NPW equal to zero.

## EXAMPLE 5-9

A piece of land may be purchased for $610,000 to be strip mined for the underlying coal. Annual net income will be $200,000 per year for ten years. At the end of the ten years the surface of the land will be restored as required by a federal law on strip mining. The cost of reclamation will be $1,500,000 more than the resale value of the land after it is restored. Using a 10% interest rate determine whether the project is desirable.

The investment opportunity may be described by the following cash flow.

| Year | Cash flow |
|------|-----------|
| 0 | −$610 thousand |
| 1–10 | +200 (per year) |
| 10 | −1500 |

$$NPW = -610 + 200(P/A,10\%,10) - 1500(P/F,10\%,10)$$
$$= -610 + 200(6.145) - 1500(0.3855)$$
$$= -610 + 1229 - 578 = +41$$

Since NPW is positive, the project is desirable.

*EXAMPLE 5-10*

Two pieces of construction equipment are being analyzed.

| Year | Alternative A | Alternative B |
|------|---------------|---------------|
| 0 | −$2000 | −$1500 |
| 1 | +1000 | +700 |
| 2 | +850 | +300 |
| 3 | +700 | +300 |
| 4 | +550 | +300 |
| 5 | +400 | +300 |
| 6 | +400 | +400 |
| 7 | +400 | +500 |
| 8 | +400 | +600 |

Based on an 8% interest rate, which alternative should be selected?

*Alternative A*

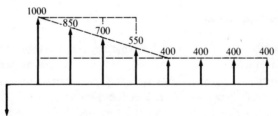

PW of benefits $= 400(P/A,8\%,8) + 600(P/A,8\%,4) - 150(P/G,8\%,4)$
$= 400(5.747) + 600(3.312) - 150(4.650) = 3588.50$

PW of cost $= 2000$

Net present worth $= 3588.50 - 2000 = +1588.50$

*Alternative B*

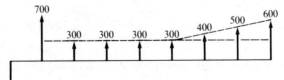

$$\text{PW of benefits} = 300(P/A,8\%,8) + (700 - 300)(P/F,8\%,1)$$
$$+ 100(P/G,8\%,4)(P/F,8\%,4)$$
$$= 300(5.747) + 400(0.9259) + 100(4.650)(0.7350)$$
$$= 2436.24$$

PW of cost $= 1500$

Net present worth $= 2436.24 - 1500 = +936.24$
    To maximize NPW, choose $A$.

## ASSUMPTIONS IN SOLVING ECONOMIC ANALYSIS PROBLEMS

In solving economic analysis problems there are a number of decisions that must be made. In this section six items will be examined.

### End-of-Year Convention

As we indicated in Chapter 4, economic analysis textbooks follow the end-of-year convention. This makes "$A$" a series of end-of-period receipts or disbursements. And in problems we generally assume all series of payments or disbursements occur at the end of the interest period. This is, of course, a very self-serving assumption for it allows us to use values from our compound interest tables without any adjustments.

A cash flow diagram of $P$, $A$, and $F$ for the end-of-period convention is as follows:

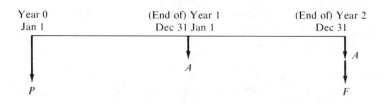

If one were to adopt a middle-of-period convention, the diagram would be:

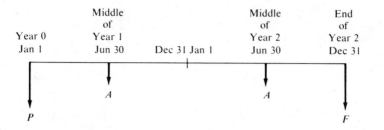

As the diagrams illustrate, only *A* shifts; *P* remains at the beginning of period and *F* at the end of period irrespective of the convention. The compound interest tables in the Appendix are based on the end-of-period convention.

### Viewpoint of Economic Analysis Studies

When we make economic analysis calculations we must proceed from a point of reference. Generally, we will want to take the point of view of a total firm when doing industrial economic analyses. Example 3-1 vividly illustrated the problem. In the example, the shipping department of a firm decided it could save money by having its printing work done by a commercial printer rather than sending it to the in-house printing department. An analysis, from the viewpoint of the shipping department revealed that this was correct. The shipping department could get for $298.50 the same printing it had been charged $397.20 for by the in-house printing department. But further analysis showed that the printing department costs would decline less than the commercial printer would charge. From the viewpoint of the firm the net result would be an increase in total cost. From Example 3-1 we see it *is* important that the viewpoint of the study be carefully considered. Selecting a narrow viewpoint, like the shipping department, may result in a suboptimal decision from the viewpoint of the firm. For this reason the viewpoint of the total firm is the normal point of reference in industrial economic analyses.

### Sunk Costs

We know that it is the *differences between alternatives* that are relevant in economic analysis. Events that have occurred in the past really have no bearing on what we should do in the future. When the judge says, "$30 fine or 3 days in jail" the events that led to these unhappy alternatives really are unimportant. What *is* important are the current and future differences between the alternatives. Past costs, like past events, have no bearing on deciding between alternatives unless the past costs somehow affect the present or future costs. In general, past costs do not affect the present or the future and so we refer to them as *sunk costs* and disregard them.

## Borrowed Money Viewpoint

In most economic analyses the various alternatives require spending money. And so it is natural to ask the source of the money. Obviously the source will vary from situation to situation. The total problem is in fact two problems. One can be called the financing (or obtaining money) problem and the other the investment (or spending money) problem. Experience has shown that the problems should be separated. When separated, the problems of obtaining money and of spending it are both logical and straight forward. Failure to separate the problems sometimes produces confusing problems and poor decision-making. The conventional assumption in economic analysis is that the required money is borrowed. With this assumption it is reasonable to use the interest rate on borrowed money as the interest rate in economic analyses. We will later examine the interest rate question in much greater depth.

## Effect of Inflation and Deflation

For the present we will assume that prices are stable. This means that a machine that costs $5000 today can be expected to cost the same amount several years hence. Inflation and deflation is a serious problem in many situations, but we will disregard it for now.

## Income Taxes

This aspect of economic analyses, like inflation-deflation, must be considered if a realistic analysis is to be done. We will defer our introduction of income taxes into economic analyses until later.

## SUMMARY

Present worth analysis is suitable for almost any economic analysis problem. But it is particularly desirable when we wish to know the present worth of future costs and benefits. And we frequently want to know the value today of such things as income producing assets, stocks, and bonds.

For present worth analysis the proper economic criteria are:

Fixed input                    Maximize the PW of benefits

Fixed output          Minimize the PW of costs

Neither input nor      Maximize (PW of benefits − PW of costs)
    output fixed           or more simply stated: Maximize NPW

To make valid comparisons we need to analyze each alternative in a problem over the same analysis period or planning horizon. If the alternatives do not have equal lives, some technique must be used to achieve a common analysis period. One method is to select an analysis period equal to the least common multiple of the alternative lives. Another method is to select an analysis period and then compute end-of-analysis-period salvage values for the alternatives.

Capitalized cost is the present worth of cost for an infinite analysis period ($n = \infty$). When $n = \infty$, the fundamental relationship between $A$, $P$, and $i$ is $A = iP$. Some form of this equation is used whenever there is a problem with an infinite analysis period.

There are a number of assumptions that are routinely made in solving economic analysis problems. They include the following.

1. Present sums $P$ are beginning of period and all series payments or disbursements $A$ and future sums $F$ occur at the end of the interest period. The compound interest tables were derived on this basis.

2. On industrial economic analyses, the appropriate point of reference from which to compute the consequences of alternatives is the total firm. Taking a narrower view of the consequences can result in suboptimal solutions.

3. Only the differences between the alternatives are relevant. Past costs are *sunk costs* and generally do not affect present or future costs. For this reason they are ignored.

4. The investment problem should be isolated from the financing problem. We generally assume that all required money is borrowed at interest rate $i$.

5. For now stable prices are assumed. The problem of inflation-deflation is deferred to Chapter 13. Similarly, income taxes are deferred to Chapter 11.

## PROBLEMS

**5-1** On February 1 the Miro Company needs to purchase some office equipment. The company is presently short of cash and expects to be short for several months. The company treasurer has indicated that he could pay for the equipment as follows.

| Date | Payment |
|------|---------|
| April 1 | $150 |
| June 1 | 300 |
| Aug 1 | 450 |
| Oct 1 | 600 |
| Dec 1 | 750 |

A local office supply firm has been contacted and they will agree to sell the equipment to the firm now and to be paid according to the treasurer's payment schedule. If interest will be charged at 3% every two months, with compounding once every two months, how much office equipment can the Miro Company buy now? (*Answer:* $2020)

**5-2** A rather wealthy man decided he would like to arrange for his descendants to be well educated. He would like each child to have $20,000 for his or her education. He plans to set up a perpetual trust fund so that six children will receive this assistance each generation. He estimates that there will be four generations per century, spaced 25 years apart. He expects the trust to be able to obtain a 4% rate of return, and the first recipients to receive the money ten years hence. How much money should he now set aside in the trust? (*Answer:* $129,700)

**5-3** How much would the owner of a building be justified in paying for a sprinkler system that will save $750 a year in insurance premiums if the system has to be replaced every 20 years, and has a salvage value equal to 10% of its initial cost? Assume money is worth 7%. (*Answer:* $8156)

**5-4** A man had to have the muffler replaced on his two-year-old car. The repairman offered two alternatives. For $30 he would install a muffler guaranteed for two years. But for $45 he would install a muffler guaranteed "for as long as you own the car." Assuming the present owner expects to keep the car for about three more years, which muffler would you advise him to have installed if you thought 20% were a suitable interest rate and the less expensive muffler would only last two years?

**5-5** A consulting engineer has been engaged to advise a town how best to proceed with the construction of a 200,000 acre-foot water supply reservoir. Since only 120,000 acre-feet of storage will be required for the next 25 years, an alternative to building the full capacity now is to build the reservoir in two stages. Initially, the reservoir could be built with 120,000 acre-feet of capacity and then 25 years hence the additional 80,000 acre-feet of capacity could be added by increasing the height of the reservoir. Estimated costs are as follows.

| *Build in two stages* | Construction cost | Annual maintenance cost |
|------------------------|-------------------|-------------------------|
| First stage: | | |
|   120,000 acre-foot reservoir | $14,200,000 | $75,000 |
| Second stage: | | |
|   Add 80,000 acre-feet of capacity, | | |
|     additional construction and | | |
|     maintenance costs | 12,600,000 | 25,000 |

| Build full capacity now | Construction cost | Annual maintenance |
|---|---|---|
| 200,000 acre-foot reservoir | $22,400,000 | $100,000 |

If interest is computed at 4%, which construction plan is preferred?

**5-6** An engineer has received two bids for an elevator to be installed in a new building. The bids, plus his evaluation of the elevators, are as follows:

| | *Bids* | | *Engineer's estimates* | |
|---|---|---|---|---|
| *Alternative* | *Installed cost* | *Service life* | *Annual operating cost including repairs* | *Salvage value at end of service life* |
| Westinghome | $45,000 | 10 yrs | $2700/yr | $3000 |
| Itis | 54,000 | 15 yrs | 2850/yr | 4500 |

The engineer will make a present worth analysis using a 10% interest rate. Prepare the analysis and determine which bid should be accepted.

**5-7** A railroad branch line is to be constructed to a missile site. It is expected the railroad line will be used for 15 years, after which the missile site will be removed and the land turned back to agricultural use. The railroad track and ties will be removed at that time. In building the railroad line either treated or untreated wood ties may be used. Treated ties have an installed cost of $6 and a ten year life; untreated ties are $4.50 with a six year life. If at the end of fifteen years the ties then in place have a remaining useful life of four years or more, they will be used by the railroad elsewhere and have an estimated salvage value of $3 each. Anytime ties are removed that are at the end of their service life, or are too close to the end of their service life to be used elsewhere, they are sold for $0.50 each. Determine the most economical plan for the initial railroad ties and their replacement for the fifteen year period. Make a present worth analysis assuming 8% interest.

**5-8** A weekly business magazine offers a one-year subscription for $10 and a three-year subscription for $20. If you thought you would read the magazine for at least the next three years, and consider 20% as a minimum rate of return, which way would you purchase the magazine, with three one-year subscriptions or a single three-year subscription? (*Answer:* Choose the three-year subscription)

**5-9** A manufacturer is considering purchasing equipment which will have the following financial effects:

| Year | Disbursements | Receipts | |
|---|---|---|---|
| 0 | $4400 | $ 0 | -4400 |
| 1 | 660 | 880 | -220 |
| 2 | 660 | 1980 | 1320 |
| 3 | 440 | 2420 | 1980 |
| 4 | 220 | 1760 | 1540 |

If money is worth 6%, should he invest in the equipment?

**5-10** Jerry Stans, a young industrial engineer, prepared an economic analysis for some equipment to replace one production worker. The analysis showed that the present worth of benefits (of employing one less production worker) just equaled the present worth of the equipment costs, based on a ten year useful life for the equipment. It was decided not to purchase the equipment. A short time later, the production workers won a new three year union contract that granted them an immediate 40¢ per hour wage increase, plus an additional 25¢ per hour wage increase in each of the two subsequent years. Assume that in all future years a 25¢ per hour per year wage increase will be granted. Jerry Stans has been asked to revise his earlier economic analysis. The present worth of benefits of replacing one production employee will now increase. Assuming an interest rate of 8%, the justifiable cost of the automation equipment (with a ten year useful life) will increase by how much? Assume the plant operates a single eight-hour shift, 250 days per year.

**5-11** The management of an electronics manufacturing firm believes it is desirable to install some automation equipment in their production facility. They believe the equipment would have a ten-year life with no salvage value at the end of ten years. The plant engineering department has surveyed the plant and suggested there are eight mutually exclusive alternatives available.

| Plan | Initial cost | Net annual benefit |
|------|-------------|--------------------|
| 1 | $265 thousands | $51 thousands |
| 2 | 220 | 39 |
| 3 | 180 | 26 |
| 4 | 100 | 15 |
| 5 | 305 | 57 |
| 6 | 130 | 23 |
| 7 | 245 | 47 |
| 8 | 165 | 33 |

If the firm expects a 10% rate of return, which alternative, if any, should they adopt? (*Answer:* Plan 1)

**5-12** The president of the E. L. Echo Corporation thought it would be appropriate for his firm to "endow a chair" in the Industrial Engineering Department of the local university; that is, he was considering making a gift to the university of sufficient money to pay the salary of one professor forever. One professor in the department would be designated the E. L. Echo Professor of Industrial Engineering, and his salary would come from the fund established by the Echo Corporation. If the professor will receive $21,000 per year, and the interest received on the endowment fund is expected to remain at 4.5%, what lump sum of money will the Echo Corporation need to provide to establish the endowment fund? (*Answer:* $466,700)

**5-13** A man who likes cherry blossoms very much would like to have an urn full of them put on his grave once each year forever after he dies. In his will he will leave a certain sum of money in the trust of a local bank to pay the florist's annual bill. How much money should be left for this purpose? Make whatever assumptions you feel are justified by the facts presented. State your assumptions, and compute a solution.

**5-14** A local symphony association offers memberships as follows:

  Continuing membership        $10 per year
  Patron lifetime membership   $250

The patron membership has been based on the symphony association's belief that it can obtain a 4% rate of return on its investment. If you believed 4% to be an appropriate rate of return, would you be willing to purchase the patron membership? Explain why or why not.

**5-15** A battery manufacturing plant has been ordered to cease discharging acidic waste liquids containing mercury into the city sewer system. As a result the firm must now adjust the pH and remove the mercury from its waste liquids. Three firms have provided quotations on the necessary equipment. An analysis of the quotations provided the following table of costs.

| Bidder | Installed cost | Annual operating cost | Annual income from mercury recovery | Salvage value |
|---|---|---|---|---|
| Foxhill Instrument | $35,000 | $8,000 | $2,000 | $20,000 |
| Quicksilver | 40,000 | 7,000 | 2,200 | 0 |
| Almaden | 100,000 | 2,000 | 3,500 | 0 |

If the installation can be expected to last 20 years and money is worth 7%, which equipment should be purchased? (*Answer:* Almaden)

**5-16** A firm is considering three mutually exclusive alternatives as part of a production improvement program. The alternatives are:

|  | A | B | C |
|---|---|---|---|
| Installed cost | $10,000 | $15,000 | $20,000 |
| Uniform annual benefit | 1,625 | 1,530 | 1,890 |
| Useful life | 10 years | 20 years | 20 years |

For each alternative the salvage value at the end of its useful life is zero. At the end of 10 years *A* could be replaced with another *A* with identical cost and benefits. The minimum attractive rate of return is 6%. Which alternative should be selected?

**5-17** A steam boiler is needed as part of the design of a new plant. The boiler can be fired either by natural gas, fuel oil, or coal. A decision must be made on which fuel to use. An analysis of the costs shows that the installed cost, with all controls, would be least for natural gas at $30,000; for fuel oil it would be $55,000 and for coal it would be $180,000. If natural gas is used rather than fuel oil, the annual fuel cost will increase by $7500. If coal is used rather than fuel oil, the annual fuel cost will be $15,000 per year less. Assuming 8% interest, a 20-year analysis period, and no salvage value, which is the most economical installation?

**5-18** An investor has carefully studied a number of companies and their common stock. From his analysis he has decided that the stocks of six firms are the best of the many he has examined. They represent about the same amount of risk and so he would like to determine the one in which to invest. He plans to keep the stock for four years, and requires a 10% minimum attractive rate of return.

| Common stock | Price/Share | Annual end-of-year dividend/Share | Estimated price at end of 4 years |
|---|---|---|---|
| Western House | $24\frac{1}{2}$ | $1.25 | $32 |
| Fine Foods | 45 | 4.50 | 45 |
| Mobile Motors | $30\frac{5}{8}$ | 0 | 42 |
| Trojan Products | 12 | 0 | 20 |
| U.S. Tire | $33\frac{3}{8}$ | 2.00 | 40 |
| Wine Products | $52\frac{1}{2}$ | 3.00 | 60 |

Which stock, if any, should the investor consider purchasing? (*Answer:* Trojan Products)

# 6
# Annual
# Cash Flow
# Analysis

This chapter is devoted to annual cash flow analysis—the second of the three major analysis techniques. As has been said previously, alternatives must be resolved into a form so they may be compared. This means we must use the equivalence concept to convert from a cash flow representing the alternative into some equivalent sum or equivalent cash flow.

In present worth analysis we resolved an alternative into an equivalent cash sum. This might have been an equivalent present worth of cost, an equivalent present worth of benefit, or an equivalent net present worth. But instead of computing equivalent present sums we could compare alternatives based on their equivalent annual cash flows. Depending on the particular situation we may wish to compute the equivalent uniform annual cost (EUAC), the equivalent uniform annual benefit (EUAB), or their difference (EUAB − EUAC).

To prepare for a discussion of annual cash flow analysis we will begin by reviewing some annual cash flow calculations, and then examine annual cash flow criteria. Following this, we will proceed with annual cash flow analysis.

## ANNUAL CASH FLOW
## CALCULATIONS

### Resolving a Present Cost to an
### Annual Cost

In prior chapters equivalence techniques were used to convert money at one point in time to some equivalent sum or series. In annual cash flow analysis the goal will be to convert money to an equivalent uniform annual cost or benefit. The simplest case would be to convert a present sum $P$ to a series of equivalent uniform end-of-period cash flows. This is illustrated in Example 6-1.

### EXAMPLE 6-1

A woman bought $1000 worth of furniture for her home. If she expects it to last ten years, what will be her equivalent uniform annual cost if interest is 7%

$P = 1000$        $A$ $A$ $A$ $A$ $A$ $A$ $A$ $A$ $A$ $A$     $n = 10$ years

$i = 7\%$

$$\text{Equivalent uniform annual cost} = P(A/P, i\%, n)$$
$$= 1000(A/P, 7\%, 10)$$
$$= \$142.40$$

### Treatment of Salvage Value

In a situation where there is a salvage value, or future value at the end of the useful life of an asset, the result is to decrease the equivalent uniform annual cost.

### EXAMPLE 6-2

The woman in Example 6-1 now believes she can resell the furniture at the end of ten years for $200. Under these circumstances what is her equivalent uniform annual cost?

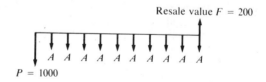

Resale value $F = 200$

$A$ $A$ $A$ $A$ $A$ $A$ $A$ $A$ $A$ $A$

$P = 1000$

For this situation the problem may be solved by any of three different calculations.

1. EUAC $= P(A/P,i\%,n) - F(A/F,i\%,n)$       (6-1)
   $= 1000(A/P,7\%,10) - 200(A/F,7\%,10)$
   $= 1000(0.1424) - 200(0.0724)$
   $= 142.40 - 14.48 = \$127.92$

This method reflects the annual cost of the cash disbursement minus the annual benefit of the future resale value.

2. The Eq. 6-1 relationship may be modified by an identity presented in Chapter 4.

$$(A/P,i\%,n) = (A/F,i\%,n) + i \qquad (6\text{-}2)$$

Substituting this into Eq. 6-1 gives:

EUAC $= P(A/F,i\%,n) + Pi - F(A/F,i\%,n)$
    $= (P - F)(A/F,i\%,n) + Pi$       (6-3)
    $= (1000 - 200)(A/F,7\%,10) + 1000(0.07)$
    $= 800(0.0724) + 70 = 57.92 + 70 = \$127.92$

This method computes the equivalent annual cost due to the unrecovered $800 when the furniture is sold and adds annual interest on the $1000 investment.

3. If the value for $(A/F,i\%,n)$ from Eq. 6-2 is substituted into Eq. 6-1, we obtain:

EUAC $= P(A/P,i\%,n) - F(A/P,i\%,n) + Fi$
    $= (P - F)(A/P,i\%,n) + Fi$       (6-4)
    $= (1000 - 200)(A/P,7\%,10) + 200(0.07)$
    $= 800(0.1424) + 14 = 113.92 + 14 = \$127.92$

This method computes the annual cost of the $800 decline in value during the ten years plus interest on the $200 tied up in the furniture as the salvage value.

Example 6-2 illustrated that when there is an initial disbursement $P$ followed by a salvage value $F$, the annual cost may be computed in three different ways.

1. EUAC $= P(A/P,i\%,n) - F(A/F,i\%,n)$
2. EUAC $= (P - F)(A/F,i\%,n) + Pi$
3. EUAC $= (P - F)(A/P,i\%,n) + Fi$

Each of the three calculations gives the same results. In practice the first and last methods are most commonly used.

*EXAMPLE 6-3*

Bill owned a car for five years. One day he wondered what his uniform annual cost for maintenance and repairs had been. He assembled the following data.

| Year | Maintenance and repair cost for year |
|------|--------------------------------------|
| 1    | $ 45                                 |
| 2    | 90                                   |
| 3    | 180                                  |
| 4    | 135                                  |
| 5    | 225                                  |

Compute the equivalent uniform annual cost (EUAC) assuming 7% interest and end-of-year disbursements. The EUAC may be computed for this irregular series of payments in two steps.

1.  Compute the present worth of cost for the five years using single payment present worth factors.
2.  With the PW of cost known, compute EUAC using the capital recovery factor.

$$\begin{aligned}
\text{PW of cost} &= 45(P/F,7\%,1) + 90(P/F,7\%,2) + 180(P/F,7\%,3) \\
&\quad + 135(P/F,7\%,4) + 225(P/F,7\%,5) \\
&= 45(0.935) + 90(0.873) + 180(0.816) + 135(0.763) \\
&\quad + 225(0.713) \\
&= \$531
\end{aligned}$$

$$\text{EUAC} = 531(A/P,7\%,5) = 531(0.244) = \$130$$

*EXAMPLE 6-4*

Bill reexamined his calculations and found that he had reversed the year three and four maintenance and repair costs in his table. The correct table is:

| Year | Maintenance and repair cost for year |
|------|--------------------------------------|
| 1    | $ 45                                 |
| 2    | 90                                   |
| 3    | 135                                  |
| 4    | 180                                  |
| 5    | 225                                  |

Recompute the EUAC.

This time the schedule of disbursements is a gradient series plus a uniform annual cost as follows:

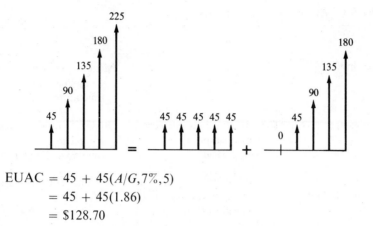

$$EUAC = 45 + 45(A/G,7\%,5)$$
$$= 45 + 45(1.86)$$
$$= \$128.70$$

(Since the timing of the Example 6-3 and 6-4 expenditures are different, we would not expect to obtain the same EUAC.)

The examples have shown four essential points concerning cash flow calculations.

1.  There is a direct relationship between the present worth of cost and the equivalent uniform annual cost. It is

    $$EUAC = (PW \text{ of cost})(A/P,i\%,n).$$

2.  In a problem, an expenditure of money increases the EUAC, while a receipt of money (like selling something for its salvage value) decreases EUAC.

3.  When there are irregular cash disbursements over the analysis period, a convenient method of solution is to first determine the PW of cost. Then using the equation in item 1 above, the EUAC may be calculated.

4.  Where there is an increasing uniform gradient, EUAC may be rapidly computed using the gradient to uniform series factor, $(A/G,i\%,n)$.

## ANNUAL CASH FLOW ANALYSIS

For economic efficiency the criteria are presented in Table 6-1. One notices immediately that the table is quite similar to the one in Chapter 5. In the case of fixed input, for example, the present worth criterion is "maximize PW of benefits" and the annual cost criterion is "maximize equivalent uniform annual benefit." It is apparent that if you are maximizing the present worth of benefits, you at the same time must be maximizing the equivalent uniform annual benefit. This is illustrated in Example 6-5.

**Table 6-1.** Annual cash flow analysis

|  | *Situation* | *Criterion* |
|---|---|---|
| Fixed input | Amount of money or other input resources are fixed | Maximize equivalent uniform annual benefits (maximize EUAB) |
| Fixed output | There is a fixed task, benefit, or other output to be accomplished | Minimize equivalent uniform annual cost (minimize EUAC) |
| Neither input nor output fixed | Neither amount of money or other inputs nor amount of the benefits or other outputs are fixed | Maximize (EUAB − EUAC) |

## EXAMPLE 6-5

A firm is considering which of two devices to install to reduce costs in a particular situation. Both devices cost $1000 and have useful lives of five years and no salvage value. Device $A$ can be expected to result in $300 savings annually. Device $B$ will provide cost savings of $400 the first year, but will decline $50 annually, making the second year savings $350, the third year savings $300, and so forth. With interest at 7%, which device should the firm purchase?

*Device A*

   EUAB = $300

*Device B*

   EUAB = $400 - 50(A/G,7\%,5) = 400 - 50(1.865) = \$306.75$

To maximize EUAB, select device $B$.

Example 6-5 was previously presented as Example 5-1 where we found:

*Device A*

   PW of benefits $= 300(P/A,7\%,5) = 300(4.100) = \$1230$

This may be converted to EUAB by multiplying by the capital recovery factor:

   EUAB $= 1230(A/P,7\%,5) = 1230(0.2439) = \$300$

*Device B*

   PW of benefits $= 400(P/A,7\%,5) - 50(P/G,7\%,5)$
   $= 400(4.100) - 50(7.647) = \$1257.65$

and hence

$$EUAB = 1257.65(A/P,7\%,5) = 1257.65(0.2439) = \$306.75$$

We see, therefore, that it is easy to convert the present worth analysis results into the annual cash flow analysis results. We could go from annual cash flow to present worth just as easily using the series present worth factor.

### EXAMPLE 6-6

Three alternatives are being considered for improving an operation on the assembly line. The cost of the equipment varies as do their annual benefits compared to the present situation. Each of the alternatives has a ten-year life and a scrap value equal to 10% of its original cost.

|  | Plan A | Plan B | Plan C |
|---|---|---|---|
| Installed cost of equipment | $15,000 | $25,000 | $33,000 |
| Material and labor savings per year | 14,000 | 9,000 | 14,000 |
| Annual operating expenses | 8,000 | 6,000 | 6,000 |
| End-of-useful life scrap value | 1,500 | 2,500 | 3,300 |

If interest is 8%, which alternative, if any, should be adopted?

Since neither installed cost nor output benefits are fixed, the economic criterion is to maximize (EUAB − EUAC).

|  | Plan A | Plan B | Plan C | Do nothing |
|---|---|---|---|---|
| *Equivalent uniform annual benefit* (EUAB): |  |  |  |  |
| Material and labor per year | $14,000 | $9,000 | $14,000 | $0 |
| Scrap value $(A/F,8\%,10)$ | 104 | 172 | 228 | 0 |
| EUAB = | $14,104 | $9,172 | $14,228 | $0 |
| *Equivalent uniform annual cost* (EUAC): |  |  |  |  |
| Installed cost $(A/P,8\%,10)$ | $ 2,235 | $3,725 | $ 4,917 | $0 |
| Annual operating expenses | 8,000 | 6,000 | 6,000 | 0 |
| EUAC = | $10,235 | $9,725 | $10,917 | $0 |
| (EUAB − EUAC) = | $ 3,869 | −$553 | $ 3,311 | $0 |

Based on our criterion of maximize (EUAB − EUAC), plan *A* is the best of the four alternatives. We note, however, that since the do-nothing alternative has (EUAB − EUAC) = 0, it is a more desirable alternative than plan *B*.

## ANALYSIS PERIOD

In the last chapter we saw that the analysis period was an important consideration in computing present worth comparisons. It was essential that a common analysis period be used for each alternative. In annual cash flow comparisons we again have the analysis period question. Example 6-7 will help in examining the problem.

### EXAMPLE 6-7

Two pumps are being considered for purchase. If interest is 7%, which pump should be bought?

|  | Pump A | Pump B |
|---|---|---|
| Initial cost | $7000 | $5000 |
| End-of-useful life salvage value | 1500 | 1000 |
| Useful life | 12 yrs | 6 yrs |

The annual cost for 12 years of pump $A$:
Using Eq. 6-4

$$EUAC = (P - F)(A/P,i\%,n) + Fi$$
$$= (7000 - 1500)(A/P,7\%,12) + 1500(0.07)$$
$$= 5500(0.1259) + 105 = \$797$$

Now compute the annual cost for six years of pump $B$:

$$EUAC = (5000 - 1000)(A/P,7\%,6) + 1000(0.07)$$
$$= 4000(0.2098) + 70 = \$909$$

For a common analysis period of twelve years we need to replace pump $B$ at the end of its six year useful life. If we assume that another pump $B'$ can be obtained, with the same $5000 initial cost, $1000 salvage value and six year life, the cash flow will be as follows.

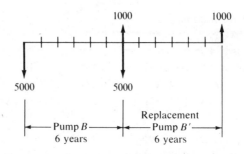

For the 12 year analysis period the annual cost for pump $B$:

$$\begin{aligned}
\text{EUAC} &= [5000 - 1000(P/F,7\%,6) + 5000(P/F,7\%,6) - 1000(P/F,7\%,12)] \\
&\quad \times [A/P,7\%,12] \\
&= [5000 - 1000(0.6663) + 5000(0.6663) - 1000(0.4440)][0.1259] \\
&= [5000 - 666 + 3331 - 444][0.1259] \\
&= (7211)(0.1259) = \$909
\end{aligned}$$

For $B$ the annual cost for the six-year analysis period is the same as the annual cost for the twelve-year analysis period. This is not a surprising conclusion when one recognizes that the annual cost of the first six-year period is repeated in the second six-year period. By assuming that the shorter lived equipment is replaced by equipment with identical economic consequences, we have avoided the analysis period problem. Select pump $A$.

### Analysis Period Equal to Alternative Lives

When the analysis period for an economy study coincides with the useful lives for each alternative we have an ideal situation which causes no difficulties. The economy study is based on this analysis period.

### Analysis Period a Common Multiple of Alternative Lives

When the analysis period is a common multiple of the alternative lives (for example, in Example 6-7 the analysis period was twelve years with six and twelve year alternative lives), a "replacement with an identical item with the same costs, performance, and so forth," is frequently assumed. This means that when an alternative has reached the end of its useful life it is assumed to be replaced with an identical item. As shown in Example 6-7, the result is that the EUAC for pump $B$ with a six year useful life is equal to the EUAC for the entire analysis period based on pump $B$ plus the replacement pump $B'$.

Under these circumstances of identical replacement it is appropriate to compare the annual cash flows computed for alternatives based on their own service lives. In Example 6-7 the annual cost for pump $A$, based on its twelve year service life, was compared with the annual cost for pump $B$, based on its six year service life.

## Analysis Period Is a Continuing Requirement

Many times the economic analysis is to determine how to provide for a more or less continuing requirement. One might need to pump water from a well as a continuing requirement. There is no distinct analysis period. In this situation the analysis period is assumed to be long but undefined. If, for example, we had a continuing requirement to pump water and alternative pump useful lives were seven and eleven years, respectively, what should we do? The customary assumption is that the annual cash flow based on a seven year life of one pump may be compared to the annual cash flow based on the eleven year life of the other pump. And this is done without much concern that the least common multiple of the seven and eleven year lives is 77 years. This comparison of differing lived alternatives assumes identical replacement (with identical costs, performance, and so forth) when an alternative reaches the end of its useful life. Example 6-8 illustrates the situation.

### EXAMPLE 6-8

Pump $B$ in Example 6-7 is now believed to have a nine year useful life. Assuming the same initial cost and salvage value, compare it with pump $A$ using the same 7% interest rate. If we assume that the need for $A$ or $B$ will exist for some continuing period, the comparison of annual costs for the unequal lives is an acceptable technique. For twelve years of pump $A$:

$$EUAC = (7000 - 1500)(A/P, 7\%, 12) + 1500(0.07) = \$797$$

For nine years of pump $B$:

$$EUAC = (5000 - 1000)(A/P, 7\%, 9) + 1000(0.07) = \$684$$

For minimum EUAC, select pump $B$.

### Infinite Analysis Period

At times we have an alternative with a limited (finite) useful life in an infinite analysis period situation. The equivalent uniform annual cost may be computed for the limited life. The assumption of identical replacement (replacements have identical costs, performance, and so forth) is often appropriate. Based on this assumption, the same EUAC occurs for each replacement of the limited life alternative. The EUAC for the infinite analysis period is therefore equal to the EUAC computed for the limited life. With identical replacement,

$$\underset{\substack{\text{for infinite} \\ \text{analysis period}}}{EUAC} = \underset{\substack{\text{for limited} \\ \text{life } n}}{EUAC}$$

A somewhat different situation occurs when there is an alternative with an infinite life in a problem with an infinite analysis period.

$$\text{EUAC}_{\substack{\text{for infinite} \\ \text{analysis period}}} = P(A/P,i\%,\infty) + \text{any other costs}$$

When $n = \infty$, we have $A = Pi$ and hence $(A/P,i\%,\infty)$ equals $i$.

$$\text{EUAC}_{\substack{\text{for infinite} \\ \text{analysis period}}} = Pi + \text{any other costs}$$

## EXAMPLE 6-9

In the construction of the aqueduct to expand the water supply of a city there are two alternatives for a particular portion of the aqueduct. Either a tunnel may be constructed through a mountain or a pipeline laid to go around the mountain. If there is a permanent need for the aqueduct, should the tunnel or the pipeline be selected for this particular portion of the aqueduct? Assume a 6% interest rate.

|  | *Tunnel through mountain* | *Pipeline around mountain* |
|---|---|---|
| Initial cost | $5.5 million | $5 million |
| Maintenance | 0 | 0 |
| Useful life | permanent | 50 yrs |
| Salvage value | 0 | 0 |

*Pipeline*

$$\text{EUAC} = \$5 \text{ million}(A/P,6\%,50)$$
$$= \$5 \text{ million}(0.0634) = \$317,000$$

*Tunnel*

For the tunnel, with its permanent life, we want $(A/P,6\%,\infty)$. For an infinite life the capital recovery is simply interest on the invested capital. So $(A/P,6\%,\infty) = i$

$$\text{EUAC} = Pi = \$5.5 \text{ million}(0.06)$$
$$= \$330,000$$

For fixed output, minimize EUAC. Select the pipeline.

The difference in annual cost between a long life and an infinite life is small unless an unusually low interest rate is used. Here the tunnel was assumed to be permanent. At 6% the annual cost = $5.5 million × 0.06 = $330,000. What would be the annual cost if an 85-year life were assumed for the tunnel?

$$\text{EUAC} = \$5.5 \text{ million}(A/P,6\%,85)$$
$$= \$5.5 \text{ million}(0.0604) = \$332,000$$

The difference in time between 85 years and infinity is a great one indeed, yet the difference in annual costs is slight.

### Some Other Analysis Period

The analysis period in a particular problem may be something other than one of the four we have so far described. It may be equal to the life of the shorter lived alternative, the life of the longer lived alternative, or something entirely different. One must carefully examine the consequences of each alternative throughout the analysis period and in addition see what differences there might be in salvage values, and so forth, at the end of the analysis period.

### SUMMARY

Annual cash flow analysis is the second of the three major methods of resolving alternatives into comparable values. When an alternative has an initial cost $P$ and salvage value $F$ there are three ways of computing the equivalent uniform annual cost:

$$\text{EUAC} = P(A/P,i\%,n) - F(A/F,i\%,n)$$
$$\text{EUAC} = (P - F)(A/F,i\%,n) + Pi$$
$$\text{EUAC} = (P - F)(A/P,i\%,n) + Fi$$

All three equations give the same answer.

The relationship between the present worth of cost and the equivalent uniform annual cost is

$$\text{EUAC} = (\text{PW of cost})(A/P,i\%,n)$$

The three annual cash flow criteria are

| | |
|---|---|
| For fixed input | Maximize EUAB |
| For fixed output | Minimize EUAC |
| Neither input nor output fixed | Maximize (EUAB − EUAC) |

In present worth analysis there must be a common analysis period. Annual cash flow analysis, however, allows some flexibility provided the necessary assumptions are suitable in the situation being studied. The analysis period may be different from the lives of the alternatives, and a valid cash flow analysis made, provided the following two criteria are met.

1.  When an alternative has reached the end of its useful life it is assumed to be replaced by an identical replacement (with the same costs, performance, and so forth).
2.  The analysis period is a common multiple of the useful lives of the alternatives, or there is a continuing or perpetual requirement for the selected alternative.

If both these conditions do not apply, then it is necessary to make a detailed study of the consequences of the various alternatives over the entire analysis period with particular attention to the difference between the alternatives at the end of the analysis period.

There is very little numerical difference between a long lived alternative and a perpetual alternative. As the value of $n$ increases, the capital recovery factor approaches $i$. At the limit

$$(A/P,i\%,\infty) = i$$

## PROBLEMS

**6-1**  A certain industrial firm desires an economic analysis to determine which of two different machines should be purchased. Each machine is capable of performing the same task in a given amount of time. Assume the minimum attractive return is 6%. The following data are to be used in this analysis:

|  | Machine X | Machine Y |
|---|---|---|
| First cost | $5000 | $8000 |
| Estimated life | 5 yrs | 12 yrs |
| Salvage value | 0 | $2000 |
| Annual maintenance cost | 0 | $ 150 |

Which machine would you choose? Base your answer on annual cost. (*Answer:* $X = \$1187$; $Y = \$986$)

**6-2**  An electronics firm invested $60,000 in a precision inspection device. It cost $4000 to operate and maintain in the first year, and $3000 in each of the subsequent years. At the end of four years the firm changed their inspection procedure, eliminating the need for the device. The purchasing agent was very fortunate in being able to sell the inspection device for the $60,000 that had originally been paid for it. The plant manager asks you to compute the equivalent uniform annual cost of the device during the four years it was used. Assume interest at 10% per year. (*Answer:* $9287)

**6-3**  A firm is about to begin pilot plant operation on a process it has developed. One item of optional equipment that could be obtained is a heat exchanger unit. The company finds that one can be obtained now at the very favorable price of $30,000, and that this unit can be used in other company operations. It is estimated that the heat exchanger unit will be worth $35,000 at the end of eight years. This seemingly high salvage value is due primarily to the fact that the $30,000 purchase price is really a rare bargain. If the firm believes 15% is an appropriate rate of return, what annual benefit is needed to justify the purchase of the heat exchanger unit? (*Answer:* $4135)

6-4 The maintenance foreman of a plant in reviewing his records found that a large press had the following maintenance cost record for the last five years:

| | |
|---|---|
| 5 years ago | $600 |
| 4 years ago | 700 |
| 3 years ago | 800 |
| 2 years ago | 900 |
| Last year | 1000 |

After consulting with a lubrication specialist, he changed the preventive maintenance schedule. He believes that this year maintenance will be $900 and will decrease $100 a year in each of the following four years. If his estimate of the future is correct, what will be the equivalent uniform annual maintenance cost for the ten-year period? Assume interest at 8%. (*Answer:* $756)

6-5 A firm purchased some equipment at a very favorable price of $30,000. The equipment resulted in an annual net saving of $1000 per year during the eight years it was used. At the end of eight years the equipment was sold for $40,000. Assuming interest at 8%, did the equipment purchase prove to be desirable?

6-6 A manufacturer is contemplating replacing a production machine tool. The new machine would cost $3700, have a life of four years, have no salvage value, and save him $500 per year in direct labor costs and $200 per year indirect labor costs. The existing machine tool was purchased four years ago at a cost of $4000. It will last four more years and have no salvage value at the end of that time. It could be sold now for $1000 cash. Assume money is worth 8% and that the difference in taxes, insurance, and so forth, for the two alternatives is negligible. Determine whether or not the new machine should be purchased.

6-7 Two possible routes for a power line are under study. Data on the routes are as follows.

| | *Around the lake* | *Under the lake* |
|---|---|---|
| Length | 15 miles | 5 miles |
| First cost | $5000/mile | $25,000/mile |
| Maintenance | $200/mile/yr | $400/mile/yr |
| Useful life | 15 yrs | 15 yrs |
| Salvage value | $3000/mile | $5000/mile |
| Yearly power loss | $500/mile | $500/mile |
| Annual property taxes | 2% of first cost | 2% of first cost |

If 7% interest is used, should the power line be routed around the lake or under the lake? (*Answer:* Around the lake)

6-8 Steve Lowe must pay his property taxes in two equal installments on December 1 and April 1. The two payments are for taxes for the fiscal year that begins on July 1 and ends the following June 30. Steve purchased a home on September 1. He estimates the annual property taxes will be $850 per year. Assuming the annual property taxes remain at $850 per year for the next several years, Steve plans to open a savings account and to make uniform monthly deposits the first of each month. The account is to be used to pay the taxes when they are due. To begin the account Steve will deposit a lump sum that is equivalent to the monthly payments that will not have been made for the first year's taxes. The savings account pays 9% interest, compounded monthly and payable quarterly (March 31, June 30, September 30 and December 31). How much money should Steve put into the account when he opens it on September 1? What uniform monthly deposit should he make from that time on? A careful *exact* solution is expected. (*Answer:* Initial deposit $350.28; monthly deposit $69.02)

**6-9** An oil refinery finds that it is now necessary to process its waste liquids in a costly treating process before discharging them into a nearby stream. The engineering department estimates that the waste liquid processing will cost $30,000 at the end of the first year. By making process and plant alterations, it is estimated that the waste treatment cost will decline $3000 each year. As an alternate a specialized firm, Hydro-Clean, Inc., has offered a contract to process the waste liquids for the ten years for a fixed price of $15,000 per year, payable at the end of each year. Either way there should be no need for waste treatment after ten years. If the refinery manager considers 8% a suitable interest rate, should he accept the Hydro-Clean offer or not?

**6-10** Bill Anderson buys an automobile every two years as follows: initially he pays a downpayment of $800 on a $3500 car. The balance is paid in 24 equal monthly payments with interest at 12%. When he has made the last payment on the loan, he trades in the two-year old car for $800 on a new $3500 car, and the cycle begins over again. Doug Jones decided on a different purchase plan. He felt he would be better off if he paid $3500 cash for a new car. Then he would make a monthly deposit in a savings account so that at the end of two years he would have $2700 in the account. The $2700 plus the $800 trade-in value of the car will allow Doug to replace his two-year old car by paying $3500 for a new one. The bank pays 4% interest, compounded quarterly.

(a) What is Bill Anderson's monthly payment to pay off the loan on the car? $127.17/\text{mon}$

(b) After he purchased the new car for cash, how much should Doug Jones deposit in his savings account monthly to have sufficient money for the next car two years hence? $325.85$

$\frac{8}{108.63}$ =

(c) Why is Doug Jones' monthly savings account deposit smaller than Bill Anderson's payment?

**6-11** Claude James, a salesman, needs a new car for use in his business. He expects to be promoted to a supervisory job at the end of three years, and so his concern now is to have a car for the three years he expects to be "on the road." The company will reimburse their salesmen each month at the rate of 14¢ per mile driven. Claude has decided to drive a low priced automobile. He finds, however, that there are three different ways of obtaining the automobile.

(a) Purchase for cash. The price is $3800.

(b) Lease the car. The monthly charge is $90 on a 36 month lease, payable at the end of each month. At the end of the three-year period the car is returned to the leasing company.

(c) Lease the car with option to purchase it at the end of the lease. Pay $100 a month for 36 months. At the end of that time, Claude could purchase the car, if he chooses, for $1100.

Claude believes he should use a 12% interest rate in determining which alternative to select. If the car could be sold for $1500 at the end of three years, which method should he use to obtain it?

**6-12** A college student has been looking for a new tire for his car and has located the following alternatives.

| Tire guaranty | Price per tire |
|---------------|----------------|
| 12 mo         | $15.95         |
| 24            | 24.95          |
| 36            | 34.95          |
| 48            | 44.95          |

If the student feels that the guaranty period is a good estimate of the tire life and that a 10% interest rate is appropriate, which tire should he buy?

**6-13** A suburban taxi company is considering buying taxis with diesel engines instead of gasoline engines. The cars average 30,000 miles a year with a useful life of three years for the taxi with the gas engine and four years for the diesel taxi. Other comparative information is as follows.

|                                   | Diesel | Gasoline |
|-----------------------------------|--------|----------|
| Vehicle cost                      | $5000  | $4000    |
| Fuel cost per gallon              | 0.65   | 0.75     |
| Mileage in miles per gallon       | 25     | 17       |
| Annual repairs                    | 300    | 200      |
| Annual insurance premium          | 500    | 500      |
| End of useful life resale value   | 500    | 700      |

Determine the more economical choice if interest rate is 6%.

**6-14** When he started work on his 22nd birthday, Wayne Morris decided to invest money each month with the objective of becoming a millionaire by the time he reaches his 65th birthday. If he expects his investments to yield 18% per annum, compounded monthly, how much should he invest each month? (*Answer:* $6.92 a month)

# 7
# Rate of Return Analysis

The third of the three major analysis methods is rate of return. In this chapter we will examine three aspects of rate of return. First, we will begin by describing the meaning of rate of return. Second, the calculation of rate of return will be illustrated. Third, rate of return analysis problems will be presented. In an appendix to this chapter we will describe difficulties sometimes encountered in attempting to compute an unknown interest rate for certain kinds of cash flows.

## RATE OF RETURN

In Chapter 4 we examined four plans to repay $5000 in five years with interest at 6% (Table 4-1). In each of the four plans the amount loaned ($5000) and the loan duration (five years) was the same. Yet the total interest paid to the lender varied from $900 to $1691 depending on the loan repayment plan. We saw however that the lender received 6% interest each year on the amount of money actually owed. And at the end of five years, the principal and interest payments exactly repaid the $5000 debt with interest at 6%. We say the lender received a 6% rate of return.

Rate of return may be defined as the interest rate paid on the unpaid balance of a *loan* such that the payment schedule makes the unpaid loan balance equal to zero when the final payment is made.

Instead of lending money we might invest $5000 in a machine tool with a five year useful life and an equivalent uniform annual benefit of $1187. An appropriate question is, what rate of return would we receive on this investment? The cash flow would be as follows.

| Year | Cash flow |
|------|-----------|
| 0 | −$5000 |
| 1 | +1187 |
| 2 | +1187 |
| 3 | +1187 |
| 4 | +1187 |
| 5 | +1187 |

We would recognize the cash flow as plan 3 of Table 4-1. We know that five payments of $1187 are equivalent to a present sum of $5000 when interest is 6%. Therefore, the rate of return on this investment is 6%. Stated in terms of an investment we may define rate of return as follows:

> Rate of return is the interest rate earned on the unrecovered *investment* such that the payment schedule makes the unrecovered investment equal to zero at the end of the life of the investment.

It must be understood that the 6% rate of return does not mean an annual return of 6% on the $5000 investment, or $300 in each of the five years. Instead, each $1187 payment represents a 6% return on the unrecovered investment plus the partial return of the investment. This may be tabulated as follows

| Year | Cash flow | Unrecovered investment at beginning of year | 6% return on unrecovered investment | Investment repayment at end of year | Unrecovered investment at end of year |
|------|-----------|---------------------------------------------|--------------------------------------|--------------------------------------|---------------------------------------|
| 0 | −$5000 | | | | |
| 1 | +1187 | $5000 | $300 | $887 | $4113 |
| 2 | +1187 | 4113 | 247 | 940 | 3173 |
| 3 | +1187 | 3173 | 190 | 997 | 2176 |
| 4 | +1187 | 2176 | 131 | 1056 | 1120 |
| 5 | +1187 | 1120 | 67 | 1120 | 0 |
| | | | $935 | $5000 | |

This cash flow represents a situation where the $5000 investment has benefits that produce a 6% rate of return. But in the five year period the total return is only $935, far less than $300 per year for five years. The reason, we can see, is because rate of return is defined as the interest rate earned on the unrecovered investment.

Although the two definitions of rate of return are stated differently, one in terms of a loan and the other in terms of an investment, there is only one fundamental concept being described. It is that the rate of return is the interest rate at which the benefits are equivalent to the costs.

## CALCULATION OF RATE OF RETURN

To calculate a rate of return on an investment we must convert the various consequences of the investment into a cash flow. Then we will solve the cash flow for the unknown value of $i$. This value of $i$ is the rate of return. Five forms of the cash flow equation are:

$$\text{PW of benefits} - \text{PW of costs} = 0 \qquad (7\text{-}1)$$

$$\frac{\text{PW of benefits}}{\text{PW of costs}} = 1 \qquad (7\text{-}2)$$

$$\text{Net present worth} = 0 \qquad (7\text{-}3)$$

$$\text{EUAB} - \text{EUAC} = 0 \qquad (7\text{-}4)$$

$$\text{PW of costs} = \text{PW of benefits} \qquad (7\text{-}5)$$

The five equations represent the same concept in different forms. They can relate costs and benefits with rate of return $i$ as the only unknown. The calculation of rate of return is illustrated by the following examples.

### *EXAMPLE 7-1*

A $8200 investment returned $2000 per year over a five year useful life. What was the rate of return on the investment? Using Eq. 7-2:

$$\frac{\text{PW of benefits}}{\text{PW of costs}} = 1 \qquad \frac{2000(P/A,i\%,5)}{8200} = 1$$

Rewriting the equation we see that

$$(P/A,i\%,5) = \frac{8200}{2000} = 4.1$$

Then look at the interest tables for the value of $i$ where $(P/A,i\%,5) = 4.1$. If no tabulated value of $i$ gives this value, we will then find values on either side of the desired value (4.1) and interpolate to find the rate of return $i$.

From interest tables we find:

| $i$ | $(P/A,i\%,5)$ |
|-----|---------------|
| 6% | 4.212 |
| 7% | 4.100 |
| 8% | 3.993 |

In this example no interpolation is needed as the rate of return for this investment is exactly 7%.

*EXAMPLE 7-2*

An investment resulted in the following cash flow. Compute the rate of return.

| Year | Cash flow |
|------|-----------|
| 0 | − $700 |
| 1 | + 100 |
| 2 | + 175 |
| 3 | + 250 |
| 4 | + 325 |

EUAB − EUAC = 0

$100 + 75(A/G,i\%,4) - 700(A/P,i\%,4) = 0$

In this situation we have two different interest factors in the equation. We will not be able to solve it as easily as Example 7-1. Since there is no convenient direct method of solution, we will solve the equation by trial and error. Try $i = 5\%$.

EUAB − EUAC = 0

$100 + 75(A/G,5\%,4) - 700(A/P,5\%,4)$

$100 + 75(1.439) - 700(0.282)$

At $i = 5\%$    EUAB − EUAC = 208 − 197 = +11

The EUAC is too low. If the interest rate is increased, EUAC will increase. Try $i = 8\%$.

EUAB − EUAC = 0

$100 + 75(A/G,8\%,4) - 700(A/P,8\%,4)$

$100 + 75(1.404) - 700(0.302)$

At $i = 8\%$    EUAB − EUAC = 205 − 211 = −6

This time the EUAC is too large. We see that the true rate of return is between 5% and 8%. Try $i = 7\%$.

EUAB − EUAC = 0

$100 + 75(A/G,7\%,4) - 700(A/P,7\%,4)$

$100 + 75(1.416) - 700(0.295)$

At $i = 7\%$    EUAB − EUAC = 206 − 206 = 0

The rate of return is 7%.

*EXAMPLE 7-3*

Given the cash flow below, calculate the rate of return on the investment.

| Year | Cash flow |
|------|-----------|
| 0 | −$100 |
| 1 | +20 |
| 2 | +30 |
| 3 | +20 |
| 4 | +40 |
| 5 | +40 |

Using NPW = 0, try $i = 10\%$.

$$NPW = -100 + 20(P/F,10\%,1) + 30(P/F,10\%,2) + 20(P/F,10\%,3)$$
$$+ 40(P/F,10\%,4) + 40(P/F,10\%,5)$$
$$= -100 + 20(0.9091) + 30(0.8264) + 20(0.7513) + 40(0.6830)$$
$$+ 40(0.6209)$$
$$= -100 + 18.18 + 24.79 + 15.03 + 27.32 + 24.84$$
$$= -100 + 110.16$$
$$= +10.16$$

The trial interest rate $i$ is too low. Select a second trial $i = 15\%$.

$$NPW = -100 + 20(0.8696) + 30(0.7561) + 20(0.6575) + 40(0.5718)$$
$$+ 40(0.4972)$$
$$= -100 + 17.39 + 22.68 + 13.15 + 22.87 + 19.89$$
$$= -100 + 95.98$$
$$= -4.02$$

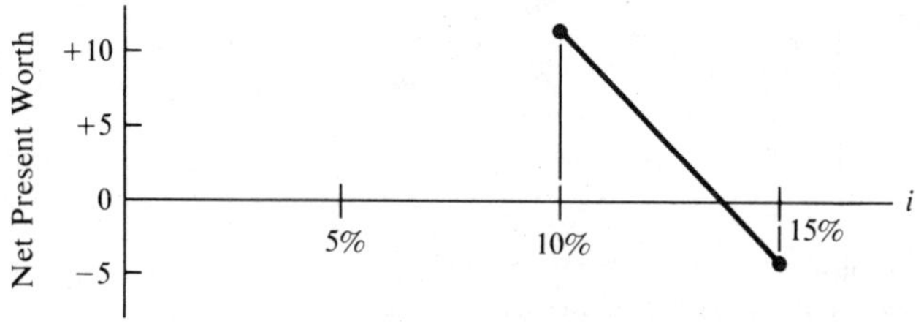

**Figure 7-1.** Plot of NPW vs interest rate $i$.

These two points are plotted in Figure 7-1. By linear interpolation we compute the rate of return as follows:

$$i = 10\% + (15\% - 10\%)\left(\frac{10.16}{10.16 + 4.02}\right) = 13\tfrac{1}{2}\%$$

We can prove that the rate of return is very close to $13\frac{1}{2}\%$ by showing that the unrecovered investment is very close to zero at the end of the life of the investment.

| Year | Cash flow | Unrecovered investment at beginning of year | $13\frac{1}{2}\%$ return on unrecovered investment | Investment repayment at end of year | Unrecovered investment at end of year |
|---|---|---|---|---|---|
| 0 | −$100 | | | | |
| 1 | +20 | $100.0 | $13.5 | $6.5 | $93.5 |
| 2 | +30 | 93.5 | 12.6 | 17.4 | 76.1 |
| 3 | +20 | 76.1 | 10.3 | 9.7 | 66.4 |
| 4 | +40 | 66.4 | 8.9 | 31.1 | 35.3 |
| 5 | +40 | 35.3 | 4.8 | 35.2 | 0.1* |

*This small unrecovered investment indicates that the rate of return is slightly less than $13\frac{1}{2}\%$.

If in Figure 7-1 NPW had been computed for a broader range of values of $i$, Figure 7-2 would have been obtained. From the figure it is apparent that the error resulting from linear interpolation increases as the interpolation width increases.

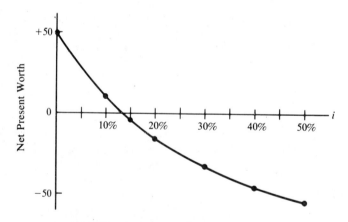

**Figure 7-2.**  Replot of NPW vs interest rate $i$ over a larger range of values.

### Plot of NPW vs $i$

Figure 7-2 represents a very important graph. The plot of NPW vs $i$ can be an important source of information. A cash flow representing an investment followed by benefits from the investment would have a NPW vs $i$ plot (we will call it a NPW plot for convenience) in the form of Figure 7-3.

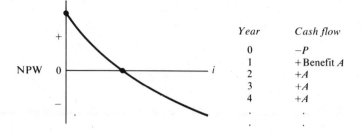

| Year | Cash flow |
|------|-----------|
| 0    | $-P$      |
| 1    | $+$ Benefit $A$ |
| 2    | $+A$      |
| 3    | $+A$      |
| 4    | $+A$      |
| .    | .         |
| .    | .         |

**Figure 7-3.** Typical NPW plot for an
investment.

If, on the other hand, borrowed money was involved, the NPW plot would appear as in Figure 7-4. This form of cash flow typically results when one is a borrower of money. In such a case the usual pattern is, there is a receipt of borrowed money early in the time period with a later repayment of an equal sum plus payment of interest on the borrowed money. In all cases where interest is charged the NPW at 0% will be negative.

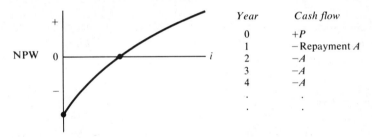

| Year | Cash flow |
|------|-----------|
| 0    | $+P$      |
| 1    | $-$ Repayment $A$ |
| 2    | $-A$      |
| 3    | $-A$      |
| 4    | $-A$      |
| .    | .         |
| .    | .         |

**Figure 7-4.** Typical NPW plot for borrowed
money.

How do we determine the interest rate paid by the borrower in this situation? Typically we would write an equation (such as PW of income = PW of disbursements) and solve for the unknown $i$. Is the resulting $i$ positive or negative from the borrower's point of view? If the lender said he was receiving, say $+11\%$ on the debt, it seems reasonable to say the borrower is faced with $-11\%$ interest. But this is not the way interest is discussed. Rather, interest is referred to in absolute terms without associating a sign to it. A banker says he pays 5% interest on savings accounts and charges 11% on personal loans.

Thus we implicitly recognize interest as a charge for the use of money and a receipt for letting others use our money. In determining the interest rate in a particular situation we solve for a single unsigned value of it. We then view this value of $i$ in the customary way, that is, either as a charge for borrowing money, or a receipt for lending money.

### EXAMPLE 7-4

A new corporate bond was initially sold by a stockbroker to an investor for $1000. The issuing corporation promised to pay to the bondholder $40 interest on the $1000 face value of the bond every six months, and to repay the $1000 at the end of ten years. After one year the bond was sold by the original buyer for $950.

(a) What rate of return did the original buyer receive on his investment?

(b) What rate of return can the new buyer (paying $950) expect to receive if he keeps the bond for its remaining nine-year life?

(a)

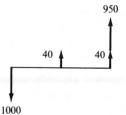

Since $40 is received each six months, we will solve the problem using a six-month interest period.

Let PW of cost = PW of benefits.

$$1000 = 40(P/A,i\%,2) + 950(P/F,i\%,2)$$

Try $i = 1\frac{1}{2}\%$.

$$1000 = 40(1.956) + 950(0.9707) = 78.24 + 922.17 = 1000.41$$

The interest rate per six months is very close to $1\frac{1}{2}\%$. This means the nominal (annual) interest rate is $2 \times 1.5\% = 3\%$. The effective interest rate $= (1 + 0.015)^2 - 1 = 3.02\%$.

(b)

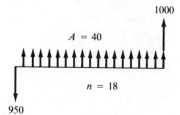

We have the same $40 semi-annual interest payments. For six-month interest periods:

$$950 = 40(P/A,i\%,18) + 1000(P/F,i\%,18)$$

Try $i = 5\%$.

$$950 = 40(11.690) + 1000(0.4155) = 467.60 + 415.50 = 883.10$$

The PW of benefits is too low. Try a lower interest rate. Try $i = 4\%$.

$$950 = 40(12.659) + 1000(0.4936) = 506.36 + 493.60 = 999.96$$

The value of $i$ is between 4% and 5%. By interpolation

$$i = 4\% + (1\%)\left(\frac{999.96 - 950.00}{999.96 - 883.10}\right) = 4.43\%$$

The nominal interest rate is $2 \times 4.43\% = 8.86\%$. The effective interest rate is $(1 + 0.0443)^2 - 1 = 9.05\%$.

## RATE OF RETURN ANALYSIS

Rate of return analysis is probably the most frequently used exact analysis technique in industry. Although there are infrequent problems in computing rate of return, its major advantage outweighs the occasional difficulty. The major advantage is that we can compute a single figure of merit that is readily understood. Consider these statements:

1. The net present worth on the project is $32,000.
2. The equivalent uniform annual net benefit is $2800.
3. The project will produce a 23% rate of return.

While none of the statements tells the complete story, the third one gives a measure of desirability of the project in terms that are widely understood. It is this acceptance by engineers and businessmen alike of rate of return that has promoted its more frequent use than present worth or annual cash flow methods.

There is another advantage to rate of return analysis. In both present worth and annual cash flow calculations one must select an interest rate for use in the calculations—this may be a difficult and controversial item. In rate of return analysis no interest rate is introduced into the calculations (except as described in the appendix of this chapter). Instead we compute a rate of return (more accurately called internal rate of return) from the cash flow. To decide how to proceed, the calculated rate of return is compared with a preselected minimum attractive rate of return (MARR). This is the same value of $i$ used for present worth and annual cash flow analysis.

When there are two alternatives, rate of return analysis is performed by computing the incremental rate of return ($\Delta$ROR) on the difference between the alternatives. If the incremental rate of return is greater than or equal to the minimum attractive rate of return, choose the higher cost alternative. If the incremental rate of return is less than the minimum attractive rate of return, choose the lower cost alternative.

| Situation | Decision |
|---|---|
| ΔROR ⩾ MARR | Choose the higher cost alternative |
| ΔROR < MARR | Choose the lower cost alternative |

Rate of return analysis is illustrated by a group of examples.

### EXAMPLE 7-5

If an electromagnet is installed on the input conveyor of a coal processing plant, it will pick up scrap metal in the coal. The removal of this
metal will save an estimated $1200 per year in machinery damage being
caused by metal. The electromagnetic equipment has an estimated useful life of five years and no salvage value. Two suppliers have been
contacted. Leaseco will provide the equipment in return for three
beginning-of-year annual payments of $1000 each. Saleco will provide
the equipment for $2783. If the minimum attractive rate of return
(MARR) is 10%, which supplier should be selected?

Since both suppliers will provide equipment with the same useful life
and benefits, this is a fixed output situation. In rate of return analysis
the method of solution is to examine the differences between the
alternatives.

| Year | Leaseco | Saleco | Difference between alternatives Saleco-Leaseco |
|---|---|---|---|
| 0 | −$1000 | −$2783 | −$1783 |
| 1 | $\begin{cases} -1000 \\ +1200 \end{cases}$ | +1200 | +1000 |
| 2 | $\begin{cases} -1000 \\ +1200 \end{cases}$ | +1200 | +1000 |
| 3 | +1200 | +1200 | 0 |
| 4 | +1200 | +1200 | 0 |
| 5 | +1200 | +1200 | 0 |

Compute the net present worth (NPW) at various interest rates on the
increment of investment represented by the difference between the
alternatives.

| Year n | Cash flow Saleco-Leaseco | PW† at 0% | PW† at 8% | PW† at 20% | PW† at ∞% |
|---|---|---|---|---|---|
| 0 | −$1783 | −$1783 | −$1783 | −$1783 | −$1783 |
| 1 | +1000 | +1000 | +926 | +833 | 0 |
| 2 | +1000 | +1000 | +857 | +694 | 0 |
| 3 | 0 | 0 | 0 | 0 | 0 |

| Year | Cash flow | PW† | PW† | PW† | PW† |
|---|---|---|---|---|---|
| n | Saleco-Leaseco | at 0% | at 8% | at 20% | at ∞% |
| 4 | 0 | 0 | 0 | 0 | 0 |
| 5 | 0 | 0 | 0 | 0 | 0 |
| | NPW = | +217 | 0 | −256 | −1783 |

†Each year the cash flow is multiplied by $(P/F,i\%,n)$.

At 0%: $(P/F,0\%,n) = 1$ for all values of $n$

At ∞%: $(P/F,\infty\%,0) = 1$

$(P/F,\infty\%,n) = 0$ for all other values of $n$

These data are plotted in Figure 7-5. From the figure we see that NPW = 0 at $i = 8\%$.

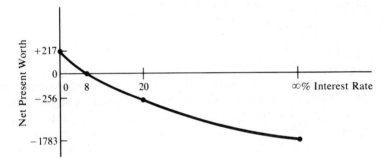

**Figure 7-5.** NPW plot for Example 7-5.

Thus the incremental rate of return (ΔROR) of selecting Saleco rather than Leaseco is 8%. This is less than the 10% MARR. Select Leaseco.

*EXAMPLE 7-6*

You are given the choice of selecting one of two mutually exclusive alternatives. The alternatives are as follows.

| Year | Alternative 1 | Alternative 2 |
|---|---|---|
| 0 | −$10 | −$20 |
| 1 | +15 | +28 |

Any money not invested here may be invested elsewhere at the minimum attractive rate of return (MARR) of 6%. If you can only choose one alternative one time, which one would you select?

We will select the lesser cost alternative 1, unless we find the additional cost of alternative 2 produces sufficient additional benefits that we would prefer alternative 2. If we consider alternative 2 in relation to alternative 1 then,

Higher cost alternative 2 = lower cost alternative 1 + differences
between them

or

Differences between them = higher cost alternative 2 − lower cost
alternative 1

The choice between the two alternatives reduces down to an examination
of the differences between them. We can compute the rate of return on
the differences between the alternatives. Writing the alternatives again.

| Year | Alternative 1 | Alternative 2 | Alternative 2 − Alternative 1 |
|------|---------------|---------------|-------------------------------|
| 0 | −$10 | −$20 | −$20 − (−$10) = −$10 |
| 1 | +$15 | +$28 | +$28 − (+$15) = +$13 |

PW of cost = PW of benefit

$$10 = 13(P/F,i\%,1)$$

$$(P/F,i\%,1) = \frac{10}{13} = 0.7692$$

One can see that if $10 increases to $13 in one year the interest rate must
be 30%. The interest tables confirm this conclusion. The 30% rate of
return on the difference between the alternatives is far higher than the
6% MARR. The additional $10 investment to obtain alternative 2 is
superior to investing the $10 elsewhere at 6%. To obtain this desirable
increment of investment, with its 30% rate of return, alternative 2 is
selected.

To understand more about Example 7-6, compute the rate of return
for each alternative.

*Alternative 1*

PW of cost = PW of benefit

$$\$10 = \$15(P/F,i\%,1)$$

$$(P/F,i\%,1) = \frac{10}{15} = 0.667$$

From the interest tables: rate of return = 50%.

*Alternative 2*

PW of cost = PW of benefit

$$\$20 = \$28(P/F,i\%,1)$$

$$(P/F,i\%,1) = \frac{20}{28} = 0.7143$$

From the interest tables: rate of return = 40%.

One is tempted to select alternative 1, based on these rate of return computations. We have already seen, however, that this is not the correct solution. Solve the problem again, this time using present worth analysis.

*Present Worth Analysis*

> *Alternative 1*
>
> $$NPW = -10 + 15(P/F,6\%,1) = -10 + 15(0.943) = +\$4.15$$
>
> *Alternative 2*
>
> $$NPW = -20 + 28(P/F,6\%,1) = -20 + 28(0.943) = +\$6.40$$

Alternative 1 has a 50% rate of return and a NPW (at the 6% MARR) of +$4.15. Alternative 2 has a 40% rate of return on a larger investment, with the result that its NPW (at the 6% MARR) is +$6.40. Our economic criterion is to maximize the return, rather than the rate of return. To maximize NPW, select alternative 2. This agrees with the rate of return analysis on the differences between the alternatives.

## EXAMPLE 7-7

A firm is considering which of two devices to install to reduce costs in a particular situation. Both devices cost $1000 and have useful lives of five years and no salvage value. Device $A$ can be expected to result in $300 savings annually. Device $B$ will provide cost savings of $400 the first year, but will decline $50 annually, making the second year savings $350, the third year savings $300, and so forth. For a 7% minimum attractive rate of return, which device should the firm purchase? This problem has been solved by present worth analysis (Example 5-1) and annual cost analysis (Example 6-5). This time we will use rate of return analysis. The example has fixed input ($1000) and differing outputs (savings).

| Year | Device A | Device B | Difference between alternatives Device A-Device B |
|------|----------|----------|---------------------------------------------------|
| 0 | −$1000 | −$1000 | $0 |
| 1 | +300 | +400 | −100 |
| 2 | +300 | +350 | −50 |
| 3 | +300 | +300 | 0 |
| 4 | +300 | +250 | +50 |
| 5 | +300 | +200 | +100 |

For the difference between the alternatives, write a single equation with $i$ as the only unknown.

EUAC = EUAB

$[100(P/F,i\%,1) + 50(P/F,i\%,2)](A/P,i\%,5)$
$$= [50(F/P,i\%,1) + 100](A/F,i\%,5)$$

The equation is cumbersome, but need not be solved. Instead, we observe that the sum of the costs (−100 and −50) equals the sum of the benefits (+50 and +100). This indicates that 0% is the incremental rate of return on the $A - B$ increment of investment. This is less than the 7% MARR, therefore the increment is undesirable. Reject Device $A$ and choose Device $B$.

**Analysis Period**

In discussing present worth analysis and annual cash flow analysis an important consideration was the analysis period. This is also true in rate of return analysis. For two alternatives the method of solution is to examine the differences between the alternatives. The examination must necessarily cover the selected analysis period. An assumption that an alternative can be replaced with one of identical costs and performance appears dubious at best. For now we can only suggest that the assumptions made should reflect one's perception of the future as accurately as possible.

Example 7-8 is a problem where the analysis period is a common multiple of the alternative service lives, and where identical replacement is assumed. It will illustrate an analysis of the differences between the alternatives over the analysis period.

*EXAMPLE 7-8*

Two machines are being considered for purchase. If the minimum attractive rate of return (minimum required interest rate) is 10%, which machine should be bought?

|                                   | Machine X | Machine Y |
|-----------------------------------|-----------|-----------|
| Initial cost                      | $200      | $700      |
| Uniform annual benefit            | 95        | 120       |
| End of useful life salvage value  | 50        | 150       |
| Useful life                       | 6 years   | 12 years  |

The solution is based on a 12-year analysis period and a replacement machine $X$ that is identical to the present machine $X$. The cash flow for the differences between the alternatives is as follows.

| Year | Machine X | Machine Y | Machine Y − Machine X |
|------|-----------|-----------|-----------------------|
| 0    | − $200    | − $700    | − $500                |
| 1    | + 95      | + 120     | + 25                  |
| 2    | + 95      | + 120     | + 25                  |

| Year | Machine X | Machine Y | Machine Y − Machine X |
|------|-----------|-----------|------------------------|
| 3 | +95 | +120 | +25 |
| 4 | +95 | +120 | +25 |
| 5 | +95 | +120 | +25 |
| 6 | $\left\{\begin{array}{l} +95 \\ +50 \\ -200 \end{array}\right.$ | +120 | +25 <br> +150 |
| 7 | +95 | +120 | +25 |
| 8 | +95 | +120 | +25 |
| 9 | +95 | +120 | +25 |
| 10 | +95 | +120 | +25 |
| 11 | +95 | +120 | +25 |
| 12 | $\left\{\begin{array}{l} +95 \\ +50 \end{array}\right.$ | +120 <br> +150 | +25 <br> +100 |

PW of cost = PW of benefits

$$500 = 25(P/A,i\%,12) + 150(P/F,i\%,6) + 100(P/F,i\%,12)$$

The sum of the benefits over the 12 years is \$550 which is only a little greater than the \$500 additional cost. This indicates that the rate of return is quite low. Try $i = 1\%$.

$$500 = 25(11.255) + 150(0.942) + 100(0.887)$$
$$= 281 + 141 + 89 = 511$$

The interest rate is too low. Try $i = 1\tfrac{1}{2}\%$.

$$500 = 25(10.908) + 150(0.914) + 100(0.836)$$
$$= 273 + 137 + 84 = 494$$

The rate of return is about 1.3%.

The rate of return on the difference between $Y$ and $X$ is 1.3%, far below the 10% minimum attractive rate of return. The additional investment to obtain $Y$ yields an unsatisfactory rate of return, therefore $X$ is the preferred alternative.

## SUMMARY

Rate of return may be defined as the interest rate paid on the unpaid balance of a loan such that the loan is exactly repaid by the schedule of payments. On an investment, rate of return is the interest rate earned on the unrecovered investment such that the payment schedule makes the unrecovered investment equal to zero at the end of the life of the investment.

Although the two definitions of rate of return are stated differently, there is only one fundamental concept being described. It is that the rate of return is the interest rate at which the benefits are equivalent to the costs.

There are a variety of ways of writing the cash flow equation in which the rate of return *i* may be the single unknown. Five of them are:

PW of benefits − PW of costs = 0

$$\frac{\text{PW of benefits}}{\text{PW of costs}} = 1$$

NPW = 0

EUAB − EUAC = 0

PW of costs = PW of benefits

*Rate of Return Analysis.* Rate of return analysis is the most frequently used method in industry as the resulting rate of return is readily understood. Also, the difficulties in selecting a suitable interest rate to use in the calculations are avoided.

*Criteria.*

*Two alternatives*

Compute the incremental rate of return (ΔROR) on the difference between the alternatives.

If ΔROR ⩾ MARR, choose the higher cost alternative.

If ΔROR < MARR, choose the lower cost alternative.

*Three or more alternatives*

Incremental analysis is needed. See Chapter 8.

## PROBLEMS

**7-1**   Peter Minuit bought an island from the Manhattoes Indians in 1626 for $24 worth of glass beads and trinkets. The 1976 estimate of the value of land on this island was $9 billion. What rate of return would the Indians have received if they had retained title to the island rather than selling it for $24.

**7-2**   A man buys a corporate bond from a bond brokerage house for $925. The bond has a face value of $1000 and pays 4% of its face value each year. If the bond will be paid off at the end of ten years, what rate of return will the man receive? (*Answer:* 4.97%)

**7-3**   A well known industrial firm has issued $1000 bonds that carry a 4% nominal interest rate paid semi-annually. The bonds mature 20 years from now, at which time the industrial firm will redeem them for $1000 plus the terminal semi-annual interest payment. From the financial pages of your newspaper you learn that the bonds may be purchased for $715 each ($710 for the bond plus a $5 sales commission). What nominal rate of return would you receive if you purchased the bond now and held it to maturity 20 years hence? (*Answer:* 6.6%)

**7-4** One aspect of obtaining a college education is the prospect of improved future earnings, compared to non-college graduates. Sharon Shay estimates that a college education has a $28,000 equivalent cost at graduation. She believes the benefits of her education will occur throughout 40 years of employment. She thinks she will have a $1000 per year higher income during the first 10 years out of college, compared to a non-college graduate. During the subsequent 10 years she projects an annual income that is $3000 per year higher. During the last 20 years of employment she estimates an annual salary that is $6000 above the level of the non-college graduate. Assuming her estimates are correct, what rate of return will she receive as a result of her investment in a college education?

**7-5** An investor purchased a one-acre lot on the outskirts of a city for $3500 cash. Each year he paid $80 of property taxes. At the end of four years he sold the lot. After deducting his selling expenses the investor received $4400. What rate of return did he receive on his investment? (*Answer:* 3.8%)

**7-6** A popular reader's digest offers a lifetime subscription to the magazine for $100. Such a subscription may be given as a gift to an infant at birth (the parents can read it in those early years), or taken out by an individual for himself. Normally, the magazine costs $6.45 per year. Knowledgeable people say it probably will continue indefinitely at this $6.45 rate. What rate of return would be obtained if a life subscription were purchased for an infant rather than paying $6.45 per year beginning immediately? You may make any reasonable assumptions, but the compound interest factors must be *correctly* used.

**7-7** On April 2, 1976 an engineer bought a $1000 bond of an American airline for $875. The bond pays 6% on its principal amount of $1000, half in each of its April 1 and October 1 semi-annual payments, and will repay the $1000 principal sum on October 1, 1989. What nominal rate of return will the engineer receive from the bond if he holds it to its maturity on October 1, 1989? (*Answer:* 7.5%)

**7-8** The cash price of a machine tool is $3500. The dealer is willing to accept a $1200 down payment and 24 end-of-month monthly payments of $110 each. At what effective interest rate are these terms equivalent? (*Answer:* 14.4%)

**7-9** A local bank makes automobile loans. It charges 4% per year in the following manner: If $3600 is borrowed to be repaid over a three-year period, the bank interest charge is $3600 × 0.04 × 3 years = $432. The bank deducts the $432 of interest from the $3600 loan and gives the customer $3168 in cash. The customer must repay the loan by paying 1/36 of $3600 or $100 the end of each month for 36 months. What nominal interest rate is the bank actually charging for this loan?

**7-10** Upon graduation every engineer must decide whether or not to go on to graduate school. Estimate the costs of going full time to the university to obtain a Master of Science degree. Then estimate the resulting costs and benefits. Combine the various consequences into a cash flow table and compute the rate of return.

**7-11** In his uncle's will, Bill is to choose one of two alternatives:

*Alternative 1.* $2000 cash

*Alternative 2.* $150 cash now plus $100 per month for 20 months beginning the first day of next month

(a) At what rate of return are the two alternatives equivalent?

(b) If Bill thinks the rate of return in (a) is too low, which alternative should he select?

**7-12** A man buys a table saw at a local store for $175. He may either pay cash for it, or pay $35 now and $12.64 a month for 12 months beginning 30 days hence. If the man chooses the time payment plan, what is the nominal interest rate he will be charged? *(Answer: 15%)*

**7-13** In January, 1973 an investor purchased a convertible debenture bond issued by the XLA Corporation. The bond cost $1000, and paid $60 per year interest in annual payments on December 31. Under the convertible feature of the bond, it could be converted into 20 shares of common stock by tendering the bond together with $400 cash. The day after the investor received the December 31, 1975 interest payment he submitted the bond together with $400 to the XLA Corporation. In return he received the 20 shares of common stock. The common stock paid no dividends. On December 31, 1977, the investor sold the stock for $1740 terminating his five year investment in XLA Corporation. What rate of return did he receive?

**7-14** A man owns a corner lot. He must decide which of several alternatives to select in trying to obtain a desirable return on his investment. After much study and calculation, he decides that the two best alternatives are:

|                         | Build gas station | Build soft ice cream stand |
|-------------------------|-------------------|----------------------------|
| First cost              | $80,000           | $120,000                   |
| Annual property taxes   | 3000              | 5000                       |
| Annual income          | 11,000            | 16,000                     |
| Life of building        | 20 yrs            | 20 yrs                     |
| Salvage value           | 0                 | 0                          |

If the owner wants a minimum attractive rate of return on his investment of 6%, which of the two alternatives would you recommend to him?

**7-15** Two alternatives are as follows

| Year | Alternative A | Alternative B |
|------|---------------|---------------|
| 0    | −$2000        | −$2800        |
| 1    | +800          | +1100         |
| 2    | +800          | +1100         |
| 3    | +800          | +1100         |

If 5% is considered the minimum attractive rate of return, which alternative should be selected?

**7-16** The Southern Guru Copper Company operates a large mine in a South American country. A legislator in the National Assembly said in a speech that most of the capital for the mining operation was provided by loans from the World Bank with the result that Southern Guru had only $500,000 of its own money actually invested in the property. The cash flow for the mine was as follows:

| Year | Cash flow |
|------|-----------|
| 0 | $0.5 million investment |
| 1 | 3.5 million profit |
| 2 | 0.9 million profit |
| 3 | 3.9 million profit |
| 4 | 8.6 million profit |
| 5 | 4.3 million profit |
| 6 | 3.1 million profit |
| 7 | 6.1 million profit |

64%

The legislator divided the $30.4 million total profit by the $0.5 million investment. This produced, he said, a 6080% rate of return on the investment. Southern Guru claims their actual rate of return is much lower. They ask you to compute their rate of return.

# Appendix 7-A
## Difficulties
## in Solving for an
## Unknown Interest Rate

Occasionally we encounter a cash flow that cannot be solved for a single positive interest rate. In this appendix we will attempt to indicate a way to resolve the difficulty. Example 7A-1 illustrates the situation.

### EXAMPLE 7A-1

The Going Aircraft Company has an opportunity to supply four large airplanes to Interair, a foreign airline. Interair will pay $19 million when the contract is signed and $10 million one year later. Going estimates its second and third year net cash flows at $50 million each when the planes are being produced. Interair will take delivery of the planes during year four and agrees to pay $20 million at the end of that year and the $60 million balance at the end of year five. Compute the rate of return on this project. Computation of NPW at various interest rates using single payment present worth factors* is presented.

| Year | Cash flow | 0% | 10% | 20% | 40% | 50% |
|------|-----------|------|--------|--------|--------|--------|
| 0 | +$19 | +$19 | +$19 | +$19 | +$19 | +$19 |
| 1 | +10 | +10 | +9.1 | +8.3 | +7.1 | +6.7 |
| 2 | −50 | −50 | −41.3 | −34.7 | −25.5 | −22.2 |
| 3 | −50 | −50 | −37.6 | −28.9 | −18.2 | −14.8 |
| 4 | +20 | +20 | +13.7 | +9.6 | +5.2 | +4.0 |
| 5 | +60 | +60 | +37.3 | +24.1 | +11.2 | +7.9 |
|   | NPW = | +$9 | +$0.2 | −$2.6 | −$1.2 | +$0.6 |

The NPW plot for this cash flow is represented in Figure 7A-1. From the NPW plot we see this cash flow produces *two* points at which NPW = 0.

*For example, for year 2 and $i = 10\%$ : PW $= -50(P/F,10\%,2) = -50(0.826) = -41.3$.

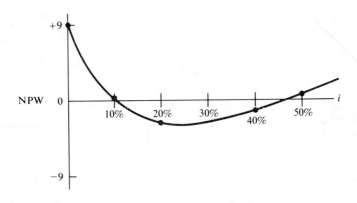

**Figure 7A-1.** NPW plot.

Thus there are *two* positive rates of return, one at about 10.1% and the other at about 47%

Example 7A-1 produced unexpected and undesirable results. We want to know when we may expect this kind of result, what it means, and how we may resolve the difficulty. In the next section we will see that the solution of an economic analysis problem is really the solution of a mathematical equation.

## Converting a Cash Flow to a Mathematical Equation

| Year | Cash flow |
|------|-----------|
| 0    | $-P$      |
| 1    | $+A_1$    |
| 2    | $+A_2$    |
| .    | .         |
| .    | .         |
| .    | .         |
| $n$  | $+A_n$    |

Setting NPW = PW of benefits minus PW of cost = 0, and using single payment present worth factors, the cash flow may be rewritten as

$$+A_1(1 + i)^{-1} + A_2(1 + i)^{-2} + \cdots + A_n(1 + i)^{-n} - P = 0 \qquad (7A\text{-}1)$$

If we let $X = (1 + i)^{-1}$, then Eq. 7A-1 may be written

$$+A_1 X + A_2 X^2 + \cdots + A_n X^n - P = 0 \qquad (7A\text{-}2)$$

Rearranging terms,

$$A_n X^n + \cdots + A_2 X^2 + A_1 X - P = 0 \qquad\qquad (7A\text{-}3)$$

Eq. 7A-3 is an $n$th order polynomial to which Descartes' Rule may be applied. The rule is

> If a polynomial with real coefficients has $m$ sign changes, then the number of positive roots will be $m - 2k$, where $k$ is a positive integer or zero ($k = 0,1,2,3,\ldots$).

A sign change is where successive nonzero terms, written according to descending powers of $X$, have different signs (that is, change from $+$ to $-$ or vice versa). Descartes' Rule means that the number of positive roots of the polynomial cannot exceed $m$, the number of sign changes; the number of positive roots must either be equal to $m$ or less by an even integer.

Descartes' Rule for polynomials gives the following.

| Number of sign changes, $m$ | Number of positive values of $X$ |
|:---:|:---:|
| 0 | 0 |
| 1 | 1 |
| 2 | 2 or 0 |
| 3 | 3 or 1 |
| 4 | 4, 2 , or 0 |

But the Eq. 7A-3 polynomial is not the equation that represents the economic analysis problem. By substituting $(1 + i)^{-1}$ in place of $X$ we return to Eq. 7A-1. We see from the relationship $X = (1 + i)^{-1}$ that a positive value of $X$ does not insure a positive value of $i$. In fact, whenever $X$ is greater than one, $i$ is negative. Thus in a particular situation when $m$ equals two sign changes, there would either be two or zero positive values of $X$. But this means there could be two, one, or zero positive values of $i$ for the corresponding economic analysis equation with two sign changes. From this we can form a rule of signs.

## Cash Flow Rule of Signs

There may be as many positive values of $i$ as there are sign changes in the cash flow.

Sign changes are computed the same way as for Descartes' Rule. A sign change is where successive nonzero values in the cash flow have different signs (that is, change from $+$ to $-$, or vice versa). A zero cash flow is ignored. The five cash flows in Table 7A-1 illustrate the counting of the number of sign changes.

**Table 7A-1.** Sign changes in cash flows

| | Cash flow | | | | |
|---|---|---|---|---|---|
| Year | A | B | C | D | E |
| 0 | +$100 | −$100 | −$100 | +$50 | +$50 |
| 1 | +10 | +10 | 0 | +40 | −50 |
| 2 | +50 | +50 | +50 | −100 | +50 |
| 3 | +20 | +20 | 0 | +10 | −10 |
| 4 | +40 | +40 | +80 | +10 | −30 |
| No. of sign changes | 0 | 1 | 1 | 2 | 3 |

The cash flow rule of signs indicates the following possibilities:

| Number of sign changes, m | Number of positive values of i |
|---|---|
| 0 | 0 |
| 1 | 1 or 0 |
| 2 | 2, 1, or 0 |
| 3 | 3, 2, 1, or 0 |

Thus there are three possibilities that need examination: zero sign changes, one sign change, and more than one sign change.

## Zero Sign Changes

There are two situations that produce zero sign changes in a cash flow. Either all terms have a positive sign, representing receipts, or all terms have a negative sign, reflecting a series of disbursements.

The first case would be like walking into a store and finding oneself the millionth customer and hence the recipient of money and gifts. This would be the utopian something-for-nothing situation. There is no value of $i$ that can be computed, for there are no disbursements to offset the receipts.

The second case represents the less happy situation of disbursements without any compensating receipts. It would be like the periodic purchase of Irish Sweepstakes tickets—but never winning anything. There is no value of $i$ that would reflect this economic situation.

## One Sign Change

Unless there are rather unusual circumstances one sign change is the normal cash flow pattern. Given one sign change it is extremely likely that we have a situation where there will be a single positive value of $i$.

There is no positive value of $i$ whenever *in an investment situation* the

subsequent benefits do not equal the magnitude of the investment. An
example would be as follows.

| Year | Cash flow |
|:----:|:---------:|
| 0 | −$50 |
| 1 | +20 |
| 2 | +20 |

Also, there is no positive value of $i$ whenever *in a borrowing situation* the
subsequent repayments do not equal the magnitude of the borrowed money.

| Year | Cash flow |
|:----:|:---------:|
| 0 | +$50 |
| 1 | −20 |
| 2 | −20 |

These circumstances can be readily identified and hence present no con-
fusion or particular problem.

**Two or More Sign Changes**

When there are two or more sign changes in the cash flow we know that
there are several possibilities concerning the number of positive values of $i$.
Probably the greatest danger in this situation is to fail to recognize the
multiple possibilities and to solve for a value of $i$. Then one might incor-
rectly (or correctly) assume that the value of $i$ that is obtained is the only
positive value of $i$. In this multiple sign change situation one approach is
to prepare a NPW plot like Figure 7A-1. This may be a tedious procedure
but it graphically portrays the exact situation.

**What the Difficulties Mean**

If there is a single positive value of $i$ we have no problem. On the other
hand a situation with no positive value of $i$ or multiple positive values
represents a situation that may be attractive, unattractive, or confusing.
Where there are multiple values of $i$, none of them should be considered a
suitable measure of the rate of return or attractiveness of the cash flow.
We classify a cash flow as an investment situation if we put money into
a project and benefits of the project come back to us. The rate of return

tells us something about the attractiveness of the project. But what was the situation in Example 7A-1?

| Year | Cash flow |
|------|-----------|
| 0 | +$19 |
| 1 | +10 |
| 2 | −50 |
| 3 | −50 |
| 4 | +20 |
| 5 | +60 |

At the beginning money is generated and flows out of the project. In years 2 and 3 money is spent on production, followed by receipts in the final two years. We know that 10.1% is a value of $i$ at which NPW = 0. What does this mean? If the initial outflows from the project are invested in some *external investment* at 10.1% and then their compound amount returned to the project in year 2, the project, or internal investment, will have $i = 10.1\%$. Similarly, if the initial outflows from the project can be put into an external investment at 47% then the project or internal investment will also show a 47% interest rate. This sounds unbelievable so let's demonstrate both situations.

For $i = 10.1\%$:

| Year | Cash flow |
|------|-----------|
| 0 | +$19 invest for 2 years in an *external investment* at 10.1%  $F = +19(F/P,10.1\%,2)$  $F = +19(1.21) = +23$ |
| 1 | +10 invest for 1 year in an *external investment* at 10.1%  $F = +10(F/P,10.1\%,1)$  $F = +10(1.101) = +11$ |
| 2 | +11 return to internal investment  +23 return to internal investment  −50 |
| 3 | −50 |
| 4 | +20 |
| 5 | +60 |

When the external investment phase is completed the cash flow for the internal investment is as follows.

| Year | Cash flow |
|------|-----------|
| 0 | $ 0 |
| 1 | 0 |
| 2 | − 16 |
| 3 | − 50 |
| 4 | + 20 |
| 5 | + 60 |

We see that for the internal investment there is one sign change so we can solve for the internal value of $i$ directly. It will be 10.1% if when we use that value NPW = 0.

$$NPW = -16(P/F,10.1\%,2) - 50(P/F,10.1\%,3) + 20(P/F,10.1\%,4)$$
$$+ 60(P/F,10.1\%,5)$$
$$= -16(0.825) - 50(0.749) + 20(0.681) + 60(0.618)$$
$$= -13.2 - 37.5 + 13.6 + 37.1$$
$$= 0$$

Similarly for $i = 47\%$:

### Cash flow

| | |
|---|---|
| 0 | + $19 invest for 2 years in an *external investment* at 47% |
| | $F = +19(F/P,47\%,2) = +19(2.161) = 41.1$ ┐ |
| 1 | + 10 invest for 1 year in an *external investment* at 47% |
| | $F = +10(F/P,47\%,1) = +10(1.47) = +14.7$ ┐ |
| | ⎧ + 14.7 return to project ← |
| 2 | ⎨ + 41.1 return to project ← |
| | ⎩ − 50 |
| 3 | − 50 |
| 4 | + 20 |
| 5 | + 60 |

When the cash outflows to the external investment are returned, the transformed cash flow is tabulated below.

| Year | Cash flow |
|------|-----------|
| 0 | $ 0 |
| 1 | 0 |
| 2 | + 5.8 |
| 3 | − 50 |
| 4 | + 20 |
| 5 | + 60 |

There is the need for external investment of the 5.8 at year 2 that will be needed in the project at year 3.

$$F = +5.8(F/P,47\%,1) = 5.8(1.47) = +8.5$$

When the 8.5 is returned to the project in year 3 the resulting net required investment in year 3 is $-50 + 8.5 = -41.5$. Now the internal investment cash flow is as follows.

| Year | Cash flow |
|------|-----------|
| 0 | $ 0 |
| 1 | 0 |
| 2 | 0 |
| 3 | $-41.5$ |
| 4 | $+20$ |
| 5 | $+60$ |

Solving for the rate of return on the internal investment

PW of costs = PW of benefits

At year 3    $41.5 = 20(P/F,i\%,1) + 60(P/F,i\%,2)$    Try $i = 47\%$
$41.5 = 20(0.680) + 60(0.464)$
$41.5 = 13.6 + 27.9$
$41.5 = 41.5$

From the computations we have seen that the two positive interest rates (10.1% and 47%) require that the internal investment and the external investment both earn the same interest rate. Thus, if one were prepared to agree that the appropriate interest rate for external investments is 47% then we would have to agree that the resulting interest rate on the internal investment is 47%. But if a suitable interest rate on external investments (say putting the money in a savings account) were only 6%, then what is the rate of return on the internal investment? None of the calculations we have made so far tell us. But in the next section we will see that this is a practical approach for solving the multiple interest rate problem.

**External Interest Rate**

From the discussion of the meaning of multiple interest rates we see an important general situation.

Solving a cash flow for an unknown interest rate means that money in any required external investment is assumed to earn the same interest rate as money invested in the internal investment.

This occurs irrespective of whether there is only one or more than one positive interest rate.

There can be no particular reason why we would assume that external investments earn the same rate of return as internal investments. The required external investment may be of short duration, like a year or two. And as will be discussed later, it is very likely that the rate of return available on a capital investment in the business is two or three times the rate of return available on short duration external deposits or other external investments of money.

Thus two interest rates are reasonable, one on the internal investment and one on the temporary external investment. By separating the interest rates we have also provided the means for resolving any difficulties that arise from multiple positive rates of return. What we desire is to determine the rate of return on the internal investment assuming a realistic value for the rate of return available on the external investment. With the external interest rate we can compute the effect of any required external investments. The results can be introduced back into the cash flow in the same manner as was done in our detailed examination of the external investment assumed in Example 7A-1. Example 7A-1 will now be presented again, but this time with a preselected external interest rate of 6%.

### EXAMPLE 7A-2

Take the Example 7A-1 cash flow and assume that any money held outside of the project earns 6% interest (that is, the external interest rate = 6%).

| Year | Cash flow |
|------|-----------|
| 0    | +$19      |
| 1    | +10       |
| 2    | −50       |
| 3    | −50       |
| 4    | +20       |
| 5    | +60       |

At both year 0 and year 1 there is a flow of money resulting from the advance payments before the aircraft are manufactured. The money will be needed later to help pay the production costs. If the external interest rate is 6%, the +19 (million dollars) will be invested externally for two years and the $+10$ for one year. Their compound amount at the end of year 2 will be:

$$
\begin{aligned}
\text{Compound amount at end of year 2} &= +19(F/P,6\%,2) + 10(F/P,6\%,1) \\
&= +19(1.124) + 10(1.06) \\
&= +21.4 + 10.6 \\
&= +32
\end{aligned}
$$

When this amount is returned to the project the net cash flow for year 2 becomes $-50 + 32 = -18$. The resulting cash flow for the project is now presented.

*Computation of NPW at various*
*interest rates using single payment*
*present worth factors*

| Year | Cash flow | $0\%$ | $8\%$ | $10\%$ |
|------|-----------|-------|-------|--------|
| 0 | $ 0 | $ 0 | $ 0 | $ 0 |
| 1 | 0 | 0 | 0 | 0 |
| 2 | $-18$ | $-18$ | $-15.4$ | $-14.9$ |
| 3 | $-50$ | $-50$ | $-39.7$ | $-37.6$ |
| 4 | $+20$ | $+20$ | $+14.7$ | $+13.7$ |
| 5 | $+60$ | $+60$ | $+40.8$ | $+37.3$ |
| | NPW $=$ | $+\$12$ | $+\$0.4$ | $-\$1.5$ |

The cash flow has one sign change indicating there is either zero or one positive interest rate. We have located a point where NPW $= 0$ at

$$
i = 8\% + 2\%\left(\frac{0.4}{1.5 + 0.4}\right) = 8\% + 2\%(0.21) = 8.4\%
$$

Thus we have identified the single positive root for the cash flow. Assuming an external interest rate of 6%, the rate of return on the Interair plane contract is 8.4%.

In Example 7A-2 we accomplished two tasks.

1. A realistic interest rate was used to find equivalent sums when money must be invested externally. This external interest rate should reflect the rate on external investment opportunities and therefore be independent of the rate of return on any particular internal investment.

2. Through the use of an external interest rate the number of sign changes in the cash flow was reduced to one, insuring that there would not be multiple positive rates of return.

### Resolving Multiple Rate of
### Return Problems

Example 7A-1 contained a cash flow that produced two positive rates of return. Yet, on closer examination we saw a cash flow for external investment in the initial part of the problem. The outflow of cash was invested at the external interest rate until it was needed in the project. It was returned to the project at the end of year 2. In so doing the number of sign changes in the cash flow was reduced from two to one. A single positive rate of return was then computed. In this way the multiple rate of return difficulty was resolved. The general approach for resolving cash flows with multiple positive rates of return is to use an external interest rate to reduce the number of sign changes to one. When we use an external interest rate to adjust a cash flow, the external interest rate will have an impact on the resulting cash flow and hence on the project or internal rate of return. To keep the sensitivity of the internal rate of return to the external interest rate as small as possible, the cash flow adjustments should be kept to a minimum. In this way we can overcome the multiple rate of return problem but do it without severe changes to the cash flow.

### SUMMARY

In certain rare situations we find that solution of a cash flow equation results in more than one positive rate of return. This is possible by the Cash Flow Rule of Signs. A sign change is where successive nonzero values in the cash flow have different signs (that is, change from + to −, or vice versa).

Zero sign changes indicates no rate of return for the cash flow entries are either all disbursements or they are all receipts.

One sign change is the usual situation and a single positive rate of return generally results. There will, however, be no rate of return whenever loan repayments are less than the loan or an investment fails to return benefits equal to the investment.

Multiple sign changes may result in multiple positive rates of return. The difficulty is not that it will happen, but that the analyst may not recognize that the cash flow has multiple sign changes and may have multiple positive rates of return. When they occur, none of the multiple rates of return are a suitable measure of the economic desirability of the project represented by the cash flow.

Multiple positive rates of return indicate a project that at some time has money invested outside the project. Since investments outside the project may earn interest at a different rate from the internal project rate of return, an external rate should be selected. This approach leaves the rate of return on the money actually invested in the project as the single unknown. The number of sign changes are thereby reduced to one, eliminating the possibility of multiple positive rates of return.

## PROBLEMS

**7A-1**  The owner of a walnut orchard wished to enjoy some of his wealth and yet not sell the orchard for ten years. He negotiated an agreement with the Omega Insurance Company as follows. Omega would pay the owner $4000 per year in 20 equal annual payments beginning immediately. At the end of the tenth year, the owner is obligated to sell the orchard and with the proceeds pay Omega $75,000 at that time. Omega will, of course, continue the $4000 annual payments to the retired orchardist for nine more years. What interest rate was used in devising the agreement between Omega and the orchardist?

**7A-2**  A group of businessmen formed a partnership to buy and race an Indianapolis type racing car. They agreed to pay an individual $50,000 for the car and associated equipment. The payment was to be in a lump sum at the end of the year. In what must be "beginner's luck," the group won a major race the first week and won a total of $80,000. The rest of the first year was not so good with the result that at the end of the first year, the group had to pay out $35,000 for expenses plus the $50,000 for the car and equipment. The second year was a poor one. The group paid $70,000 at the end of the second year to clear up the racing debts. During the third and fourth year, racing income just equalled costs. Thus, when the group was approached by a prospective buyer for the car, the group readily accepted $80,000 cash which was paid at the end of the fourth year. What rate of return did the group obtain from their racing venture?

**7A-3**  A student organization at the beginning of the Fall quarter purchased and operated a soft drink vending machine as a means of helping finance the organization's activities. The machine cost $75. The vending machine was installed at a gasoline station near the university. The student organization pays $75 every three months to the station owner in return for the right to keep their vending machine at the station. During the year the student organization owned the machine, they received the following quarterly income from it, before making the $75 quarterly payment to the station owner:

|  | *Income* |
|---|---|
| Fall quarter | $150 |
| Winter quarter | 25 |
| Spring quarter | 125 |
| Summer quarter | 150 |

At the end of one year, the student group resold the machine for $50. Determine the quarterly cash flow, and then determine the nominal annual rate of return the organization received on their investment.  (*Answer:* 172%)

**7A-4**  Given the following cash flow:

| Year | Cash flow |
|------|-----------|
| 0 | −$500 |
| 1 | +2000 |
| 2 | −1200 |
| 3 | −300 |

Determine the rate of return on the internal investment. If necessary, assume external investments earn 6% interest.   (*Answer:* 20.2%)

**7A-5**  Given the following cash flow:

| Year | Cash flow |
|------|-----------|
| 0 | −$500 |
| 1 | +200 |
| 2 | −500 |
| 3 | +1200 |

Determine the rate of return on the internal investment. If necessary, assume external investments earn 6% interest.   (*Answer:* 19.6%)

**7A-6**  Given the following cash flow:

| Year | Cash flow |
|------|-----------|
| 0 | −$100 |
| 1 | +360 |
| 2 | −570 |
| 3 | +360 |

Determine the rate of return on the internal investment. If necessary, assume external investments earn 6% interest.

# 8
# Incremental
# Analysis

We have previously seen how to solve problems by each of the three major methods, with one exception. For three or more alternatives, no rate of return solution was given. The reason is that under these circumstances incremental analysis is required and it had not been discussed. This chapter will show how to solve that problem.

Incremental analysis can be defined as the examination of the differences between alternatives. By so doing we emphasize the fact that in choosing between alternatives we are really deciding whether or not the differential costs are justified by the differential benefits.

In retrospect we see that the simplest form of incremental analysis was presented in the rate of return chapter. We were doing incremental analysis by the rate of return evaluation of the differences between two alternatives. We recognized that the two alternatives could be related as follows:

Higher cost alternative = lower cost alternative
+ differences between them

In this chapter we will see that incremental analysis can be examined either graphically or numerically. We will begin by looking at graphical representations of problems. Then we will proceed with numerical solutions of rate of return problems. Finally, we will see that a graphical representation may be useful in examining problems whether using incremental analysis or not.

## GRAPHICAL SOLUTIONS

In the last chapter we examined problems with two alternatives. Our method of solution represented a form of incremental analysis. A graphical review of that situation will help to introduce incremental analysis.

### EXAMPLE 8-1

This is a review of Example 7-6. There were two mutually exclusive alternatives:

| Year | Alternative 1 | Alternative 2 |
|------|---------------|---------------|
| 0    | −$10          | −$20          |
| 1    | +15           | +28           |

If 6% interest is assumed, which alternative should be selected? For this problem we will plot the two alternatives on a PW of benefits vs PW of cost graph.

*Alternative 1*

PW of cost = $10
PW of benefit = $15$(P/F,6\%,1)$ = 15(0.943) = $14.15

*Alternative 2*

PW of cost = $20
PW of benefit = $28$(P/F,6\%,1)$ = 28(0.943) = $26.40

The alternatives are plotted in Figure 8-1. Figure 8-1 looks very simple and yet it tells us a great deal about the situation.

On a graph of PW of benefits vs PW of cost (for convenience we will call it a benefit-cost graph) there will be a line where NPW = 0. Where the scales used on the two axes are identical, as in this case, the resulting line will be at a 45° angle. If unequal scales are used the line will be at some other angle. For the chosen interest rate (6% in this example) this NPW = 0 line divides the graph into an area of desirable alternatives and undesirable alternatives. To the left (or above) the line is desirable, while to the right (or below) the line is undesirable. We see that to the left of the line, PW of benefits exceeds the PW of cost or we could say NPW is positive. To the right of the line PW of benefits is less than PW of cost hence NPW is negative.

In this example both alternatives are to the left of the NPW = 0 line. Therefore both alternatives will have a rate of return greater than

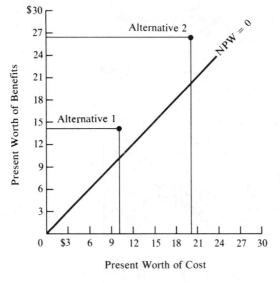

**Figure 8-1.** PW of benefits vs PW of cost
graph.

the 6% interest rate used in constructing the graph. In fact, other rate of return lines could also be computed and plotted on the graph *for this special case of a one year analysis period.* We must emphasize at the outset that the additional rate of return lines shown in Figure 8-2 can be plotted only for this special situation. For analysis periods greater than one year the NPW = 0 line is the only line that can be accurately drawn. The graphical results in Figure 8-2 agree with the calculations made in Example 7-6, where the rates of return for the two alternatives were 50% and 40% respectively. Figure 8-2 shows that the slope of a line on the graph represents a particular rate of return for this special case of a one year analysis period. Between the origin and alternative 1 the slope represents a 50% rate of return while from the origin to alternative 2 the slope represents a 40% rate of return. Also from Figures 8-1 and 8-2 we see more clearly the relationship:

Higher cost alternative 2 = lower cost alternative 1
+ differences between them

The differences between the alternatives are represented by a line as shown in Figure 8-3. Viewed in this manner we clearly see that alternative 2 may be considered two separate increments of investment. The

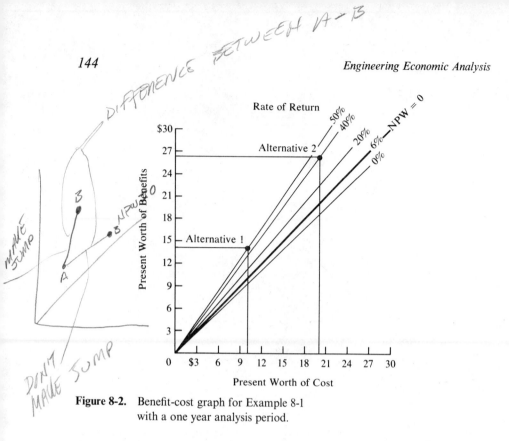

**Figure 8-2.** Benefit-cost graph for Example 8-1
with a one year analysis period.

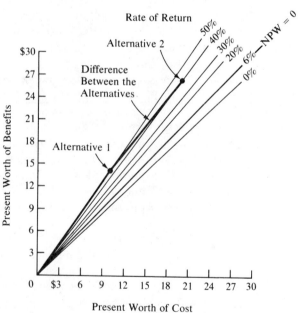

**Figure 8-3.** Benefit-cost graph for alternatives
with a one year analysis period.

first increment is alternative 1 and the second one is the difference between the alternatives. Thus we will select alternative 2 if the difference between the alternatives is a desirable increment of investment. Since the slope of the line represents a rate of return, we see that the increment is desirable if the slope of the increment is greater than the slope of the 6% line that corresponds to NPW = 0. We can readily see that the slope is greater and hence the increment of investment is attractive. In fact a careful examination shows that the "difference between the alternatives" line has the same slope as the 30% rate of return line. We can say, therefore, that the incremental rate of return from selecting alternative 2 rather than alternative 1 is 30%. (This is the same, of course, as we computed in Example 7-6.) We conclude that alternative 2 is the preferred alternative.

## EXAMPLE 8-2

Solve Example 7-8 by means of a benefit-cost graph. Two machines are being considered for purchase. If the minimum attractive rate of return is 10%, which machine should be bought?

|  | Machine X | Machine Y |
|---|---|---|
| Initial cost | $200 | $700 |
| Uniform annual benefit | 95 | 120 |
| End of useful life salvage value | 50 | 150 |
| Useful life | 6 years | 12 years |

Using a 12-year analysis period

*Machine X*

$$\text{PW of cost*} = 200 + (200 - 50)(P/F,10\%,6) - 50(P/F,10\%,12)$$
$$= 200 + 150(0.5645) - 50(0.3186) = 269$$
$$\text{PW of benefit} = 95(P/A,10\%,12) = 95(6.814) = 647$$

*Machine Y*

$$\text{PW of cost*} = 700 - 150(P/F,10\%,12) = 700 - 150(0.3186) = 652$$
$$\text{PW of benefit} = 120(P/A,10\%,12) = 120(6.814) = 818$$

---

*Salvage value is considered a reduction in cost rather than a benefit.

The two alternatives are plotted in Figure 8-4. In Figure 8-4 we see that the increment *Y-X* has a slope much less than the 10% rate of return line. The rate of return on the increment of investment is less than 10%, hence the increment is undesirable. This means that machine *X* should be selected rather than machine *Y*.

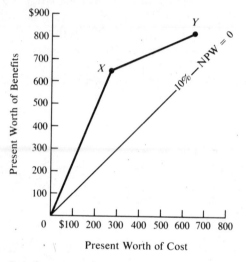

**Figure 8-4.** Benefit-cost graph.

The two example problems show us some aspects of incremental analysis. We will now proceed to examine multiple alternative problems. We can solve multiple alternative problems by present worth and annual cash flow analysis without any difficulties. Rate of return analysis requires that for two alternatives the differences between them must be examined to see whether or not they are desirable. Obviously if we can choose between two alternatives, then by a successive examination we can choose between multiple alternatives. Figure 8-5 illustrates the method.

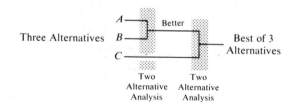

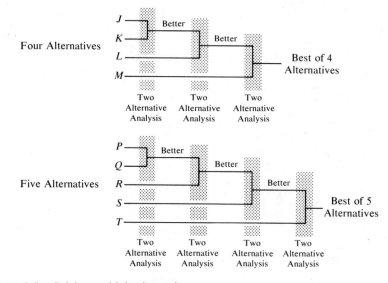

**Figure 8-5.** Solving multiple alternative problems by successive two alternative analyses.

## EXAMPLE 8-3

Given the three mutually exclusive alternatives below, each has a 20-year life and no salvage value. If the minimum attractive rate of return is 6%, which alternative should be selected?

|                        | A      | B      | C      |
|------------------------|--------|--------|--------|
| Initial cost           | $2000  | $4000  | $5000  |
| Uniform annual benefit | 410    | 639    | 700    |

At 6% the PW of benefits = uniform annual benefit × series present worth factor.

PW of benefits = uniform annual benefit $(P/A,6\%,20)$

PW of benefits for Alternative $A$ = \$410(11.47) = \$4703
Alternative $B$ = \$639(11.47) = \$7329
Alternative $C$ = \$700(11.47) = \$8029

Figure 8-6 is a plot of the situation. From Figure 8-6 we can see that the slope of the line from the origin to $A$ is greater than the 6% line (NPW = 0). Thus the rate of return for $A$ is greater than 6%. For the

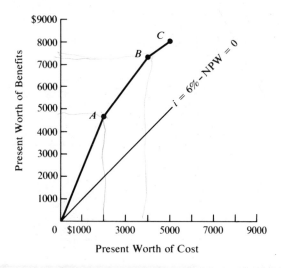

**Figure 8-6.**  Benefit-cost graph for Example 8-3.

increment of additional cost of *B* over *A* the slope of the line (*B-A*) is greater than the 6% line. This indicates that the rate of return on the increment of investment also exceeds 6%. But the slope of the increment (*C-B*) indicates its rate of return is less than 6% and hence undesirable. We conclude that the *A* investment is satisfactory as well as the *B-A* increment. Therefore *B* is satisfactory. The *C-B* increment is unsatisfactory so *C* is undesirable compared to *B*. Our decision is to select alternative *B*.

## EXAMPLE 8-4

Further study of the three alternatives of Example 8-3 reveals that the alternative *A* uniform annual benefit was overstated. It is now projected to be 122 rather than 410. Replot the benefit-cost graph for this changed situation.

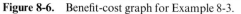

    Alternative *A'*     PW of benefits = 122(*P/A*,6%,20)
                                         = 122(11.47) = 1399

Figure 8-7 shows the revised plot of the three alternatives. The graph shows that the revised *A'* is no longer desirable. We see that it has a rate of return less than 6%. Now we wish to examine *B*. Should we compare it to the do-nothing alternative (which is represented by the origin), or as a *B-A'* increment over *A'*? Graphically, should we examine the line 0-*B* or *B-A'*? Since *A'* is an undesirable alternative it should be discarded

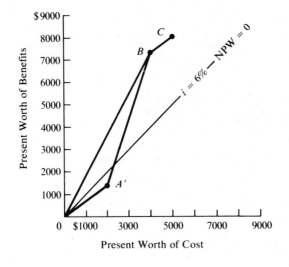

**Figure 8-7.** Benefit-cost graph for Example 8-4.

and not considered further. Thus ignoring $A'$, we should consider $B$, rather than do-nothing, which is the line $0$-$B$. Alternative $B$ is preferred over the do-nothing alternative since it has a rate of return greater than 6%. Then the increment $C$-$B$ is examined and, as we saw previously, it is an undesirable increment of investment. The decision to select $B$ has not changed. (This should be no surprise. If an inferior $A$ has become an even less attractive $A'$, we still would select the superior alternative $B$.)

The graphical solution of the four example problems has helped to visualize the mechanics of incremental analysis. While problems could be solved this way, in practice they are solved mathematically rather than graphically. We will now proceed to solve problems mathematically by incremental rate of return analysis.

## INCREMENTAL RATE OF RETURN ANALYSIS

To illustrate incremental rate of return analysis we will solve the examples again by mathematical rather than graphical methods.

*EXAMPLE 8-5*

Solve Example 8-1 mathematically. With two mutually exclusive alternatives and a 6% minimum attractive rate of return (MARR), which alternative should be selected?

| Year | Alternative 1 | Alternative 2 |
|------|---------------|---------------|
| 0    | $-$10$        | $-$20$        |
| 1    | $+15$         | $+28$         |

Examine the differences between the alternatives.

| Year | Alternative 2 $-$ Alternative 1 |
|------|----------------------------------|
| 0    | $-20 - (-10) = -10$              |
| 1    | $+28 - (+15) = +13$              |

Incremental rate of return

$$10 = 13(P/F,i\%,1)$$

$$(P/F,i\%,1) = \frac{10}{13} = 0.769$$

$$\Delta ROR = 30\%$$

The incremental rate of return ($\Delta ROR$) is greater than the minimum attractive rate of return (MARR) hence we will select the alternative that gives this increment.

Higher cost alternative 2 = lower cost alternative 1

+ increment between them

Select alternative 2.

*EXAMPLE 8-6*

Recompute Example 8-3. Minimum attractive rate of return = 6%. Each alternative has a 20-year life and no salvage value.

|                        | A     | B     | C     |
|------------------------|-------|-------|-------|
| Initial cost           | $2000 | $4000 | $5000 |
| Uniform annual benefit | 410   | 639   | 700   |

A practical first step is to compute the rate of return for each alternative.

Alternative A     $2000 = 410(P/A,i\%,20)$

$$(P/A,i\%,20) = \frac{2000}{410} = 4.87 \qquad i = 20\%$$

Alternative *B*     $4000 = 639(P/A,i\%,20)$

$$(P/A,i\%,20) = \frac{4000}{639} = 6.26 \quad i = 15\%$$

Alternative *C*     $5000 = 700(P/A,i\%,20)$

$$(P/A,i\%,20) = \frac{5000}{700} = 7.14$$

The rate of return is between 12% and 15%

$$i = 12\% + \left[\frac{7.47 - 7.14}{7.47 - 6.26}\right](3\%) = 12.8\%$$

At this point we would reject any alternative that failed to meet the minimum attractive rate of return (MARR) criterion of 6%. All three alternatives exceed the MARR in this example.

Next we will see that the alternatives are arranged in order of increasing PW of cost. Then we can examine the increments between the alternatives.

|  | *A* | *B* | *C* |
|---|---|---|---|
| Initial cost | $2000 | $4000 | $5000 |
| Uniform annual benefit | 410 | 639 | 700 |
| Rate of return | 20% | 15% | 12.8% |

|  | *Increment* *B-A* | *Increment* *C-B* |
|---|---|---|
| Incremental cost | $2000 | $1000 |
| Incremental uniform annual benefit | 229 | 61 |
| Incremental rate of return: |  |  |

$2000 = 229(P/A,i\%,20)$

$$(P/A,i\%,20) = \frac{2000}{229}$$

ΔROR
9.6%

$1000 = 61(P/A,i\%,20)$

$$(P/A,i\%,20) = \frac{1000}{61}$$

ΔROR
2.0%

The *B-A* increment is satisfactory, therefore *B* is preferred over *A*. The *C-B* increment has an unsatisfactory 2% rate of return therefore *B* is preferred over *C*. Conclusion: Select alternative *B*.

*LE 8-7*

imple 8-4. Alternative *A* in the previous example was believed
...~ an overstated benefit. The new situation for *A* (we will again
call it *A′*) is a uniform annual benefit of 122. Compute the rate of return
for *A′*.

$$2000 = 122(P/A,i\%,20) \qquad (P/A,i\%,20) = \frac{2000}{122} = 16.4 \qquad i = 2\%$$

This time *A′* has a rate of return less than the MARR of 6%. Alternative
*A′* will be rejected and the problem now becomes one of selecting the
best from between *B* and *C*. From Example 8-6 we saw that the increment *C-B* had a ΔROR of 2% and it tóo was undesirable. Thus we
again select alternative *B*.

## EXAMPLE 8-8

The following information is provided for five mutually exclusive alternatives that have 20-year useful lives. If the minimum attractive rate of
return is 6%, which alternative should be selected?

|  | *A* | *B* | *C* | *D* | *E* |
|---|---|---|---|---|---|
| Cost | $4000 | $2000 | $6000 | $1000 | $9000 |
| Uniform annual benefit | 639 | 410 | 761 | 117 | 785 |
| PW of benefit† | 7330 | 4700 | 8730 | 1340 | 9000 |
| Rate of return | 15% | 20% | 11% | 10% | 6% |

†PW of benefit = uniform annual benefit$(P/A,6\%,20)$ = 11.47(uniform annual benefit).
These values will be used later to plot a PW of cost vs PW of benefit curve.

We see that the rate of return for each alternative equals or exceeds the
MARR, therefore no alternatives are rejected at this point. Next we will
rearrange the alternatives to put them in order of increasing PW of cost.

|  | *D* | *B* | *A* | *C* | *E* |
|---|---|---|---|---|---|
| Cost | $1000 | $2000 | $4000 | $6000 | $9000 |
| Uniform annual benefit | 117 | 410 | 639 | 761 | 785 |
| Rate of return | 10% | 20% | 15% | 11% | 6% |

|  | *Increment* *B-D* | *Increment* *A-B* | *Increment* *C-A* |
|---|---|---|---|
| ΔCost | $1000 | $2000 | $2000 |
| ΔAnnual benefit | 293 | 229 | 122 |
| ΔRate of return | 29% | 10% | 2% |

Beginning with the analysis of the increment *B-D* we compute an incremental rate of return of 29%. Alternative *B* is thus preferred to *D* and *D* may be discarded at this point. The incremental rate of return for *A-B* is also satisfactory with *A* retained and *B* now discarded. The *C-A* increment has a rate of return less than the MARR. Therefore *C* is discarded and *A* continues to be retained. At this point we have examined four alternatives (*D, B, A, C*) and retained *A* after discarding the other three. Now we must decide whether *A* or *E* is the superior alternative. The increment we will examine is *E-A*. (Note that the increment *E-C* would have no particular meaning for we have already discarded *C*.)

|  | *Increment* *E-A* |
|---|---|
| ΔCost | $5000 |
| ΔAnnual benefit | 146 |

Over the 20-year useful life the total benefits (20 × 146 = 2920) are less than the cost. There is no rate of return on this increment (or one might say the ΔROR < 0%). This is an unsatisfactory increment so *E* is discarded. Alternative *A* is the best of the five alternatives.

The benefit-cost graph (Figure 8-8) illustrates an interesting situation. All five alternatives have rates of return equal to or greater than the minimum attractive rate of return (MARR) of 6%. Yet on detailed examination we see that alternatives *C* and *E* contain increments of investment that are unsatisfactory. Even though *C* has an 11% rate of return, it is unsatisfactory when compared to alternative projects. Also noteworthy is the fact that the project with the greatest rate of return (alternative *B*) is not the best alternative—rather, it is alternative *A*. This is because the proper economic criterion in this situation is to accept all separable increments of investment that have a rate of return greater than the 6% MARR. We found a desirable *A-B* increment with a 10% incremental rate of return. A relationship between alternatives *A* and *B*, and the computed rates of return are:

Higher cost alternative *A* =   alternative *B*   = differences between *A* and *B*
       15% rate of return      20% rate of return       10% rate of return

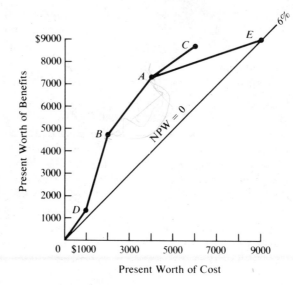

**Figure 8-8.**    Benefit-cost graph for Example 8-8
data.

By selecting *A* we have, in effect, acquired a 20% rate of return on $2000 and a 10% rate of return on an additional $2000. Both of these are desirable. Taken together as *A*, the result is a 15% rate of return on a $4000 invest-ment. This is economically preferable to a 20% rate of return on a $2000 investment, assuming we seek to invest money whenever we can identify an investment opportunity that meets our 6% minimum attractive rate of return criterion. This implies that we have sufficient money to accept all investment opportunities that come to our attention where the MARR is exceeded. This abundant supply of money is considered appropriate in most industrial analyses, but it is not likely to be valid for individuals. The selection of an appropriate minimum attractive rate of return is discussed in Chapter 15. The conclusion, based on the MARR assumption above, is to select alternative *A*.

### Elements in Incremental Rate of Return Analysis

The incremental analysis procedure has several steps.

1.  Compute the rate of return for each alternative. Generally there is a do-nothing alternative or at least one alternative whose rate of return exceeds the minimum attractive rate of return. Under these

circumstances we may reject any alternative whose ROR < MARR. (This step is not essential, but it immediately rejects unacceptable alternatives. One must insure, however, in any case that the lowest cost alternative has a rate of return ≥ MARR.)

2. Rank the remaining alternatives in their order of increasing PW of cost. If any higher cost alternative has a rate of return greater than MARR, and greater than the rate of return of a lower cost alternative, then the lower cost alternative may be immediately rejected.

3. Make a two-alternative analysis of the two lowest cost alternatives. Compute the incremental rate of return (ΔROR) on the cash flow representing the differences between the alternatives. If ΔROR ≥ MARR, retain the higher cost alternative and reject the lower cost alternative. If ΔROR < MARR, retain the lower cost alternative and reject the higher cost alternative.

4. Take the preferred alternative from step 3. Consider the next higher cost alternative and proceed with another two alternative comparison.

5. Continue until all alternatives have been examined and the best of the multiple alternatives has been identified.

## Incremental Analysis Where There Are Unlimited Alternatives

These are situations where the possible alternatives are a more or less continuous function. For example, an analysis to determine the economical height of a dam represents a situation where the number or alternatives could be infinite. If the alternatives were limited, however, to heights in even feet, the number of alternatives would still be large and have many of the qualities of a continuous function. Consider the following example.

### EXAMPLE 8-9

A careful analysis has been made of the consequences of constructing a dam in the Blue Canyon. It would be feasible to construct a dam at this site with a height anywhere from 200 to 500 feet. Using a 4% minimum attractive rate of return and a 75-year life the various data were used to construct Figure 8-9. Note particularly that the dam heights are plotted on the x-axis along with the associated PW of cost. What height of dam should be constructed? Five points have been labelled on Figure 8-9 to aid in the discussion. Dam heights below point *A* have a PW of cost > PW of benefit hence the rate of return is less than MARR and we would not build a dam of these heights. In the region of point *B* an increment of additional PW of cost (ΔC) produces a larger increment

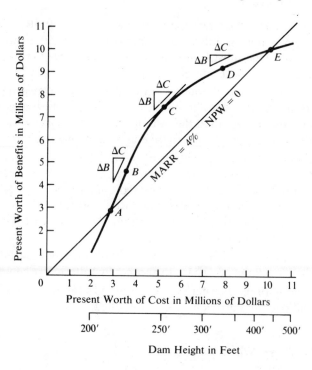

**Figure 8-9.**  Benefit-cost graph for Example 8-9.

of PW of benefit ($\Delta B$). These are therefore desirable increments of additional investment and hence dam height.

At point $D$ (and also at point $E$) the reverse is true. An increment of additional investment ($\Delta C$) produces a smaller increment of PW of benefit ($\Delta B$). This is undesirable. We do not want these increments and so the dam should not be built to these heights. At point $C$ we are at the point where $\Delta B = \Delta C$. Lower dam heights have desirable increments of investment and higher dam heights have unfavorable increments of investment. The optimal dam height therefore is where $\Delta B = \Delta C$. On the figure this corresponds to a height of approximately 250 feet. By this time you probably have recognized that another way of defining the point where $\Delta B = \Delta C$ is to describe it as the point where the slope of the curve equals the slope of the NPW = 0 line.

The techniques for solving discrete alternatives or continuous function alternatives are really the same. We proceed by adding increments whenever $\Delta ROR \geqslant MARR$ and discarding increments when $\Delta ROR < MARR$.

## PRESENT WORTH ANALYSIS WITH
## BENEFIT-COST GRAPHS

Any of the example problems presented so far in this chapter could be solved by the present worth method. The benefit-cost graphs we introduced here can be used to graphically solve problems by present worth analysis.

For neither fixed input nor fixed output the criterion in present worth analysis is to maximize NPW. In the Example 8-1 problem we had the case of two alternatives and the MARR equal to 6%.

| Year | Alternative 1 | Alternative 2 |
|------|---------------|---------------|
| 0 | −$10 | −$20 |
| 1 | +15 | +28 |

In Example 8-1 we computed

| | | |
|---|---|---|
| PW of cost | $10 | $20 |
| PW of benefit | 14.5 | 26.42 |

These points are plotted in Figure 8-10. Looking at Figure 8-10 we see that the NPW = 0 line is at 45° since identical scales were used on both axes. The point for alternate 2 is plotted at the coordinates (PW of cost, PW of benefits). We drop a vertical line from alternative 2 to the diagonal NPW = 0 line. The coordinates of any point on the graph are (PW of

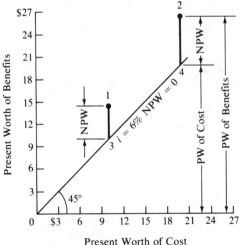

**Figure 8-10.** Benefit-cost graph for PW analysis.

cost, PW of benefits), but along the NPW = 0 line (45° line) the $x$ and $y$ coordinates are equal. Thus the coordinates of point 4 are also (PW of cost, PW of cost). Since

$$NPW = PW \text{ of benefits} - PW \text{ of cost}$$

the vertical distance from point 4 to alternative 2 represents NPW. Similarly, the vertical distance between point 3 and alternative 1 presents NPW for alternative 1. Since the criterion is to maximize NPW we select alternative 2 with larger NPW. The same technique is used in situations where there are multiple alternatives or continuous alternatives. In Example 8-8 there were five alternatives. Figure 8-11 shows a plot of the five alternatives. We can see that $A$ has the greatest NPW and therefore is the preferred alternative.

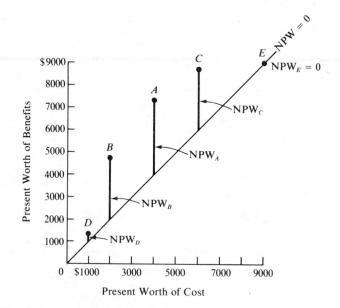

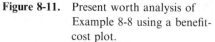

**Figure 8-11.**   Present worth analysis of
Example 8-8 using a benefit-
cost plot.

## CHOOSING AN ANALYSIS
## METHOD

At this point we have examined in detail the three major economic analysis techniques: present worth analysis, annual cash flow analysis, and rate of return analysis. A practical question is, which method should be used for a particular problem? While the obvious answer is to use the method requiring the least computations, there are a number of factors that may affect the decision.

1. Unless the minimum attractive rate of return (minimum required interest rate for invested money) is known, neither present worth analysis nor annual cash flow analysis can be done.

2. Present worth analysis and annual cash flow analysis often require far less computations than rate of return analysis.

3. In some situations a rate of return analysis is easier to explain to people unfamiliar with economic analysis. At other times, an annual cash flow analysis may be easier to explain.

4. Business enterprises generally adopt one, or at most two, analysis techniques for broad categories of problems. If you work for a corporation, and the policy manual specifies rate of return analysis, you would appear to have no choice in the matter.

Since one may not always be able to choose the analysis technique computationally best suited to the problem, this book illustrates how to use each of the three methods in all feasible situations. And ironically, the most difficult method—rate of return analysis—is the most frequently used by engineers in industry.

## SUMMARY

Incremental analysis is based on a systematic examination of each separable increment of investment. For each increment the computed incremental rate of return ($\Delta$ROR) is compared to the minimum attractive rate of return (MARR). An increment is economically desirable whenever $\Delta$ROR $\geqslant$ MARR.

A graph of the PW of benefits vs PW of cost (called a benefit-cost graph) can be an effective way to examine two alternatives by incremental analysis. And since multiple alternative incremental analysis is done by the successive analysis of two alternatives, benefit-cost graphs can be used to solve multiple alternative and continuous function alternatives as easily as two alternative problems.

The important steps in incremental rate of return analysis are:

1. Compute the rate of return for each alternative. Reject any alternatives where the rate of return is less than the minimum attractive rate of return.

2. Rank the remaining alternatives in their order of increasing PW of cost. If any higher cost alternative has a rate of return greater than

the rate of return of a lower cost alternative, then the lower cost alternative may be immediately rejected.

3. Compute the ΔROR on the differences between the two lowest cost alternatives. If the ΔROR ⩾ MARR, the increment is desirable. The higher cost alternative should be retained and the lower cost alternative rejected. The opposite decision is made when ΔROR < MARR.

4. Take the preferred alternative from step 3. Consider the next higher cost alternative and proceed with another two alternative comparison.

5. Continue until all alternatives have been examined and the best of the multiple alternatives has been identified.

In incremental analysis we assume that all increments of investment that exceed the MARR are desirable. This assumption is an important one that probably is appropriate for businesses but not for individuals. It is discussed at length in Chapter 15.

Benefit-cost graphs, being a plot of the PW of benefits vs the PW of costs, can also be used in present worth analysis to graphically show the NPW for each alternative.

## PROBLEMS

Unless otherwise noted all problems should be solved by rate of return analysis.

8-1 A firm is considering moving its manufacturing plant from Chicago to a new location. The Industrial Engineering Department was asked to identify the various alternatives together with the costs to relocate the plant, and the benefits. They examined six likely sites, together with the do-nothing alternative of keeping the plant at its present location. Their findings are summarized below:

| Plant location | First cost | Uniform annual benefit |
|---|---|---|
| Denver | $300 thousand | $ 52 thousand |
| Dallas | 550 | 137 |
| San Antonio | 450 | 117 |
| Los Angeles | 750 | 167 |
| Cleveland | 150 | 18 |
| Atlanta | 200 | 49 |
| Chicago | 0 | 0 |

The annual benefits are expected to be constant over the eight-year analysis period. If the firm uses 10% annual interest in its economic analysis, where should the manufacturing plant be located? (*Answer:* Dallas)

**8-2** In a particular situation four mutually exclusive alternatives are being considered. Each of the alternatives costs $1300 and has no end-of-useful-life salvage value.

| Alternative | Annual benefit | Useful life | Calculated rate of return |
|---|---|---|---|
| A | $100 at end of first year; *increasing* $100 per year thereafter | 6 yrs | 12% |
| B | $50 at end of first year; *increasing* $65 per year thereafter | 8 | 10% |
| C | Annual end of year benefit = $260 | 10 | 15% |
| D | $450 at end of first year; *declining* $65 per year thereafter | 8 | 12% |

If the minimum attractive rate of return (MARR) is 8%, which alternative should be selected? (*Answer:* Alternative *C*)

**8-3** A more detailed examination of the situation in Problem 8-2 reveals that there are two additional mutually exclusive alternatives to be considered. Both cost more than the $1300 for the four original alternatives.

| Alternative | Cost | Annual end of year benefit | Useful life | Calculated rate of return |
|---|---|---|---|---|
| E | $3000 | $ 488 | 10 yrs | 10% |
| F | 5850 | 1000 | 10 | 11.2% |

If the minimum attractive rate of return (MARR) remains at 8%, which one of the six alternatives should be selected? Neither alternative *E* nor *F* has any end-of-useful-life salvage value. (*Answer:* Alternative *F*)

**8-4** The owner of a downtown parking lot has employed a civil engineering consulting firm to advise him whether or not it is economically feasible to construct an office building on the site. Bill Samuels, a newly hired civil engineer, has been assigned to make the analysis. He has assembled the following data.

| Alternative | Total investment* | Total net annual revenue from property |
|---|---|---|
| Sell parking lot | $ 0 | $ 0 |
| Keep parking lot | 200,000 | 22,000 |
| Build 1-story building | 400,000 | 60,000 |
| Build 2-story building | 555,000 | 72,000 |
| Build 3-story building | 750,000 | 100,000 |
| Build 4-story building | 875,000 | 105,000 |
| Build 5-story building | 1,000,000 | 120,000 |

*Includes the value of the land.

The analysis period is to be 15 years. For all alternatives the property has an estimated resale (salvage) value at the end of 15 years equal to the present total investment. If the minimum attractive rate of return is 10%, what recommendation should Bill make?

**8-5** An oil company plans to purchase a large piece of vacant land on the corner of two busy streets for $70,000. The company has four different types of businesses that it installs on properties of this type.

| Plan | Cost of improvements* | |
|------|----------------------|---|
| A | $ 75,000 | Conventional gas station with service facilities for lubrication, oil changes, etc |
| B | 230,000 | Automatic carwash facility with gasoline pump island in front |
| C | 30,000 | Discount gas station (no service bays) |
| D | 130,000 | Gas station with low cost quick carwash facility |

*Cost of improvements does not include the $70,000 cost of land.

In each case, the estimated useful life of the improvements is 15 years. The salvage value for each is estimated to be the $70,000 cost of the land. The net annual income, after paying all operating expenses, is projected.

| Plan | Net annual income |
|------|-------------------|
| A | $23,300 |
| B | 44,300 |
| C | 10,000 |
| D | 27,500 |

If the oil company expects a 10% rate of return on its investments, which plan (if any) should be selected?

**8-6** A firm is considering three mutually exclusive alternatives as part of a production improvement program. The alternatives are:

| | A | B | C |
|---|---|---|---|
| Installed cost | $10,000 | $15,000 | $20,000 |
| Uniform annual benefit | 1625 | 1625 | 1890 |
| Useful life | 10 yrs | 20 yrs | 20 yrs |

For each alternative the salvage value at the end of its useful life is zero. At the end of ten years alternative A could be replaced with another A with identical cost and benefits. The minimum attractive rate of return is 6%. If the analysis period is 20 years, which alternative should be selected?

**8-7** Given the following four mutually exclusive alternatives:

| | A | B | C | D |
|---|---|---|---|---|
| First cost | $75 | $50 | $50 | $85 |
| Uniform annual benefit | 16 | 12 | 10 | 17 |
| Useful life | 10 yrs | 10 yrs | 10 yrs | 10 yrs |
| End-of-useful-life salvage value | 0 | 0 | 0 | 0 |
| Computed rate of return | 16.8% | 20.2% | 15.1% | 15.1% |

If the minimum attractive rate of return (MARR) is 8%, which alternative should be selected? (*Answer: A*)

**8-8** Consider the following three mutually exclusive alternatives.

| | A | B | C |
|---|---|---|---|
| First cost | $200 | $300 | $600 |
| Uniform annual benefit | 59.7 | 77.1 | 165.2 |
| Useful life | 5 yrs | 5 yrs | 5 yrs |
| End-of-useful-life salvage value | 0 | 0 | 0 |
| Computed rate of return | 15% | 9% | 11.7% |

For what range of values of minimum attractive rate of return (MARR) is alternative C the preferred alternative? Put your answer in the following form: Alternative C is preferred when _____% ≤ MARR ≤ _____%.

# 9
# Other
# Analysis
# Techniques

In this chapter we will examine four topics. They are:

Future worth analysis
Benefit-cost ratio analysis
Payback period
Sensitivity and breakeven analysis

We will see that the first, future worth analysis, is very much like present worth analysis. Rather than dealing with "now" (present worth) analysis we will deal with "then" (future worth) analysis.

Previously we have written economic analysis relationships based on

$$\text{PW of cost} = \text{PW of benefit} \quad \text{or} \quad \text{EUAC} = \text{EUAB}$$

Instead of writing it in this form, we could define the relationship as

$$\frac{\text{PW of benefit}}{\text{PW of cost}} = 1 \quad \text{or} \quad \frac{\text{EUAB}}{\text{EUAC}} = 1$$

When economic analysis is based on these ratios, the calculations are called benefit-cost ratio analysis. Payback period is an approximate analysis technique. Although defined many ways, it is generally the time required for the cumulative benefits to equal the cumulative costs.

Since much of the data gathered concerning a problem represents a projection of future consequences, there may be considerable uncertainty regarding the accuracy of the data. But since the end result of the analysis is decision-making, an appropriate question is to what extent the variations in the data will affect the decision. When small variations in a particular estimate would change selection of the alternative, the decision is said to be sensitive to the estimate. To better evaluate the impact of particular estimates, we could compute what variation would be necessary in a particular estimate to change the decision. This is called sensitivity analysis. Closely related to this is an analysis to determine the conditions where two alternatives are equivalent. This is breakeven analysis. Thus breakeven analysis is a form of sensitivity analysis.

## FUTURE WORTH ANALYSIS

In the various economic analysis techniques we have resolved the alternatives into comparable units. In present worth analysis the comparison was made in terms of the present consequences of taking the feasible courses of action. In annual cash flow analysis the comparison was in terms of equivalent uniform annual costs (or benefits). We saw that we could easily convert from present worth to annual cash flow and vice versa. But the concept of resolving alternatives into comparable units is not restricted to a present or annual comparison. The comparison may be made at any point in time. There are many situations where we would like to know what the future situation will be, if we take some particular course of action now. This is called future worth analysis.

### EXAMPLE 9-1

Bill James, a 20-year old college student, considers himself an average cigarette smoker for he consumes about a carton a week. He wonders how much money he could accumulate by the time he reaches 65 if he quit smoking and put his cigarette money into a savings account. Cigarettes cost $4 per carton. Bill expects that a savings account would earn 5% interest, compounded semiannually. Compute Bill's future worth at age 65.

Semi-annual saving = $4/carton × 26 weeks = $104
Future worth (FW) = $A(F/A,2\frac{1}{2}\%,90)$ = 104(329.1) = $34,226

### EXAMPLE 9-2

An East Coast firm has decided to establish a second plant in Kansas City. There is a factory for sale for $850,000 that, with extensive remodeling, could be used. As an alternative the company can buy vacant land for $85,000 and have a new plant constructed on the property. Either way it will be three years before the company will be able to get the plant into production. The timing and cost of the various components for the factory are given in the cash flow table below.

| Year | Construct new plant | | Remodel available factory | |
|---|---|---|---|---|
| 0 | Buy land | $85,000 | Purchase factory | $850,000 |
| 1 | Design and initial construction costs | 200,000 | Design and re-modeling costs | 250,000 |
| 2 | Balance of con-struction costs | 1,200,000 | Additional re-modeling costs | 250,000 |
| 3 | Setup production equipment | 200,000 | Setup production equipment | 250,000 |

If interest is 8%, which alternative results in the lower equivalent cost when the firm begins production at the end of the third year?

*New plant*

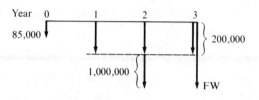

Future worth (FW) = $85,000(F/P,8\%,3) + 200,000(F/A,8\%,3)$
$+ 1,000,000(F/P,8\%,1)$
$= \$1,836,000$

*Remodel available factory*

Future worth (FW) = $850,000(F/P,8\%,3) + 250,000(F/A,8\%,3)$
$= \$1,882,000$

The total cost of remodeling the available factory ($1,600,000) is smaller than the total cost of a new plant ($1,685,000). The timing of the expenditures however, is less favorable than building the new plant. The new plant is projected to have the smaller future worth of cost, and hence is the preferred alternative.

## BENEFIT-COST RATIO ANALYSIS

At a given minimum attractive rate of return (MARR) we would consider an alternative acceptable provided

PW of benefits − PW of costs ≥ 0
EUAB − EUAC ≥ 0

At this point we also have

Benefit-cost ratio $B/C = \dfrac{\text{PW of benefits}}{\text{PW of costs}} = \dfrac{\text{EUAB}}{\text{EUAC}} \geqslant 1$

Rather than solving problems using present worth or annual cash flow analysis, we could base the calculations on the benefit-cost ratio (B/C). The criteria are presented in Table 9-1. We will illustrate B/C analysis by solving the same example problems worked by other economic analysis methods.

**Table 9-1.** Benefit-cost ratio analysis

|  | *Situation* | *Criterion* |
|---|---|---|
| Fixed input | Amount of money or other input resources are fixed | Maximize B/C |
| Fixed output | Fixed task, benefit, or other output to be accomplished | Maximize B/C |
| Neither input nor output fixed | Neither amount of money or other inputs nor amount of benefits or other outputs are fixed | *Two alternatives:* Compute incremental benefit-cost ratio ($\Delta$B/$\Delta$C) on the difference between the alternatives. If $\Delta$B/$\Delta$C $\geqslant$ 1 choose higher cost alternative; otherwise choose lower cost alternative |
|  |  | *Three of more alternatives:* Solve by benefit-cost ratio incremental analysis |

## EXAMPLE 9-3

A firm is trying to decide which of two devices to install to reduce costs in a particular situation. Both devices cost $1000 and have useful lives of five years and no salvage value. Device *A* can be expected to result in $300 savings annually. Device *B* will provide cost savings of $400 the first year, but will decline $50 annually, making the second year savings $350, the third year savings $300, and so forth. With interest at 7%, which device should the firm purchase? This problem was previously solved by present worth (Example 5-1), annual cash flow (Example 6-5) and rate of return analysis (Example 7-7).

*Device A*

PW of cost = $1000

PW of benefit = $300(P/A,7\%,5) = 300(4.10) = \$1230$

$$\mathrm{B/C} = \frac{\text{PW of benefit}}{\text{PW of cost}} = \frac{1230}{1000} = 1.23$$

*Device B*

   PW of cost = $1000

   PW of benefit = $400(P/A,7\%,5) - 50(P/G,7\%,5)$

   $= 400(4.10) - 50(7.65) = 1640 - 382 = 1258$

$$\mathrm{B/C} = \frac{\text{PW of benefit}}{\text{PW of cost}} = \frac{1258}{1000} = 1.26$$

To maximize the benefit-cost ratio, select device *B*.

*EXAMPLE 9-4*

Two machines are being considered for purchase. Assuming 10% interest, which machine should be bought?

|  | Machine X | Machine Y |
|---|---|---|
| Initial cost | $200 | $700 |
| Uniform annual benefit | 95 | 120 |
| End of useful life salvage value | 50 | 150 |
| Useful life | 6 yrs | 12 yrs |

Assuming a 12 year analysis period, the cash flow table is:

| Year | Machine X | Machine Y |
|---|---|---|
| 0 | −$200 | −$700 |
| 1–5 | +95 | +120 |
| 6 | +95 −200 +50 | +120 |
| 7–11 | +95 | +120 |
| 12 | +95 +50 | +120 +150 |

We will solve the problem using

$$\mathrm{B/C} = \frac{\text{EUAB}}{\text{EUAC}}$$

and considering the salvage value of the machines to be reductions in cost (rather than increases in benefits).

*Machine X*

$$EUAC = 200(A/P,10\%,6) - 50(A/F,10\%,6)$$
$$= 200(0.230) - 50(0.130) = 46 - 6 = \$40$$
$$EUAB = \$95$$

(Note that this assumes the replacement for the last six years has identical costs. Under these circumstances the EUAC for the first six years equals the EUAC for all twelve years.)

*Machine Y*

$$EUAC = 700(A/P,10\%,12) - 150(A/F,10\%,12)$$
$$= 700(0.147) - 150(0.047) = 103 - 7 = \$96$$
$$EUAB = \$120$$

*Machine Y − Machine X*

$$\frac{\Delta B}{\Delta C} = \frac{120 - 95}{96 - 40} = \frac{25}{56} = 0.45$$

Since the incremental benefit-cost ratio is less than one it represents an undesirable increment of investment. We therefore choose the lower cost alternative—machine $X$. If we had computed benefit-cost ratios for each machine, they would have been:

| *Machine X* | *Machine Y* |
|:---:|:---:|
| $B/C = \dfrac{95}{40} = 2.38$ | $B/C = \dfrac{120}{96} = 1.25$ |

The fact that $B/C = 1.25$ for machine $Y$ (the higher cost alternative) must not be used as the basis for suggesting that the more expensive alternative should be selected. The incremental benefit-cost ratio, $\Delta B/\Delta C$, clearly shows that $Y$ is a less desirable alternative than $X$. Also, we must not jump to the conclusion that the best alternative is always the one with the largest B/C ratio. This too, may lead to incorrect decisions as we shall see when we examine problems with three or more alternatives.

*EXAMPLE 9-5*

Given the five mutually exclusive alternatives from Example 8-8 plus an additional alternative $F$. They have 20-year useful lives and no salvage value. If the minimum attractive rate of return is 6%, which alternative should be selected?

|            | A      | B      | C      | D      | E      | F        |
|------------|--------|--------|--------|--------|--------|----------|
| Cost       | $4000  | $2000  | $6000  | $1000  | $9000  | $10,000  |
| PW of benefit | 7330 | 4700   | 8730   | 1340   | 9000   | 9500     |
| $B/C = \dfrac{\text{PW of benefit}}{\text{PW of cost}}$ | 1.83 | 2.35 | 1.46 | 1.34 | 1.00 | 0.95 |

Incremental analysis is needed to solve the problem. The steps in the solution are as follows.

1. Compute the B/C ratio for each alternative. Since there are alternatives whose $B/C \geqslant 1$, we will discard any with a $B/C < 1$. Discard alternative F.

2. Rank the remaining alternatives in order of increasing PW of cost. (Since B/C for B is greater than B/C for D, alternative D could have been discarded at this point.)

|                        | D      | B      | A      | C      | E      |
|------------------------|--------|--------|--------|--------|--------|
| Cost (= PW of cost)    | $1000  | $2000  | $4000  | $6000  | $9000  |
| PW of benefit          | 1340   | 4700   | 7330   | 8730   | 9000   |
| B/C                    | 1.34   | 2.35   | 1.83   | 1.46   | 1.00   |

|                | Increment B-D | Increment A-B | Increment C-A |
|----------------|---------------|---------------|---------------|
| ΔCost          | $1000         | $2000         | $2000         |
| ΔBenefit       | 3360          | 2630          | 1400          |
| ΔB/ΔC          | 3.36          | 1.32          | 0.70          |

3. Examine each separable increment of investment. If $\Delta B/\Delta C < 1$, the increment is not attractive. If $\Delta B/\Delta C \geqslant 1$, the increment of investment is desirable. The increments B-D and A-B are desirable. Thus of the first three alternatives D, B, and A, alternative A is the preferred alternative. The increment C-A is not attractive as $\Delta B/\Delta C = 0.70$. This indicates that of the first four alternatives (D, B, A, and C), A continues as the best of the four. Now we want to decide between A and E. This we will do by examining the increment of investment that represents the difference between these alternatives.

|                | Increment E-A |
|----------------|---------------|
| ΔCost          | $5000         |
| ΔBenefit       | 1670          |
| ΔB/ΔC          | 0.33          |

The increment is undesirable. We choose *A* as the best of the six alternatives. One should note that the best alternative in this example does not have the highest B/C ratio.

Benefit-cost ratio analysis may be graphically represented. Figure 9-1 is a graph of Example 9-5. Looking at Figure 9-1 we see that *F* has a B/C ratio < 1 and can be discarded. Alternative *D* is the starting point for examining the separable increments of investment. The slope of line *B-D* indicates a ΔB/ΔC ratio of > 1. This is also true for line *A-B*. The increment *C-A* has a slope much flatter than B/C = 1 indicating an undesirable increment of investment. Alternative *C* is therefore discarded and *A* retained. The increment *E-A* is similarly unattractive. Alternative *A* is therefore the best of the six alternatives. One should particularly note two additional things about Figure 9-1. First, even if alternatives with a B/C ratio < 1 had not been initially excluded, they would have been systematically eliminated in the incremental analysis. Since this is the case, it is not essential that the B/C ratio be computed for each alternative as an initial step in incremental analysis. Nevertheless, it seems like an orderly and logical way to approach

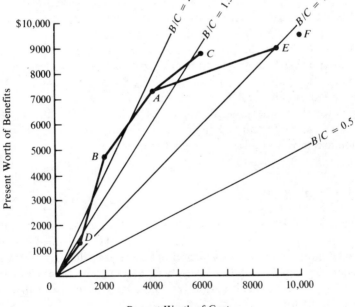

**Figure 9-1.**    Benefit-cost ratio graph of
                  Example 9-5.

a multiple alternative problem. Second, alternative $B$ had the highest B/C ratio (B/C = 2.35), but it is not the best of the six alternatives. We saw this same situation in rate of return analysis of three or more alternatives. The reason is the same in both analysis situations. We seek to maximize the total profit, not the profit rate.

### Continuous Alternatives

There are times when the feasible alternatives are a continuous function. In Chapter 8 the height of a dam was an example of this situation. It was possible to build the dam anywhere from 200 to 500 feet high. In many situations the projected capacity of an industrial plant can be varied continuously over some feasible range. In these situations we seek to add increments of investment where $\Delta B/\Delta C \geqslant 1$ and avoid increments where $\Delta B/\Delta C < 1$. The optimal size of such a project is where $\Delta B/\Delta C = 1$. Figure 9-2a shows the line of feasible alternatives with their costs and benefits. This may represent a lot of calculations to locate points through which the line passes. Figure 9-2b shows how the incremental benefit-cost ratio ($\Delta B/\Delta C$) changes as one moves along the line of feasible alternatives. In Figure 9-2b the incremental net present worth to incremental cost ratio ($\Delta NPW/\Delta C$) is also plotted. As expected we are adding increments of NPW as long as $\Delta B/\Delta C > 1$. Finally, in Figure 9-2c we see the plot of (total) NPW vs the size of the project. This three part figure demonstrates that both present worth analysis and benefit-cost ratio analysis lead to the same optimal decision. In Chapter 8 we saw that rate of return and present worth analysis led to identical decisions. Any of the exact analysis methods, present worth, annual cash flow, rate of return, or benefit-cost ratio will lead to the same decision. Benefit-cost ratio analysis is extensively used in economic analysis at all levels of government.

### PAYBACK PERIOD

Payback period is the period of time required for the profit or other benefits from an investment to equal the cost of the investment. This is the general definition for payback period but there are other definitions. Others consider depreciation of the investment and income taxes, and they too are simply called payback period. For now we will limit our discussion to the simplest form.

> Payback period is the period of time required for the profit or other savings of an investment to equal the cost of the investment.

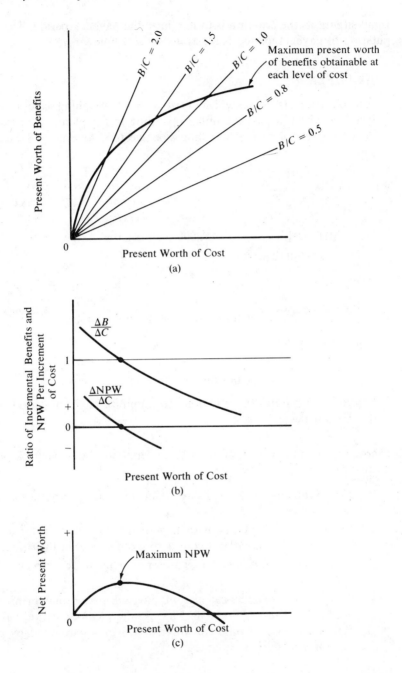

**Figure 9-2.** Selecting optimal size of project.

In all situations the criterion is to minimize the payback period. The computation of payback period is illustrated in Example 9-6.

### EXAMPLE 9-6

A firm is trying to decide which of two alternate weighing scales it should install to check a package filling operating in the plant. If both scales have a six-year life, which one should be selected? Assume an 8% interest rate.

| Alternative | Cost | Uniform annual benefit | End of useful life salvage value |
|---|---|---|---|
| Atlas scales | $2000 | $450 | $100 |
| Tom Thumb scales | 3000 | 600 | 700 |

*Atlas scales*

$$\text{Payback period} = \frac{\text{Cost}}{\text{Uniform annual benefit}} = \frac{2000}{450} = 4.4 \text{ years}$$

*Tom Thumb scales*

$$\text{Payback period} = \frac{\text{Cost}}{\text{Uniform annual benefit}} = \frac{3000}{600} = 5 \text{ years}$$

Figure 9-3 illustrates the situation. To minimize payback period, select the Atlas scales.

There are four important points to be understood about payback period calculations.

1.  This is an approximate rather than an exact, economic analysis calculation.
2.  All costs and all profits or savings of the investment prior to payback are included without considering differences in their timing.
3.  All the economic consequences beyond the payback period are completely ignored.
4.  Being an approximate calculation, payback period may or may not select the correct alternative. That is, the payback period calculations may select a different alternative from that found by exact economic analysis techniques.

This last point—that payback period may select the wrong alternative—was illustrated by Example 9-6. Using payback period the Atlas scales

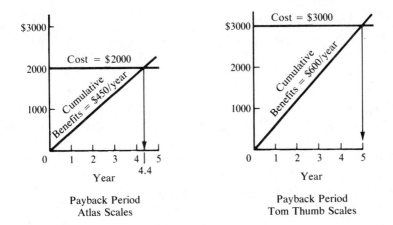

**Figure 9-3.** Payback period plots for Example 9-6.

appears to be the more attractive alternative. Yet this same problem was solved by the present worth method (Example 5-4) and there the Tom Thumb scales was the chosen alternative. A review of the problem reveals the reason for the different conclusions. The $700 salvage value at the end of six years for the Tom Thumb scales is a significant benefit. It was considered in the present worth analysis with the result that Tom Thumb scales was the more desirable alternative. The salvage value occurs after the payback period and so in the payback calculation it was ignored.

But if payback period calculations are approximate, and at times select the wrong alternative, why is the method used at all? There are two primary answers that can be given. First, the calculations can be readily made by people unfamiliar with economic analysis. One does not need to know how to use gradient factors or even to have a set of compound interest tables. Second, payback period is a readily understood concept. (Earlier we pointed out that this was also an advantage of describing the attractiveness of an alternative in the form of a rate of return.) Also, payback period does give us a useful measure. It tells us how long it will take for the cost of the investment to be recovered from the benefits of the investment. Businessmen are often very interested in this time period. A rapid return of the invested capital means that it can then be used for other purposes by the firm. But one must not confuse the speed of the return of the investment, as measured by the payback period, with economic efficiency. They are two distinctly separate concepts. One emphasizes the quickness of the return of the investment while the other considers the profitability of the investment. We can create another situation to illustrate that selecting between alternatives by the payback period criterion may result in an unwise decision.

*EXAMPLE 9-7*

A firm is purchasing production equipment for a new plant. For a particular operation two alternative machines are being considered.

|  | *Tempo machine* | *Dura machine* |
|---|---|---|
| Installed cost | $30,000 | $35,000 |
| Net annual benefit after deducting all annual expenses | $12,000 the first year, *declining* $3000 per year thereafter | $1000 the first year, *increasing* $3000 per year there-after |
| Useful life | 4 years | 8 years |

Neither machine has any salvage value. Compute the payback period for each of the alternatives.

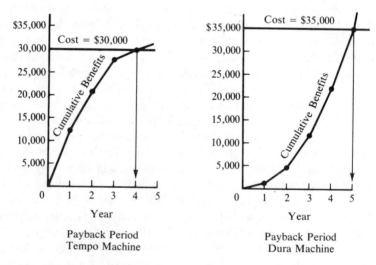

Payback Period
Tempo Machine

Payback Period
Dura Machine

**Figure 9-4.**   Payback period plots for Example 9-7.

The Tempo machine has a declining annual benefit, while the Dura has an increasing annual benefit. Figure 9-4 shows the Tempo has a four-year payback period and the Dura has a five-year payback period. To minimize the payback period the Tempo is selected. Compute the rate of return for each alternative as a check on the payback period analysis. Assume the minimum attractive rate of return is 10%.

The cash flows for the two alternatives are as follows:

| Year | Tempo machine | Dura machine |
|------|---------------|--------------|
| 0 | − $30,000 | − $35,000 |
| 1 | + 12,000 | + 1,000 |
| 2 | + 9,000 | + 4,000 |
| 3 | + 6,000 | + 7,000 |
| 4 | + 3,000 | + 10,000 |
| 5 | 0 | + 13,000 |
| 6 | 0 | + 16,000 |
| 7 | 0 | + 19,000 |
| 8 | 0 | + 22,000 |
| | $\Sigma = 0$ | $\Sigma = +57,000$ |

*Tempo machine:* Since the sum of the cash flows for the Tempo machine is zero, we see immediately that the $30,000 investment just equals the subsequent benefits. The resulting rate of return is 0%.

*Dura machine:*

$$35,000 = 1000(P/A,i\%,8) + 3000(P/G,i\%,8)$$

Try $i = 20\%$.

$$35,000 = 1000(3.84) + 3000(9.88)$$
$$= 3840 + 29,640 = 33,480$$

The 20% interest rate is too high. Try $i = 15\%$.

$$35,000 = 1000(4.49) + 3000(12.48)$$
$$= 4490 + 37,440 = 41,930$$

This time the interest rate is too low. Linear interpolation would show that the rate of return is approximately 19%.

Using an exact calculation—rate of return—it is clear that Tempo is not economically very attractive. Yet it was this alternative, and not the Dura machine, that was preferred based on the payback period calculations. On the other hand, the shorter payback period for Tempo does give a measure of the speed of the return of the investment not found in the Dura. The conclusion to be drawn is that liquidity and profitability may be two quite different criteria.

From the discussion and the examples we see that payback period can be helpful in providing a measure of the speed of the return of the investment. This might be quite important, for example, for a company that is short of working capital, or one where there are rapid changes in technology. This must not, however be confused with a careful economic analysis. We have shown that a short payback period does not always mean that the investment is desirable. Thus payback period should not be considered a suitable replacement for accurate economic analysis calculations.

## SENSITIVITY AND BREAKEVEN ANALYSIS

Sensitivity is the relative magnitude of the change in one or more elements of a problem required to change the decision. An analysis of the sensitivity of a decision to the various parameters highlights the important and significant aspects of a problem. For example, one might be concerned that the estimates for annual maintenance and future salvage value in a particular problem may vary substantially. Sensitivity analysis might indicate that the decision is insensitive to the salvage value estimate over the full range of possible values. But at the same time we might find that the decision is sensitive to changes in the annual maintenance estimate. Under these circumstances one should place greater emphasis on improving the annual maintenance estimate and less on the salvage value estimate.

As indicated at the beginning of the chapter, breakeven analysis is a form of sensitivity analysis. To illustrate the sensitivity of the decision between alternatives to particular estimates, breakeven analysis is often presented as a breakeven chart.

Sensitivity and breakeven analysis frequently are useful in engineering problems called stage construction. Should a facility be constructed now to meet its future full scale requirement, or should it be constructed in stages as the need for the increased capacity arises? Three examples of this situation are presented.

1.  Should we install a cable with 400 circuits now or a 200 circuit cable now and another 200 circuit cable later?
2.  A four inch water main is needed to serve a new area of homes. Should the four inch main be installed now or should a six inch main be installed that can later also provide an adequate water supply to adjoining areas when other homes are built?
3.  An industrial firm needs a 10,000 sq ft warehouse now and estimates that later it will need an additional 10,000 sq ft. The firm could have a 10,000 sq ft warehouse built now and later enlarged, or have the 20,000 sq ft warehouse built right away.

Examples 9-8 and 9-9 illustrate sensitivity and breakeven analysis.

## EXAMPLE 9-8

Consider the following situation where a project may be constructed to full capacity now or may be constructed in two stages.

*Construction costs*

Two stage construction
Construct first stage now                     $100,000
Construct second stage *n* years hence    120,000

Full capacity construction
Construct full capacity now                   140,000

*Other factors*

1.  All facilities will last until 40 years from now irrespective of when they are installed. At that time they will have zero salvage value.
2.  The annual cost of operation and maintenance is the same for both two stage construction and full capacity construction.
3.  Assume an 8% interest rate.

Plot a graph showing age when second stage constructed vs costs for both alternatives. Mark the breakeven point. What is the sensitivity of the decision to second stage construction 16 or more years in the future? Since we are dealing with a common analysis period, the calculations may be either annual cost or present worth. Present worth calculations appear simpler and will be used.

*Construct full capacity now*

PW of cost = $140,000

*Two stage construction*

First stage constructed now and the second stage to be constructed *n* years hence. Compute the PW of cost for several values of *n*.

$$\text{PW of cost} = 100{,}000 + 120{,}000(P/F,i\%,n)$$

| | |
|---|---|
| *n* = 5 yrs | PW = 100,000 + 120,000(0.681) = $181,700 |
| *n* = 10 | PW = 100,000 + 120,000(0.463) = 155,600 |
| *n* = 20 | PW = 100,000 + 120,000(0.214) = 125,700 |
| *n* = 30 | PW = 100,000 + 120,000(0.099) = 111,900 |

These data are plotted in the form of a breakeven chart in Figure 9-5.

Figure 9-5 portrays the PW of cost for the two alternatives. The *x*-axis variable is the time when the second stage is constructed. We see that the PW of cost for two stage construction naturally decreases as

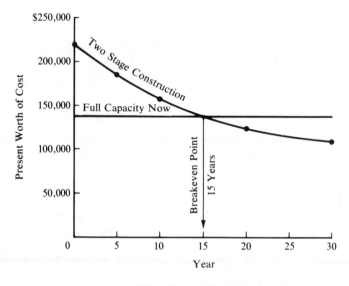

Age When Second Stage Constructed

**Figure 9-5.**  Breakeven chart for Example 9-8.

the time for the second stage is deferred. The single stage construction (full capacity now) is unaffected by the *x*-axis variable and hence is a horizontal line on the graph.

The breakeven point on the graph is the point at which both alternatives have equivalent costs. We see that if, in two stage construction, the second stage is deferred for 15 years then the PW of cost of two stage construction is equal to single stage construction. This is the breakeven point. The graph also shows that if the second stage would be needed prior to 15 years, then single stage construction, with its smaller PW of cost, would be preferred. On the other hand, if the second stage would not be required until after 15 years, two stage construction is preferred.

The decision on how to construct the project is sensitive to the age at which the second stage is needed only if the range of estimates includes 15 years. For example, if one estimated that the second stage capacity would be needed somewhere between 5 and 10 years hence, the decision is insensitive to that estimate. For any value within that range the decision would not be changed. The more economical thing to do in that situation is to build the full capacity now. But, if the second stage capacity were needed somewhere between 12 and 18 years, to take another example, the decision would be sensitive to the estimate of when the capacity would be needed.

One question posed by Example 9-8 is how sensitive the decision is to the need for the second stage at 16 years or beyond. The graph shows that the decision is insensitive. In all cases for construction on or after 16 years, two stage construction has a lower PW of cost.

### EXAMPLE 9-9

In Example 8-3 we had the following situation. Given three mutually exclusive alternatives, each with a 20-year life and no salvage value. The minimum attractive rate of return is 6%.

|                        | A      | B      | C      |
|------------------------|--------|--------|--------|
| Initial cost           | $2000  | $4000  | $5000  |
| Uniform annual benefit | 410    | 639    | 700    |

In Example 8-3 we found that alternative *B* was the preferred alternative. Here we would like to know how sensitive the decision is to our estimate of the initial cost of *B*. Obviously if *B* is preferred at an initial cost of $4000, it will continue to be preferred at any smaller initial cost. But how much higher than $4000 can the initial cost be and still have *B* the preferred alternative? The computations may be done several different ways. With neither input nor output fixed, maximizing net present worth is a suitable criterion.

*Alternative A*

$$NPW = \text{PW of benefit} - \text{PW of cost}$$
$$= 410(P/A,6\%,20) - 2000$$
$$= 410(11.47) - 2000 = 2703$$

*Alternative B*

Let $x$ = initial cost of *B*

$$NPW = 639(P/A,6\%,20) - x$$
$$= 639(11.47) - x$$
$$= 7329 - x$$

*Alternative C*

$$NPW = 700(P/A,6\%,20) - 5000$$
$$= 700(11.47) - 5000 = 3029$$

For the three alternatives we see that *B* will only maximize NPW as long as its NPW is greater than 3029.

$$3029 = 7329 - x$$
$$x = 7329 - 3029 = 4300$$

Therefore *B* is the preferred alternative if its initial cost does not exceed $4300.

Figure 9-6 is a breakeven chart for the three alternatives. Here the criterion is to maximize NPW. As a result the graph shows that *B* is preferred if its initial cost is less than $4300. At an initial cost above $4300, *C* is preferred. We have a breakeven point at $4300. When *B* has an initial cost of $4300, *B* and *C* are equally desirable.

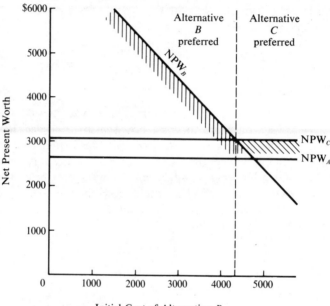

**Figure 9-6.** Breakeven chart.

Sensitivity analysis and breakeven point calculations can be very useful in identifying how different estimates affect the calculations. It must be recognized that these calculations assume all parameters except one are held constant and the sensitivity of the decision to that one variable is evaluated. Later we will look further into the problem of the impact of estimates of parameters on decision making.

## SUMMARY

In this chapter we have looked at four different analysis techniques.

*Future Worth.* When the point in time at which the comparison between alternatives will be made is at some future time, the calculation is called future worth. This is very similar to present worth which is based on the present point in time rather than a future one.

*Benefit-Cost Ratio Analysis.* This technique is based on the ratio of benefits to costs using either present worth or annual cash flow calculations. The method is graphically similar to present worth analysis. When neither input nor output is fixed, incremental benefit-cost ratios ($\Delta B/\Delta C$) are required. In this respect the method is similar to rate of return analysis. Benefit-cost ratio analysis is often used at the various levels of government.

*Payback Period.* Here we define payback as the period of time required for the profit or other savings of an investment to equal the cost of the investment. Although simple to use and simple to understand, payback is a poor analysis technique for ranking alternatives. While it provides a measure of the speed of the return of the investment, it is not an accurate measure of the profitability of the investment.

*Sensitivity and Breakeven Analysis.* These techniques are used to see how sensitive a decision is to estimates for the various parameters. Breakeven analysis is done to locate conditions where the alternatives are equivalent. This is often presented in the form of breakeven charts. Sensitivity analysis is an examination of a range of values for some parameter to determine their effect on the decision.

## PROBLEMS

**9-1** A 20-year old student decided to set aside $100 on his 21st birthday for investment. Each subsequent year through his 55th birthday he plans to increase the sum for investment on a $100 gradient. He will not set aside additional money after his 55th birthday. If he can achieve a 12% rate of return on his investment, how much will he have accrued on his 65th birthday? (*Answer:* $1,160,700)

**9-2** You have an opportunity to purchase a piece of vacant land for $30,000 cash. If you bought it, you would plan to hold the property for 15 years and then sell it at a profit. During this period you would have to pay annual property taxes of $600. You would have no income from the property. Assuming that you would want a 10% rate of return from the investment, at what net price would you have to sell it 15 years hence? (*Answer:* $144,373)

**9-3** An individual's salary is now $12,000 per year and he anticipates retiring in 30 more years. If his salary is increased by $600 each year and he deposits 10% of his yearly salary into a fund that earns 7% interest compounded annually, what will be the amount accumulated at the time of his retirement?

**9-4** A business executive is offered a management job at Generous Electric Company. They offer to give him a five-year contract which calls for a salary of $20,000 per year plus 600 shares of their stock at the end of the five years. This executive is currently employed by Fearless Bus Company and they, too, have offered him a five-year contract. It calls for a salary of $15,000 plus 100 shares of Fearless stock each year. The stock is currently worth $60 per share and pays an annual dividend of $2 per share. Assume end-of-year payments of salary and stock. Stock dividends begin one year after the stock is received. The executive believes that the value of the stock and the dividend will remain constant. If the executive considers 5% a suitable rate of return in this situation, what must the Generous Electric stock be worth per share to make the two offers equally attractive? (*Answer:* $7.40)

**9-5** Tom Jackson is preparing to buy a new car. He knows it represents a large expenditure of money, so he wants to do an analysis to see which of two cars are more economical. Alternative *A* is a U.S. built compact car. It has an initial cost of $3900 and operating costs of 9¢ per mile, excluding depreciation. Tom checked automobile resale statistics. From them he estimates the American automobile can be resold at the end of three years for $1700. Alternative *B* is a foreign built Fiasco. Its initial cost is $2900. The operating cost, also excluding depreciation, is 8¢ per mile. How low could the resale value of the Fiasco be to provide equally economical transportation? Assume Tom will drive 12,000 miles per year and considers 8% as an appropriate interest rate. (*Answer:* $49)

**9-6** A newspaper is considering purchasing locked vending machines to replace open newspaper racks for the sale of its newspapers in the downtown area. The newspaper vending machines cost $45 each. It is expected that the annual revenue from selling the same quantity of newspapers will increase $12 per vending machine. The useful life of the vending machine is unknown.

(a) To determine the sensitivity of rate of return to useful life, prepare a graph for rate of return vs useful life for lives up to eight years.

(b) If the newspaper requires a 12% rate of return, what minimum useful life must it obtain from the vending machines?

(c) What would be the rate of return if the vending machines were to last indefinitely?

**9-7** Able Plastics, an injection molding firm, has negotiated a contract with a national chain of department stores. A plastic pencil box is to be produced for a two year period. Able Plastics has never produced this item before and, therefore, requires all new dies. If the firm invests $30,000 for special removal equipment to unload the completed pencil boxes from the molding machine, one machine operator can be eliminated. This would save the firm $11,600 per year. The removal equipment has no salvage value and is not expected to be used after the two year production contract is completed. The equipment, although useless, would be serviceable for about 15 years. You have been asked to do a payback period analysis on whether or not to purchase the special removal equipment. What is the payback period? Should Able Plastics buy the removal equipment?

**9-8** A cannery is considering installing an automatic case sealing machine to replace current hand methods. If they purchase the machine for $3800 in June, at the beginning of the canning season, they will save $400 per month for the four months each year that the plant is in operation. Maintenance cost, and so forth of the case sealing machine is expected to be negligible. The case sealing machine is expected to be useful for five annual canning seasons and will have no salvage value at the end of that time. What is the payback period? Calculate the nominal rate of return based on the estimates.

**9-9** Consider three alternatives.

|  | A | B | C |
|---|---|---|---|
| First cost | $50 | $150 | $110 |
| Uniform annual benefit | 28.8 | 39.6 | 39.6 |
| Useful life* | 2 yrs | 6 yrs | 4 yrs |
| Computed rate of return | 10% | 15% | 16.4% |

*At the end of its useful life, an identical alternative (with the same cost, benefits, and useful life) may be installed.

None of the alternatives has any salvage value. If the minimum attractive rate of return is 12%, which alternative should be selected?

(a) Solve the problem by future worth analysis.

(b) Solve the problem by benefit-cost ratio analysis.

(c) Solve the problem by payback period.

(d) If the answers in parts (a), (b) and (c) differ, explain why this is the case.

**9-10** An investor is considering buying some land for $100,000 and constructing an office building on it. Three different buildings are being analysed.

|  | Building height | | |
|---|---|---|---|
|  | 2 stories | 5 stories | 10 stories |
| Cost of building (excluding cost of land) | $400,000 | $800,000 | $2,100,000 |
| Resale value* of land and building at end of 20-year analysis period | 200,000 | 300,000 | 400,000 |
| Annual rental income after deducting all operating expenses | 70,000 | 105,000 | 256,000 |

*Resale value to be considered a reduction in cost, rather than a benefit.

Using benefit-cost ratio analysis and a 8% minimum attractive rate of return, determine which alternative, if any, should be selected.

# 10
# Depreciation

We have up to this point dealt with a variety of economic analysis problems and many techniques for their solution. But in the process we have carefully avoided an important element in most economic analyses—income taxes. Now that the essential concepts have been presented, we can move to more realistic, and unfortunately more complex, situations.

This chapter and the one to follow are like Siamese twins. The requirement that we take income taxes into account in our analyses means we must understand something about the way taxes are imposed. For capital equipment we shall see that depreciation is an important component in computing income taxes. This chapter will emphasize the various aspects of depreciation, and the next chapter will illustrate how we use depreciation in income tax computations.

## Depreciation Defined

The word *depreciation* is defined as a decrease in value. This is not an entirely satisfactory definition, for *value* has several meanings. In the context of economic analysis, value may refer either to market value (the monetary value others place on property) or value to the owner. Thus, we may define depreciation in two ways: decrease in market value, or decrease in value to the owner.

## Deterioration and Obsolescence

A machine may depreciate (decline in value) because it is wearing out and no longer performing its function as well as when it was new. This is called deterioration. Many kinds of machinery require increased maintenance as they age, reflecting a slow but continuing failure of individual parts. In other types of equipment, the quality of output may decline due to wear on components and resulting poorer mating of parts. Anyone who has worked to maintain the mechanical components of an automobile will testify to deterioration due to both the failure of individual parts (such as, fan belt, muffler, or battery) and the wear on components (such as, bearings, piston rings, alternator brushes). There is another aspect of depreciation caused by obsolescence. A machine is described as obsolete when the function it performs can be done in some better manner. A machine may be in excellent working condition, yet may still be obsolete. In the 1970s, for example, there was a major shift in the construction of adding machines and calculators. Previously these business machines were complex mechanical devices, with hundreds of gears and levers. But the advancement of subminiature logic circuitry resulted in a completely different approach to adding machine and business calculator design. Electronic circuitry replaced the mechanical mechanism. Because the electronic calculators were a major improvement, mechanical calculators became obsolete in a short time. Thus, mechanical calculators, even though they had not physically deteriorated, declined in value (depreciated) rapidly.

If your automobile depreciated in the last year, that meant it had declined in market value. On the other hand, a manager who indicates a piece of machinery has depreciated may be describing a machine that has deteriorated because of use or because it has become obsolete compared to newer machinery. Both situations indicate the machine has declined in value to the owner.

The accounting profession defines depreciation in yet a third way. Depreciation, from their viewpoint, is the systematic allocation of the cost of an asset (less its salvage value) over its useful life. Thus, we now have three distinct definitions of depreciation.

1. Decline in market value of an asset.
2. Decline in value of an asset to its owner.
3. Systematic allocation of the cost of an asset (less its salvage value) over its useful life.

Although in everyday conversation we probably use depreciation to mean a decline in market value, accountants talk of depreciation as the allocation of the cost of an asset over its useful life. In determining taxable income, and hence income taxes, it is the accountant's definition of depreciation that is used in the computations. The remainder of this chapter, therefore, will deal only with this third definition of depreciation.

Almost everything seems to decline in value as time proceeds. Machines become obsolete or wear out. Automobiles literally fall apart. Buildings, too, show the heavy signs of age. But there are obvious exceptions. Manhattan Island, like land generally, has not declined in value. Rather, the market value has continued to increase. But land is not considered to be an asset subject to depreciation. Irrespective of whether the market value is going up, down, or remains unchanged, accountants consider land to be an asset that does not have a limited useful life. Therefore, its cost is not allocated over a useful life. Land is a nondepreciable asset.

### Elements of Depreciation Calculations

In the computation of depreciation, there are three components that must be considered: cost of the asset subject to depreciation, useful life, and salvage value of the asset at the end of its useful life. For any situation the cost of the asset, for depreciation purposes, will be known. It is more difficult to predict in advance the useful life of an asset. Unless the firm is familiar with a particular class of depreciable assets, the useful lives suggested by the U.S. Internal Revenue Service (IRS) are frequently used. Table 10-1 gives guideline useful lives for depreciation purposes for selected classes of assets.

The salvage value to be realized upon the disposal of the asset at the end of its useful life must be estimated. In the frequent situation where the salvage value is only a small portion of the cost of the asset, it is assumed to be zero.

## DEPRECIATION CALCULATIONS

Since depreciation is the allocation of the cost of an asset, minus its salvage value, over the useful life, it can be illustrated by a diagram, as in Figure 10-1. Our task is to arrange the annual depreciation charges to get from point $P$ to point $F$ in Figure 10-1. In Figure 10-1 the axis labelled "money" initially represents the cost of the asset at point $P$. But we will plot the curve of cost minus cumulative depreciation so the curve will reach point $F$ at the asset's end-of-useful life. We will call this declining value *book value*.

Book value = cost − depreciation charges made

Accountants would define book value as the remaining unallocated cost of an asset.

### Straight Line Depreciation

The simplest, and best known, of the various depreciation methods is straight line depreciation. In this method a constant depreciation charge is

**Table 10-1.** Guideline useful lives for depreciable assets

| *Depreciable assets used in all business activities* | *Years* |
|---|---|
| Office furniture, fixtures and equipment | 10 |
| Automobiles | 3 |
| Trucks, light general purpose | 4 |
| Buildings: | |
| Apartments and hotels | 40 |
| General factory buildings | 45 |
| Office buildings | 45 |
| Dwellings | 45 |
| Stores | 50 |
| Warehouses | 60 |

| *Depreciable assets used in the following activities* | *Years* |
|---|---|
| This category covers all depreciable assets not contained in the grouping above. Production machinery and equipment and special purpose structures are included in this category. | |
| Agriculture | 10 |
| Petroleum drilling | 6 |
| exploration | 14 |
| refining and marketing | 16 |
| Food and beverage production | 12 |
| Chemicals and allied products manufacture | 11 |
| Electrical equipment manufacturing | 12 |
| Electronic products manufacture | 8 |
| Machinery manufacture | 12 |
| Mining | 10 |
| Air transport | 6 |

made. The total amount to be depreciated, $P - F$, is divided by the useful life in years, $N$, to obtain the annual depreciation charge.

STRAIGHT LINE

$$\text{Annual depreciation charge} = \frac{1}{N}(P - F) \qquad (10\text{-}1)$$

*EXAMPLE 10-1*

| Cost of the asset, $P$ | $900 |
|---|---|
| Useful life, $N$ | 5 yrs |
| End-of-useful life salvage value, $F$ | $ 70 |

$$\text{Annual depreciation charge} = \frac{1}{N}(P - F) = \frac{1}{5}(900 - 70) = 166$$

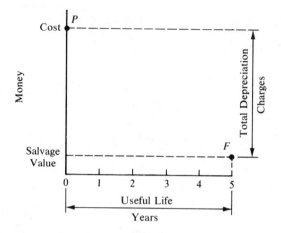

**Figure 10-1**

| Year | Book value before depreciation charge | Depreciation for year | Book value after depreciation charge |
|------|---------------------------------------|-----------------------|--------------------------------------|
| 1 | Cost = $900 | $166 | $734 |
| 2 | 734 | 166 | 568 |
| 3 | 568 | 166 | 402 |
| 4 | 402 | 166 | 236 |
| 5 | 236 | 166 | 70 = salvage value |
| | total depreciation = | $830 | |

This situation is illustrated in Figure 10-2.

There is an alternate way of computing the straight line depreciation charge in any year:

$$\text{Straight line depreciation charge for any year} = \frac{\text{Book value at beginning of year} - \text{salvage value}}{\text{Remaining useful life at beginning of year}} \quad (10\text{-}2)$$

In Example 10-1 the depreciation charge in the third year could be computed as

$$\text{Straight line depreciation for third year} = \frac{568 - 70}{3} = \$166$$

Both methods give the same result. At any time the book value of the asset would be the cost minus the depreciation to that point in time. For Example 10-1, the book value at the end of three years would be $900 − 3 ($166) = $402. In general,

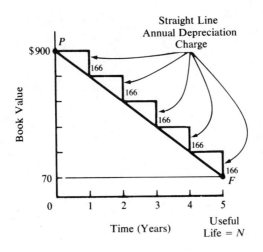

Figure 10-2.   Straight line depreciation.

$$\text{Book value end of } J\text{th year} = P - \frac{J}{N}(P - F).$$

### Sum-of-Years Digits Depreciation

Another method for allocating the cost of an asset minus salvage value over its useful life is called sum-of-years digits (SOYD) depreciation. This method results in larger than straight line depreciation charges during the early years of an asset and necessarily smaller charges as the asset nears the end of its estimated useful life. Each year the depreciation charge is computed as the remaining useful life at the beginning of the year divided by the sum of the years digits for the total useful life, with this ratio multiplied by the total amount to be depreciated $(P - F)$.

$$\begin{array}{l}\text{Sum-of-years digits deprecia-} \\ \text{tion charge for any year}\end{array} = \dfrac{\begin{array}{c}\text{Remaining useful life} \\ \text{at beginning of year}\end{array}}{\begin{array}{c}\text{Sum-of-years digits} \\ \text{for total useful life}\end{array}}(P - F) \quad (10\text{-}3)$$

where

Sum-of-years digits for total useful life $= 1 + 2 + 3 + \cdots + (N - 1) + N$

An equation for the sum-of-years digits (Sum) may be derived.

$$\text{Sum} = \quad 1 \quad + \quad 2 \quad + \cdots + (N - 1) + \quad N \qquad (10\text{-}4)$$

Write the terms in Eq. 10-4 in reverse order

$$\text{Sum} = \quad N \quad + (N-1) + \cdots + \quad 2 \quad + \quad 1 \qquad (10\text{-}5)$$

Add Eq. 10-4 and 10-5

$$2(\text{Sum}) = (N+1) + (N+1) + \cdots + (N+1) + (N+1)$$

For $N$ terms on the right side

$$2(\text{Sum}) = N(N+1)$$

$$\text{Sum-of-years digits (Sum)} = \frac{N}{2}(N+1) \qquad (10\text{-}6)$$

### EXAMPLE 10-2

Compute the sum-of-years digits (SOYD) depreciation schedule for the Example 10-1 situation.

| | |
|---|---|
| Cost of the asset, $P$ | $900 |
| Useful life, $N$ | 5 yrs |
| End-of-useful life salvage value, $F$ | $70 |

Sum-of-years digits (Sum) = $5 + 4 + 3 + 2 + 1 = 15$

$$\text{or}\quad \text{Sum} = \frac{N}{2}(N+1) = \frac{5}{2}(5+1) = 15$$

$$\text{1st year SOYD depreciation} = \frac{5}{15}(900 - 70) = \$277$$

$$\text{2nd year} = \frac{4}{15}(900 - 70) = \quad 221$$

$$\text{3rd year} = \frac{3}{15}(900 - 70) = \quad 166$$

$$\text{4th year} = \frac{2}{15}(900 - 70) = \quad 111$$

$$\text{5th year} = \frac{1}{15}(900 - 70) = \quad \underline{\quad 55}$$

$$\$830$$

These data are plotted in Figure 10-3.

## Declining Balance Depreciation

A third major method is declining balance depreciation. Here a constant depreciation rate is applied to the book value of the property. The depre-

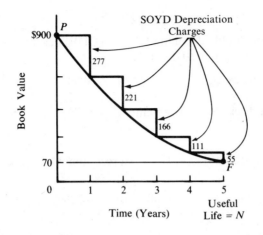

**Figure 10-3.** Sum-of-years digits
depreciation.

ciation rate allowed by the Internal Revenue Service depends on the type
of depreciable property and is stated in terms of the alternate straight line
depreciation rate which is $1/N$.

*Allowable Depreciation Rate*

Double straight line rate $\dfrac{2}{N}$    All new depreciable property except
real estate

$1\frac{1}{2}$ straight line rate $\dfrac{1.5}{N}$    All used depreciable property and new
real estate property*

$1\frac{1}{4}$ straight line rate $\dfrac{1.25}{N}$    Used rental residential property*

Since new tangible personal property (everything except real estate is called
personal property) qualifies for double the straight line rate, this method is
frequently called double declining balance depreciation, and the general
equation is

Double declining balance depreciation in any year $= \dfrac{2}{N}$ (book value)

Since book value equals cost − depreciation charges to date, double declining balance (DDB) depreciation in any year = (2/N)(cost − depreciation charges to date).

*Excludes land which may not be depreciated.

*EXAMPLE 10-3*

Compute the double declining balance (DDB) depreciation schedule for the Example 10-1 and 10-2 situation.

| | |
|---|---|
| Cost of the asset, $P$ | $900 |
| Useful life, $N$ | 5 yrs |
| End-of-useful life salvage value, $F$ | $ 70 |

$$\text{DDB depreciation} = \frac{2}{N}(P - \text{depreciation charges to date})$$

$$\text{1st year} = \frac{2}{5}(900 - 0) \quad = \$360$$

$$\text{2nd year} = \frac{2}{5}(900 - 360) = \quad 216$$

$$\text{3rd year} = \frac{2}{5}(900 - 576) = \quad 130$$

$$\text{4th year} = \frac{2}{5}(900 - 706) = \quad\quad 78$$

$$\text{5th year} = \frac{2}{5}(900 - 784) = \quad\quad 46$$
$$\overline{\hspace{1cm}\$830}$$

Figure 10-4 illustrates the situation.

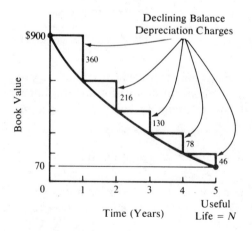

**Figure 10-4.** Declining balance depreciation.

## Declining Balance Depreciation
## in Any Year

For double declining balance the depreciation schedule is

$$\text{1st year DDB depreciation} = \frac{2}{N}(P) = \frac{2P}{N}$$

$$\text{2nd year DDB depreciation} = \frac{2}{N}\left(P - \frac{2P}{N}\right) = \frac{2P}{N}\left(1 - \frac{2}{N}\right)$$

$$\text{3rd year DDB depreciation} = \frac{2}{N}\left[P - \frac{2P}{N} - \frac{2P}{N}\left(1 - \frac{2}{N}\right)\right]$$

$$= \frac{2P}{N}\left[1 - 2\left(\frac{2}{N}\right) + \left(\frac{2}{N}\right)^2\right]$$

$$= \frac{2P}{N}\left(1 - \frac{2}{N}\right)^2$$

And in any year $n$

$$\text{DDB depreciation} = \frac{2P}{N}\left(1 - \frac{2}{N}\right)^{n-1} \tag{10-7}$$

For $1.25/N$ and $1.50/N$ declining balance depreciation, the 2 appearing in the two factors of Eq. 10-7 would be replaced by 1.25 or 1.50 as appropriate.

## Total Declining Balance Depreciation at
## End of $n$ Years

For double declining balance depreciation,

Total DDB depreciation

$$= \frac{2P}{N}\left[1 + \left(1 - \frac{2}{N}\right) + \left(1 - \frac{2}{N}\right)^2 + \cdots + \left(1 - \frac{2}{N}\right)^{n-1}\right] \tag{10-8}$$

multiply by $\left(1 - \frac{2}{N}\right)$

$(\text{Total DDB depreciation})\left(1 - \frac{2}{N}\right)$

$$= \frac{2P}{N}\left[\left(1 - \frac{2}{N}\right) + \left(1 - \frac{2}{N}\right)^2 + \cdots + \left(1 - \frac{2}{N}\right)^{n-1} + \left(1 - \frac{2}{N}\right)^n\right] \tag{10-9}$$

---

subtract Eq. 10-8 from Eq. 10-9

$$-\frac{2}{N}(\text{Total DDB depreciation}) = \frac{2P}{N}\left[-1 + \left(1 - \frac{2}{N}\right)^n\right]$$

$$\text{Total DDB depreciation} = P\left[1 - \left(1 - \frac{2}{N}\right)^n\right] \qquad (10\text{-}10)$$

(For $1.25/N$ or $1.50/N$ depreciation substitute these in Eq. 10-10 for the $2/N$ term.)

### Book Value of an Asset at End of $n$ Years

The book value at the end of $n$ years will be the cost of the asset $P$ minus the total depreciation at the end of $n$ years. For DDB depreciation,

Book value $= P -$ total DDB depreciation at end of $n$ years

$$= P - P\left[1 - \left(1 - \frac{2}{N}\right)^n\right] = P\left(1 - \frac{2}{N}\right)^n \qquad (10\text{-}11)$$

### Effect of Salvage Value on Declining Balance Depreciation

In Example 10-3 the salvage value was estimated to be $70. And we see that the calculations produced a depreciation schedule that resulted in a $70 salvage value. This is hard to explain, for the calculations did not consider our salvage value estimate. The answer is that the example problem was devised to come out this way. Since the depreciation schedule is independent of the estimated salvage value, any of the three situations in Figure 10-5 might occur in an actual situation.

In Figure 10-5a we have a relatively high salvage value with the result that the book value of the asset would decline below the estimated salvage value. While this might seem to be desirable, the IRS does not permit a taxpayer to continue to deduct depreciation charges that would drop the book value below the salvage value. By applying the IRS rule, Figure 10-5a is transformed into Figure 10-6. No depreciation charges are made that would reduce the book value below the estimated salvage value. Depreciation charges cease when the book value equals the estimated salvage value.

Figure 10-5b presents no problem, for the declining balance depreciation schedule results in a book value exactly equal to the salvage value at the end of the useful life. Obviously, this will only happen when the salvage value happens to lie on the declining balance book value curve.

Figure 10-5c represents the situation where the salvage value is beneath

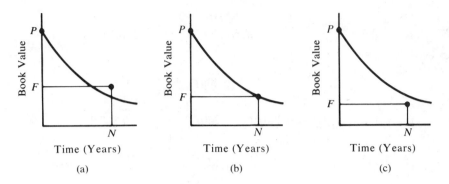

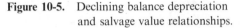

**Figure 10-5.** Declining balance depreciation and salvage value relationships.

the declining balance book value curve at the end of the depreciable life $N$. Since the declining balance book value curve is independent of the estimated salvage value, this situation could easily occur. In fact, this is *always* the case when the estimated salvage value is zero.

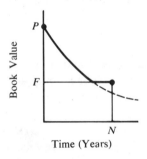

**Figure 10-6.** Declining balance depreciation terminated when salvage value is reached.

In any given declining balance depreciation situation the cost of the asset $P$, the depreciable life $N$, and the allowable depreciation rate ($1.25/N$, $1.5/N$, or $2/N$) will be known or determined from the facts. These data define the coordinates of the book value curve.* For example, for an asset with cost $P$, and depreciable life $N$ equal to five years, the double declining balance book value curve would be computed as follows.

*For an asset of cost $P$, there would be a different book value curve for each depreciable life $N$ and allowable depreciation rate ($1.25/N$, $1.5/N$, or $2/N$). Additional data on the book value curves are provided in Appendix 10-B.

| Year | DDB depreciation | | Book value |
|------|-----|-----|------------|
| 0 | | | $1.000P$ |
| 1 | $\frac{2}{5}(P - 0)$ | $= 0.400P$ | $0.600P$ |
| 2 | $\frac{2}{5}(P - 0.400P)$ | $= 0.240P$ | $0.360P$ |
| 3 | $\frac{2}{5}(P - 0.640P)$ | $= 0.144P$ | $0.216P$ |
| 4 | $\frac{2}{5}(P - 0.784P)$ | $= 0.086P$ | $0.130P$ |
| 5 | $\frac{2}{5}(P - 0.870P)$ | $= 0.052P$ | $0.078P$ |

These data are plotted in Figure 10-7. For double declining balance depreciation and a five-year depreciable life, the book value curve will decline to $0.078P$. If the estimated salvage value in this situation is less than $0.078P$, the result will look like Figure 10-5c. This is illustrated in Example 10-4.

### EXAMPLE 10-4

The cost of a new asset is $900 and its useful life is five years. If the salvage value is estimated to be $30, compute the double declining balance depreciation schedule.

| | DDB depreciation | | End-of-year book value |
|---|-----|-----|------------|
| 1st year depreciation $= \frac{2}{5}(900)$ | $= \$360$ | | $\$540$ |
| 2nd year depreciation $= \frac{2}{5}(900 - 360) =$ | 216 | | 324 |
| 3rd year depreciation $= \frac{2}{5}(900 - 576) =$ | 130 | | 194 |
| 4th year depreciation $= \frac{2}{5}(900 - 706) =$ | 78 | | 116 |
| 5th year depreciation $= \frac{2}{5}(900 - 784) =$ | 46 | | 70 |

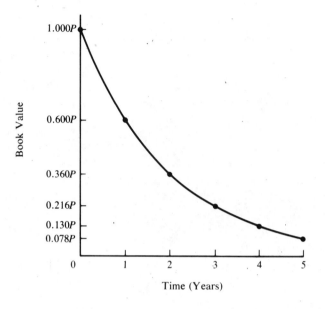

Time (Years)

**Figure 10-7.** Book value curve for double
declining balance depreciation
and five-year depreciable life.

The book value at the end of the useful life does not decline to the $30 salvage value. Since Figure 10-7 is a plot of this situation (double declining balance depreciation and five-year depreciable life), we know that the end of useful life book value will be $0.078P$.

For $P = \$900$      book value $= 0.078(900) = \$70$.

Thus the situation looks like Figure 10-5c. Two general approaches may be used to resolve the difficulty.

1.  If the asset is retained in service beyond its estimated useful life (this probably is a more frequent occurrence than one would expect), then in the subsequent years declining balance depreciation would continue to be charged until either the current estimated salvage value is reached, or until the asset is disposed of. Here the sixth year depreciation would be $(\frac{2}{5})(900 - 830) = \$28$ and seventh year depreciation would be $(\frac{2}{5})(900 - 858) = \$17$. Since the seventh year depreciation would bring the book value below the salvage value, the seventh year depreciation is decreased. The resulting DDB depreciation schedule would be

| Year | Depreciation | *End-of-year*<br>*book value* |
|------|--------------|-------------------------------|
| 1 | $360 | $540 |
| 2 | 216 | 324 |
| 3 | 130 | 194 |
| 4 | 78 | 116 |
| 5 | 46 | 70 |
| 6 | 28 | 42 |
| 7 | 12 | 30 |

2. A second alternative is to use a composite depreciation method. The rules of the IRS provide that a taxpayer may change from the declining balance method to straight line depreciation at any time during the life of an asset. In this example we could simply use straight line depreciation in the fifth year.

Since straight line depreciation in any year equals the book value minus the salvage value, all divided by the remaining useful life,

$$\frac{\text{5th year straight}}{\text{line depreciation}} = \frac{\text{Book value} - \text{salvage value}}{\text{Remaining useful life}} = \frac{116 - 30}{1} = \$86$$

Figure 10-8 illustrates the difficulty and the two possible ways of solving it.

## Declining Balance Depreciation with
## Conversion to Straight Line Depreciation

In the previous section, we encountered a situation where declining balance depreciation was not entirely satisfactory. The depreciation method failed to achieve the desired result of allocating the cost (minus salvage value) of the asset over its estimated useful life. Figure 10-5c and Example 10-4 illustrated the problem.

We saw that one way to overcome this difficulty was to adopt a composite depreciation method. Initially the asset is depreciated by the declining balance method, and then subsequently the remainder of the depreciation is computed by the straight line method. In the composite depreciation method, one must decide when to switch from declining balance to straight line depreciation. In Example 10-4 the switch was made in the last or fifth year, but the question arises, was there a better time to switch methods? In a subsequent section on selecting the preferred depreciation method, we will show that a taxpayer generally prefers to reduce the book value of an asset to its salvage value as quickly as possible. Thus we would switch

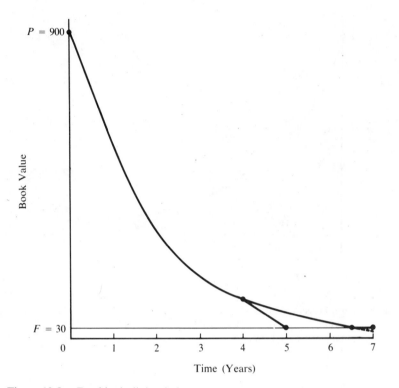

**Figure 10-8.** Double declining balance
depreciation either converted to
straight line depreciation or
continued beyond the five-year
estimated life.

from declining balance to straight line whenever straight line depreciation
results in larger depreciation charges and hence a more rapid reduction in
book value of the asset.

On this basis, the choice between the alternatives in Figure 10-8 is clear.
The conversion to straight line depreciation in year 5 is preferred over
continuing the declining balance depreciation. The same criterion (reduce
the book value of an asset to its salvage value as quickly as possible) will
be used to determine when to switch from declining balance to straight
line depreciation.

Assuming that the conversion from declining balance to straight line
depreciation could take place in any of the $N$ years, there would seem to
be $N$ possibilities. Our problem is to identify the best one.

Figure 10-9 shows the results if the conversion is made at different points
in time. We assume the most desirable depreciation schedule is the one that

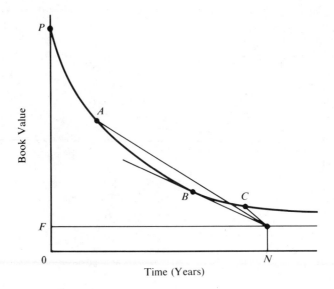

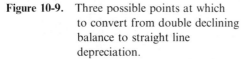

**Figure 10-9.**   Three possible points at which
to convert from double declining
balance to straight line
depreciation.

results in the most rapid decline in book value, that is, depreciates the asset
as quickly as possible. Looking at Figure 10-9 one observes that the curve
that best meets this criterion is the one that converts to straight line depreciation at point *B*. At this point the slope of the declining balance curve
equals the slope of the straight line. Selecting any other point for the conversion produces a curve that will be above our selected curve at some point
in time. Thus the point *B* conversion is best—there is no other curve that
declines as rapidly from *P* to *F*.

   Actually, the declining balance curve is like stair steps rather than a
smooth curve. For this reason the year to convert is best computed as the
point where the conversion to straight line depreciation produces an equal
or greater depreciation charge than remaining with double declining balance depreciation. Table 10-2 tabulates the conversion point for a variety
of situations.


### EXAMPLE 10-5

| | |
|---|---|
| Cost of the asset, *P* | $900 |
| Useful (and depreciable) life, *N* | 5 yrs |
| End-of-useful life salvage value, *F* | $ 30 |

**Table 10-2.** Conversion from double declining balance to straight line depreciation

Computed year *n* to change methods.

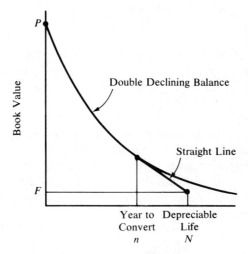

*Computed year n to change methods*

| Depreciable life of asset N (years) | Zero salvage value F = 0 | Salvage value ratio F/P | | |
|---|---|---|---|---|
| | | F/P = 0.05 | F/P = 0.10 | F/P = 0.12 |
| | *Value of n. Use straight line depreciation for year n and all subsequent years* | | | |
| 3 | 3 | | | |
| 4 | 4 | 4 | | |
| 5 | 4 | 5 | | |
| 6 | 5 | 5 | | |
| 7 | 5 | 6 | | |
| 8 | 6 | 6 | 8 | |
| 9 | 6 | 7 | 9 | |
| 10 | 7 | 7 | 9 | |
| 11 | 7 | 8 | 10 | |
| 12 | 8 | 9 | 11 | |
| 13 | 8 | 9 | 11 | |
| 14 | 9 | 10 | 12 | |
| 15 | 9 | 10 | 13 | |
| 16 | 10 | 11 | 13 | |
| 17 | 10 | 11 | 14 | |
| 18 | 11 | 12 | 15 | 18 |
| 19 | 11 | 13 | 16 | 19 |
| 20 | 12 | 13 | 16 | 19 |

*No conversion is needed*

Compute the double declining balance depreciation schedule with conversion to straight line at the most desirable time. Table 10-2 will help locate the correct conversion point.

$$F/P = 30/900 = 0.03$$

For a depreciable life of five years and $F/P$ of 0.05, the conversion point is the beginning of the fifth year. At $F = 0$ (and therefore $F/P = 0$), the conversion point is the beginning of the fourth year. Further calculations are needed to determine if the conversion point at $F/P$ of 0.03 is at four or five years.

We will proceed by computing the double declining balance depreciation for each year and also the straight line depreciation if the conversion to straight line were made in that year. To obtain the most rapid decline in book value, we will use double declining balance depreciation until we reach a point where converting to straight line results in increased depreciation.

In Example 10-4 the DDB depreciation schedule for this case was computed as:

| Year | DDB depreciation | End-of-year book value |
|------|------------------|------------------------|
| 1    | $360             | $540                   |
| 2    | 216              | 324                    |
| 3    | 130              | 194                    |
| 4    | 78               | 116                    |
| 5    | 46               | 70                     |

We can compute the straight line depreciation for any year using Eqn. 10-2.

$$\text{SL depreciation in any year} = \frac{\text{Book value at beginning of year} - \text{salvage value}}{\text{Remaining useful life at beginning of year}} \qquad (10\text{-}2)$$

| If convert to straight line depreciation in year | Straight line depreciation | DDB depreciation | Decision |
|------|------|------|------|
| 2 | $\dfrac{540 - 30}{4} = \$127.50$ | $216 | Do not convert to SL |
| 3 | $\dfrac{324 - 30}{3} = 98$ | 130 | Do not convert to SL |
| 4 | $\dfrac{194 - 30}{2} = 82$ | 78 | Convert to SL for year 4 |

The resulting depreciation schedule is:

| Year | Double declining balance with conversion to straight line depreciation |
|------|------------------------------------------------------------------------|
| 1 | $360 |
| 2 | 216 |
| 3 | 130 |
| 4 | 82 |
| 5 | 82 |
| | $870 |

## Unit of Production Depreciation

At times there may be situations where the recovery of depreciation on a particular asset is more closely related to use than to time. In these few situations (and they are rare) the depreciation in any year is:

$$\text{Depreciation in any year} = \frac{\text{Production for year}}{\text{Total lifetime production for asset}} (P - F) \qquad (10\text{-}12)$$

This method might be useful for machinery for exploiting natural resources if the resources will be exhausted before the machinery will wear out. It is not considered an acceptable method for general use in depreciating industrial equipment.

*EXAMPLE 10-6*

Cost of asset, $P$    $900
Salvage value, $F$      70

The equipment has been purchased for use in a sand and gravel pit. The sand and gravel pit will be in operation during a five-year period while a nearby airport is being reconstructed and paved. After that time the sand and gravel pit will be shut down and the equipment removed and sold. The airport reconstruction schedule calls for 40,000 cubic yards of sand and gravel on the following schedule.

| Year | Required sand and gravel |
|------|--------------------------|
| 1 | 4,000 cu yds |
| 2 | 8,000 |
| 3 | 16,000 |
| 4 | 8,000 |
| 5 | 4,000 |

Compute the unit-of-production depreciation schedule for the equipment.

The total lifetime production for the asset is 40,000 cubic yards of sand and gravel. From the airport reconstruction schedule, the first year unit-of-production depreciation would be:

First year unit-of-production depreciation $= \dfrac{4000 \text{ cu yds}}{40{,}000 \text{ cu yds}}(\$900 - \$70) = \$83$

Similar calculations for the subsequent four years gives the complete depreciation schedule.

|       | *Unit-of-production* |
| :---: | :---: |
| *Year* | *depreciation* |
| 1 | $ 83 |
| 2 | 166 |
| 3 | 332 |
| 4 | 166 |
| 5 | 83 |
|   | $830 |

It should be noted that the actual unit of production depreciation charge in any year is based on the actual production for the year rather than the scheduled production.


### Sinking Fund Depreciation

All of the depreciation methods considered so far have been based on the concept of recovering the cost of the asset (minus salvage value) over its useful life. While the results of sinking fund depreciation also equals this quantity, the underlying philosophy is different.

Sinking fund depreciation assumes that a hypothetical fund is created. Into this fund there must be deposited a uniform annual amount so that the future worth of the uniform annual deposits, earning some assumed interest rate, equals the cost of the asset minus its salvage value. Stated another way, the periodic uniform payments plus the interest earned by the imaginary fund accumulate so that the total amount $P - F$ will be in the fund by the end of the depreciable life of the asset. Example 10-7 illustrates sinking fund depreciation.


*EXAMPLE 10-7*

| | |
| :--- | :--- |
| Cost of the asset, $P$ | $900 |
| Useful life, $N$ | 5 yrs |
| End of useful life salvage value, $F$ | $ 70 |

Compute the sinking fund depreciation schedule, assuming a 6% interest rate. First, we must compute the uniform annual sinking fund deposit using the sinking fund factor

Annual deposit $= (P - F)(A/F, 6\%, 5 \text{ yrs}) = 830(0.1774) = \$147.24$

Each year $147.24 is assumed to be deposited to the sinking fund. At the end of each year 6% interest is also assumed to be added to the sinking fund. Sinking fund depreciation is the sum of the combined annual deposits.

| Year | Balance in sinking fund at beginning of year | 6% interest on balance | Uniform sinking fund deposit | Sinking fund depreciation |
|------|------|------|------|------|
| 1 | $ 0 | $ 0 | $147.24 | $147.24 |
| 2 | 147.24 | 8.83 | 147.24 | 156.07 |
| 3 | 303.31 | 18.20 | 147.24 | 165.44 |
| 4 | 468.75 | 28.12 | 147.24 | 175.36 |
| 5 | 644.11 | 38.65 | 147.24 | 185.89 |
| | | $93.80 | $736.20 | $830.00 = P − F |

## COMPARISON OF DEPRECIATION METHODS

In the previous sections six different depreciation methods* have been discussed and examples presented. Although there are variations in the details, each method allocates the cost of the asset over its productive lifetime. A combined graph of the depreciation schedules calculated in the five examples where cost $P = 900$ and salvage value $F = 70$ is shown in Figure 10-10. One must be careful about drawing conclusions based on this figure. We know, of course, that straight line depreciation always produces a uniform annual depreciation that can be characterized as a straight line. Also, we know that the sinking fund depreciation schedule will be above and the sum-of-years digits and double declining balance depreciation schedules will be below straight line depreciation. We cannot make any valid statement concerning the relationship between sum-of-years digits

*This is *not* an all-inclusive list. Another composite method allowed is DDB with conversion to SOYD. Further, the general rule is that a taxpayer may choose any reasonable method (subject to certain restrictions) of computing depreciation so long as the method is used consistently.

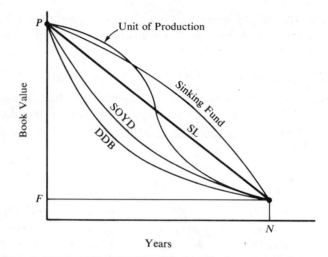

**Figure 10-10.** Five depreciation methods.

depreciation and double declining balance depreciation. Sometimes the relationship will be as shown in Figure 10-10 and sometimes the situation will be reversed with the double declining balance depreciation above sum-of-years digits depreciation. Further, double declining balance depreciation may look like any of the three situations of Figure 10-5. Where we find the case of Figure 10-5c, the sixth method (declining balance with conversion to straight line depreciation) discussed in the chapter, but not included in Figure 10-10, would be applicable. Assuming that we acquire an asset, and that we may depreciate it by any of the several methods presented, the question arises: which method should we select? We will now attempt to resolve that question.

### Selecting the Preferred Depreciation Method

The methods of computing depreciation presented in this chapter are all used at one time or another. The problem is a matter of selecting the method that best suits the objectives of the firm. In the situation where the firm pays income taxes, and the income tax rate is expected to remain constant over the depreciable life of the asset, we can readily determine the preferred depreciation method. For a firm that pays income taxes each year, depreciation is a deduction from taxable income. The result is that the greater the depreciation deduction, the less the taxable income and hence taxes for the year. Depreciation methods generally result in the same total depreciation deductions, hence it is the *timing* of the deductions that characterize the different methods. Immediate tax savings are more valuable than tax savings some years in the future due to the time value of money. For this reason a firm prefers to depreciate its assets to their salvage value as rapidly as possible.

We used this criterion to determine whether to use the declining balance method or the composite method of declining balance with conversion to straight line depreciation. But in determining the preferred depreciation method from among the several described in this chapter, the situation may not always be this simple. We will encounter situations where one depreciation method produces larger depreciation charges (and therefore a lower book value) in the early years and another depreciation method is more attractive in subsequent years. Figure 10-11 illustrates this problem. In this situation we will select the method with the largest present worth of depreciation charges. By applying this criterion we will be selecting from the two alternatives the one whose present worth of additional depreciation charges (during a part of the depreciable life of the asset) exceeds the present worth of the reduced depreciation charges (at another part of the life). Example 10-8 illustrates the situation.

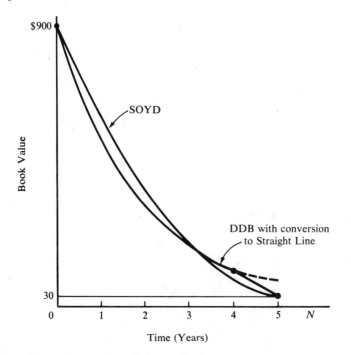

**Figure 10-11.** Alternate depreciation methods for the same situation.

*EXAMPLE 10-8*

| | |
|---|---|
| Cost of the asset, $P$ | $900 |
| Useful life, $N$ | 5 yrs |
| End-of-useful life salvage value | $ 30 |

Determine whether double declining balance or sum-of-years digits depreciation is preferred if interest is 6%.

*DDB Depreciation*

In Example 10-5 the DDB depreciation schedule was computed for these values. There it was found that DDB with conversion to straight line was the preferred form of declining balance depreciation. Below is the depreciation schedule computed in Example 10-5.

| Year | DDB with conversion to straight line depreciation | End-of-year book value |
|---|---|---|
| 1 | $360 | $540 |
| 2 | 216 | 324 |
| 3 | 130 | 194 |
| 4 | 82 | 112 |
| 5 | 82 | 30 |
| | $870 | |

*SOYD depreciation*

Sum-of-years digits (Sum) $= \dfrac{N}{2}(N + 1) = \dfrac{5}{2}(5 + 1) = 15$

| Year | SOYD depreciation | End-of-year book value |
|---|---|---|
| 1 | $\dfrac{5}{15}(900 - 30) = \$290$ | $610 |
| 2 | $\dfrac{4}{15}(900 - 30) = \ 232$ | 378 |
| 3 | $\dfrac{3}{15}(900 - 30) = \ 174$ | 204 |
| 4 | $\dfrac{2}{15}(900 - 30) = \ 116$ | 88 |
| 5 | $\dfrac{1}{15}(900 - 30) = \ 58$ | 30 |
| | $870 | |

These two depreciation schedules have been plotted in Figure 10-11. The DDB depreciation initially is larger than the SOYD depreciation. This situation is reversed beginning in the second year. Figure 10-11 shows that by year 4 the cumulative effect makes the book value for SOYD depreciation smaller than if DDB depreciation had been used. Thus while DDB depreciation is greater initially, the cumulative advantage has disappeared by year 4 when SOYD depreciation is superior.

As both depreciation schedules accomplish the same task of depreciating the asset from its cost to the expected salvage value (or a total of $900 − 30 = $870), the two depreciation schedules are equally desirable if the time value of money is ignored. The present worth of the DDB depreciation at 0% interest is $870. The present worth of the SOYD depreciation at 0% interest is also $870. But when the time value of money is considered, the two depreciation schedules will no longer be equally desirable. In Chapter 11 we will see more clearly how a depreciation schedule effects the cash flow of a profitable tax-paying firm. For now we must accept that the criterion is to select the depreciation schedule with the larger present worth of depreciation charges. Using a 6% interest rate and single payment present worth factors, the calculations are as follows:

| Year | DDB with conversion to straight line depreciation | | Single payment present worth factor $(P/F,6\%,year)$ | | Present worth of DDB depreciation |
|------|------|---|------|---|------|
| 1 | $360 | × | 0.943 | = | $339 |
| 2 | 216 | × | 0.890 | = | 192 |
| 3 | 130 | × | 0.840 | = | 109 |
| 4 | 82 | × | 0.792 | = | 65 |
| 5 | 82 | × | 0.747 | = | 61 |
| | | | | $\Sigma$PW = | $766 |

| Year | SOYD depreciation | | Single payment present worth factor $(P/F,6\%,year)$ | | Present worth of SOYD depreciation |
|------|------|---|------|---|------|
| 1 | $290 | × | 0.943 | = | $273 |
| 2 | 232 | × | 0.890 | = | 206 |
| 3 | 174 | × | 0.840 | = | 146 |
| 4 | 116 | × | 0.792 | = | 92 |
| 5 | 58 | × | 0.747 | = | 43 |
| | | | | $\Sigma$PW = | $760 |

From the calculations we see that double declining balance (with conversion to straight line) depreciation has a larger present worth of the depreciation schedule. It is therefore preferred over the SOYD depreciation in this situation.

We have examined the five principal depreciation methods.

Straight line
Sum-of-years digits
Double declining balance
Double declining balance with conversion to straight line
Sinking fund

If the time value of money is ignored there is no significant difference between the depreciation methods for they accomplish the same task—to systematically decrease book value from cost *P* to salvage value *F*.

But in many situations the choice of a depreciation method may have an impact on the reported earnings of a firm. The reported profits of the various domestic airlines, for example, are effected by the way the aircraft are depreciated. If an accelerated depreciation method (declining balance or sum-of-years digits) is used, this would result in much larger depreciation charges when aircraft are new and smaller depreciation charges as the aircraft become older. The management of an airline may not find this desirable if it has an adverse effect on earnings reported to the shareholders.

But there is still a third way to view the selection of a depreciation method. Using time value of money concepts we can select the depreciation method that provides the best economic consequences.

For economic efficiency we can determine the preferred depreciation method for any particular situation. The computations are greatly simplified if the firm is a profitable one that pays income taxes each year, and the income tax rate remains constant. An appropriate interest rate must be selected for making the necessary present worth calculations. As previously noted, the preferred depreciation method is sensitive to the interest rate used. Table 10-3 lists the preferred depreciation method for a particular set of conditions.

## PERIODIC CHANGES IN U.S. DEPRECIATION REGULATIONS

The depreciation regulations of the United States, as they affect the computation of federal taxes, are not static. Rather they are periodically adjusted according to the economic and political situation in the country. There have not been sweeping changes in the basic methods of computing

# Table 10-3. Preferred depreciation method

| Methods Compared: | Straight line |
|---|---|
| | Sum-of-years digits |
| | Double declining balance |
| | Double declining balance with conversion to ⁚ |

| Conditions: | 1. | The corporation is profitable and pays federa each year. |
|---|---|---|
| | 2. | The income tax rate remains constant throughout the depreciable life of the asset. |
| | 3. | An interest rate of 7% is used in comparing alternatives. |

| Depreciable life of asset N (years) | Zero salvage value F = 0 | Salvage value ratio F/P | | |
|---|---|---|---|---|
| | | F/P = 0.05 | F/P = 0.10 | F/P = 0.12 |
| 1 2 | **Straight line depreciation\*** | | | |
| 3 | 3 | | | |
| 4 | 4 | 4 | | |
| 5 | 4 | 5 | | |
| 6 | | 5 | *Double declining balance* | |
| 7 | | 6 | | |
| 8 | | 6 | 8 | |
| 9 | *Sum-of-years digits depreciation* | | 9 | |
| 10 | | | 9 | |
| 11 | | | 10 | |
| 12 | | | 11 | |
| 13 | | | 11 | |
| 14 | | | 12 | |
| 15 | | | 13 | |
| 16 | | | 13 | |
| 17 | | | 14 | |
| 18 | | | 15 | 18 |
| 19 | | | 16 | 19 |
| 20 | | | | 19 |

In the shaded area use double declining balance with conversion to straight line depreciation for the year indicated and all subsequent years.

\*IRS requires that straight line depreciation be used for assets with a depreciable life of less than 3 years.

depreciation. Instead some details have been changed. By examining these changes we no doubt can anticipate similar changes in the future. Three topics will be considered here. They are additional first year depreciation, amortization, and changes in depreciable lives. In the next chapter on taxes we will consider the closely related topic of investment tax credit.

## Additional First Year Depreciation

A taxpayer may elect to deduct 20% of the cost of qualifying property as additional first year depreciation in addition to the depreciation computed by the regular method. Thus the taxpayer may substantially increase his first year depreciation in this manner. There are limitations. First, the additional first year depreciation is limited to $10,000 of property for an individual or corporation in any year. Second, qualifying property means tangible personal property for use in a business or production of income. Finally, the property must have a useful life of at least six years.

## Amortization

From time to time the Congress has provided an incentive to encourage certain forms of investment. This inducement has been to allow the writing-off of the value of certain assets in a period far shorter than their useful lives. This amortization deduction takes the place of depreciation. In each of the three examples below, the amortization period was specified as 60 months.

*Emergency Facilities.* These were facilities constructed for use in support of military defense activities.

*Grain Storage Facilities.* At one time the government issued "certificates of necessity" to encourage the construction of grain storage facilities.

*Pollution Control Facilities.* Between 1969 and 1975 pollution control facilities installed in plants constructed prior to 1969 were eligible for 60 month amortization.

## Changes in Depreciable Lives

Estimation of the depreciable life of particular assets or groups of assets is subject to considerable variation. Unless a taxpayer has had experience with the useful life of a particular kind of asset, the practical result has been that federal guideline lives are adopted. Over the years there has been a series of guidelines. The Internal Revenue Service Bulletin F (1942) was followed by IRS publication 456, Depreciation Guidelines and Rules (1962) with its Revenue Procedure 62-21. This was in turn followed by the Asset Depreciation Range (ADR) System (1971). The data in Table 10-1 were taken from this publication.

## SINGLE ASSET VS MULTIPLE ASSET DEPRECIATION ACCOUNTING

There are two distinct ways of establishing the accounting records on capital equipment. Either each piece of equipment may be treated as a single item with its own depreciation schedule or a number of individual items may be lumped together and treated as a group.

If one were to buy 15 light pickup trucks for use by a construction company, the guideline useful life (Table 10-1) would be four years. But it seems likely that not all of the pickups would last for the full four years. Due to accidents, particularly hard useage, or whatever, several probably would be disposed of ("retired") prior to the four-year projected life. Several more might be retired at the end of four years, and some might be kept for one or more additional years. This would seem to be a normal and reasonable situation. If we were to plot a survivor curve for the 15 trucks, it might look something like Figure 10-12.

The survivor curve data may be replotted to produce a retirement-frequency curve. The results, shown in Figure 10-13, indicate that while there

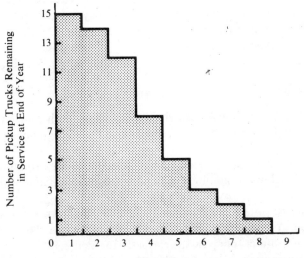

**Figure 10-12.** Survivor curve.

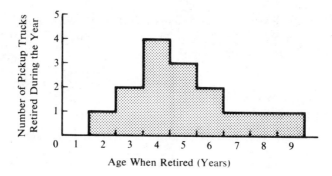

**Figure 10-13.** Retirement curve.

were retirements before the four-year estimated life, and after the four-year life, the greatest number were retired in the year that they became four years old. The data used for Figures 10-12 and 10-13 were hypothetical, but represent a realistic pattern.

Recognizing that some type of retirement frequency curve similar to Figure 10-13 exists for any group of assets, then an early retirement (that is, disposal before reaching the projected useful life) seems no more unusual that a retirement sometime beyond the projected useful life.

In multiple asset accounting the depreciation schedule recognizes this mortality dispersion of assets. Instead of a fixed predetermined schedule as has been described throughout this chapter, multiple asset depreciation is based on the cost of the assets remaining in service. Depreciation continues on the remaining assets for the years beyond the estimated mean useful life. The book value, of course, may never be less than the salvage value of the assets that continue in service. The technique of multiple asset depreciation accounting is illustrated by the following example.

*EXAMPLE 10-9*

The 15 trucks previously described cost $3000 each and have a $400 resale value whenever they are retired. Use the IRS guideline useful life (Table 10-1) of four years and assume that the pickup trucks are disposed of as indicated in Figure 10-13. Compute the depreciation schedule for the 15 pickups assuming straight line depreciation and multiple asset accounting.

It must be noted that Figures 10-12 and 10-13 represent a situation where the 15 trucks provide 60 truck-years of service. Thus the average truck useful life is four years, or exactly the depreciable life to be used in the depreciation calculations. As a result the depreciation will continue until the last truck is retired. If the trucks had survived longer, and provided more than 60 truck-years of service, the depreciation charges would continue only until the book value equals the salvage value of the trucks remaining in service.

Total cost $P = 15 \times 3000 = \$45,000$
Salvage value $F = 15 \times 400 = 6000$
Useful life $N = 4$ years
Straight line depreciation $= (1/N)(P - F) = (1/4)(P - F)$

| End of year | Number of trucks in service at end of year | | Cost of trucks in service P | Salvage value of trucks in service F | Straight line depreciation (1/4)(P − F) |
|---|---|---|---|---|---|
| 1 | | 15 | $45,000 | $6000 | $9750 |
| | Year 2 retirements | −1 | −3,000 | −400 | |
| 2 | | 14 | 42,000 | 5600 | 9100 |
| | Year 3 retirements | −2 | −6,000 | −800 | |
| 3 | | 12 | 36,000 | 4800 | 7800 |
| | Year 4 retirements | −4 | −12,000 | −1600 | |
| 4 | | 8 | 24,000 | 3200 | 5200 |
| | Year 5 retirements | −3 | −9,000 | −1200 | |
| 5 | | 5 | 15,000 | 2000 | 3250 |
| | Year 6 retirements | −2 | −6,000 | −800 | |
| 6 | | 3 | 9,000 | 1200 | 1950 |
| | Year 7 retirements | −1 | −3,000 | −400 | |
| 7 | | 2 | 6,000 | 800 | 1300 |
| | Year 8 retirements | −1 | −3,000 | −400 | |
| 8 | | 1 | 3,000 | 400 | 650 |
| | Year 9 retirements | −1 | −3,000 | −400 | |
| 9 | | 0 | 0 | 0 | 0 |

Total depreciation $39,000

## DEPLETION

Depletion is the exhaustion of natural resources as a result of their removal. Since depletion covers mineral properties, oil and gas wells, and standing timber, removal may take the form of digging up metallic or non-metallic minerals, producing petroleum or natural gas from wells, or cutting down trees.

The reason for depletion is essentially the same as the reason for depreciation. The owner of natural resources is consuming his capital investment as the natural resources are being removed and sold. Thus a portion of the gross income should be considered a return of the capital investment. The calculation of the depletion allowance is different from depreciation as there are two distinct methods of calculating depletion—cost depletion and percentage depletion. Except for standing timber and most oil and gas wells, depletion is calculated by both methods and the larger value is taken as depletion for the year. For standing timber and most oil and gas wells, only cost depletion is permissible.

**Cost Depletion**

In our calculation of depreciation we generally took the cost, useful life, and salvage value and used one of several methods to apportion the cost (minus salvage value) over the useful life. In the less frequent case where the asset is used at fluctuating rates, we might use the unit-of-production method of depreciation. For mines, oil wells, and standing timber, fluctuating production rates are the usual situation. Thus cost depletion is computed in the same manner as unit-of-production depreciation. The elements of the calculation are listed.

1. Cost of the property
2. Estimation of the number of recoverable units (tons of ore, cubic yards of gravel, barrels of oil, million cubic feet of natural gas, thousand board feet of timber, etc.)
3. Salvage value, if any, of the property

*EXAMPLE 10-10*

A small lumber company bought a timber tract for $35,000 of which $5000 was the value of the land and $30,000 was the value of the estimated $1\frac{1}{2}$ million board feet of standing timber. The first year the company cut 100,000 board feet of timber on the tract. What was the depletion allowance for the year?

$$\frac{\text{The depletion allowance per thousand}}{\text{board feet of timber}} = \frac{\$35,000 - \$5000}{1500 \text{ thousand board feet}}$$

$$= \$20 \text{ per thousand board feet}$$

The depletion allowance for the year would be
100 thousand bd ft × $20 per thousand bd ft = $2000

## Percentage Depletion

Percentage depletion is an alternate method of calculating the depletion allowance for mineral property and some oil or gas wells. (In 1975 percentage depletion was eliminated for all but the smallest petroleum companies.) The allowance is a certain percentage of the gross income from the

**Table 10-4.** Percentage depletion allowance for selected items

| Type of deposit | Percent |
| --- | --- |
| Lead, zinc, nickel, sulphur, uranium, and some oil and gas wells* | 22 |
| Gold, silver, copper, refractory and fire clay | 15 |
| Coal and sodium chloride | 10 |
| Sand, gravel, stone, clam and oyster shells, brick and tile clay | 5 |
| Most other minerals and metallic ores | 14 |

*The percentage depletion allowance for major oil and gas producers has been eliminated. For small producers the allowance is scheduled to decrease from 22% in 1980 to 15% in 1984.

property during the year. This is an entirely different concept from depreciation. Depreciation was the allocation of cost over the useful life. Percentage depletion, on the other hand, is an annual allowance of a percentage of the gross income from the property. Since percentage depletion is computed on the income rather than the cost of the property, the total depletion on a property *may exceed the cost of the property.* In computing the allowable percentage depletion on a property in any year, there is one major limitation on the amount of percentage depletion. The percentage depletion in any year is limited to not more than 50% of the taxable income from the property, computed without the deduction for depletion. The percentage depletion calculations may be illustrated by an example.

### EXAMPLE 10-11

A coal mine has a gross income of $250,000 for the year. Mining expenses equal $210,000. Compute the allowable percentage depletion deduction.

From Table 10-4, coal has a 10% depletion allowance. The percentage depletion deduction is computed from gross mining income. Then the taxable income must be computed. The allowable percentage depletion deduction is limited to the computed percentage depletion or 50% of taxable income, whichever is smaller.

*Computed percentage depletion*

| | |
|---|---|
| Gross income from mine | $250,000 |
| Depletion percentage | 10% |
| Computed percentage depletion | $ 25,000 |

*Taxable income limitation*

| | |
|---|---|
| Gross income from mine | $250,000 |
| Less: expenses other than depletion | 210,000 |
| Taxable income from mine | 40,000 |
| Deduction limitation | 50% |
| Taxable income limitation | $ 20,000 |

Since the taxable income limitation ($20,000) is less than the computed percentage depletion ($25,000), the allowable percentage depletion deduction is $20,000.

As previously stated, on mineral property and some oil and gas wells, the depletion deduction is based on either cost or percentage depletion. Each year depletion is computed by both methods and the allowable depletion deduction is the larger of the two amounts.

## SUMMARY

To prepare for the income tax effect on economic analysis it is necessary to first understand the concept of depreciation. The word *depreciation* is not readily defined. We know it means a decrease in value, but this leads us to three distinct definitions of depreciation.

1. Decline in market value of an asset.
2. Decline in value of an asset to its owner.
3. Systematic allocation of the cost of an asset (less its salvage value) over its useful life.

While the first two definitions are used in everyday discussions, it is the third or accountant's definition that is used in tax computations. For that reason the remainder of the chapter was based on allocation of cost over the useful life of an asset. Book value is the remaining unallocated cost of an asset or:

Book value = cost − depreciation charges made

Six depreciation methods for the systematic allocation of cost were presented.

1. Straight line
2. Sum-of-years digits
3. Declining balance
4. Declining balance with conversion to straight line
5. Unit of production
6. Sinking fund

Of the six methods, only the declining balance method presents any computational problems. At times the declining balance method does not achieve the desired result of allocation of the cost over the useful life of the asset. This difficulty is overcome by converting from declining balance to straight line depreciation part way through the useful life of the asset.

The different depreciation methods have an impact on the taxable income of a firm in any year. Assuming the objective is to depreciate an asset to its salvage value as rapidly as possible, then certain depreciation methods are preferred over the others. Five of the six depreciation methods are mathematical relationships and generalizations can be made about them. The sixth, unit of production, is not, and so the attractiveness of this method can only be determined from the facts of the specific situation. Of the five mathematical methods, sinking fund depreciation depreciates an asset least rapidly. Straight line depreciation is in turn less rapid than the three so-called accelerated methods: declining balance, declining balance with conversion to straight line, and sum-of-years digits. Depending on the details of a specific situation, one of these three methods can be expected to be the preferred depreciation method if the objective is to depreciate an asset as rapidly as possible. In some situations one must compute the present worth of each of the alternative depreciation schedules to properly judge which method is preferred. Periodically changes are made in the way depreciation calculations are done. Three frequent changes have been additional first year depreciation, amortization of assets over a shortened period, and changes in allowable depreciable lives.

The accounting records for capital assets may be kept either by individual asset (item depreciation accounting) or by groups of assets (group depreciation accounting). Item depreciation accounting seems easier to compute. Group depreciation accounting, however, is more realistic in its treatment of the expected useful lives of each of the assets in the group.

Depletion is the exhaustion of natural resources like minerals, oil and gas wells, and standing timber. The owner of the natural resources is consuming his investment as the natural resources are removed and sold. This concept is quite similar to unit of production depreciation and results in a computation called cost depletion. For minerals and some oil and gas wells there is an alternate calculation called percentage depletion. Percentage depletion has the unusual characteristic that in the exhaustion of the natural resources the total allowable depletion deductions may exceed the invested cost.

## PROBLEMS

**10-1** A new pickup truck can be obtained for $3700. At the end of its useful life it can be sold for $100. Using the Internal Revenue Service guideline for useful life, compute the depreciation schedule for this asset using the following.

   (a) Straight line depreciation.

   (b) Sum-of-years digits depreciation.

   (c) Double declining balance depreciation.

   (d) Double declining balance depreciation with conversion to straight line depreciation.

**10-2** The company treasurer is uncertain which of four depreciation methods is more desirable for the firm to use for a newly installed punch press. Its installed cost is $50,000, with an estimated salvage value of $8000 at the end of a ten year useful life. The firm uses an 8% interest rate in its economic analysis. Compute the depreciation schedule for the punch press for each of the four depreciation methods listed below.

   (a) Double declining balance

   (b) Straight line

   (c) SOYD

   (d) Sinking fund depreciation

   Which method would you recommend that the treasurer adopt? Explain. (*Answer:* Double declining balance)

**10-3** The RX Drug Company has just purchased a capsulating machine for $76,000. The plant engineer estimates the machine has a useful life of eight years and that it could be sold at the end of that time for $4000.

   (a) Compute the depreciation schedule for the machine by each of the following depreciation methods.

      1. Straight line depreciation.

      2. Sum-of-years digits depreciation.

      3. Double declining balance depreciation.

   (b) The Controller for RX Drug Company believes that double declining balance with conversion to straight line depreciation may be a more desirable depreciation method. Compute the depreciation schedule for double declining balance depreciation with conversion to straight line depreciation at the desirable point.

**10-4** A new machine tool is being purchased for $16,000 and is expected to have a $1000 salvage value at the end of its eight year useful life. Compute the double declining balance depreciation schedule for this capital asset. It may be desirable for this profitable firm to use double declining balance depreciation with conversion to straight line depreciation. If it is desirable, compute the depreciation schedule, making the conversion from double declining balance depreciation to straight line depreciation at the optimum time. Tabulate the resulting depreciation schedule.

**10-5** A large profitable corporation purchased a small jet plane for use of the firm's executives. The plane cost $840,000, has a six-year depreciable life and an estimated $50,000 resale value at the end of six years. Determine the preferred depreciation policy and compute the depreciation for each of the six years. If needed, use a 6% interest rate.

**10-6** When a major highway was to be constructed nearby, a farmer realized that a dry streambed running through his property might be valuable as a source of sand and gravel. He shipped samples of the sand and gravel to a testing laboratory and learned that the material met the requirements for certain low grade fill material for the highway. The farmer contacted the highway construction contractor. The contractor offered to pay 65¢ per cubic yard for 45,000 cubic yards of sand and gravel to be scooped out of the streambed. The contractor would build a haulage road to transport the sand and gravel, and would use his own equipment to load and haul the material. All the activity would take place during a single summer. The farmer hired an engineering student for the summer to watch the sand and gravel loading operation and to count the truckloads of material hauled away. For this the student received $2500. When the summer was over, the farmer calculated the results of his venture. He estimated that two acres of streambed had been stripped of the sand and gravel. The 640 acre farm had cost him $300 per acre and the farmer felt the property had not changed in value. He knew that there had been no use for the sand and gravel prior to the construction of the highway, and he could foresee no future use for any of the remaining 50,000 cubic yards of sand and gravel. Determine the farmer's depletion allowance. (*Answer:* $1462.50)

**10-7** Mr. H. Salt purchased a $\frac{1}{8}$ interest in a producing oil well for $45,000. Recoverable oil reserves for the well were estimated at that time at 120,000 barrels, $\frac{1}{8}$ of which represented Mr. Salt's share of the reserves. During the subsequent year, Mr. Salt received $12,000 as his $\frac{1}{8}$ share of the gross income from the sale of 8000 barrels of oil. From this amount he had to pay $3000 as his share of the expense of producing the oil. Compute Mr. Salt's depletion allowance for the year. (*Answer:* $3000)

**10-8** A heavy construction firm has been awarded a contract to build a large concrete dam. It is expected that a total of eight years will be required to complete the work. The firm will buy $600,000 worth of special equipment for the job. During the preparation of the job cost estimate, the following utilization schedule was computed for the special equipment:

|      | Utilization       |
| Year | (hours/year)      |
|------|-------------------|
| 1    | 6000              |
| 2    | 4000              |
| 3    | 4000              |
| 4    | 1600              |
| 5    | 800               |
| 6    | 800               |
| 7    | 2200              |
| 8    | 2200              |

At the end of the job it is estimated that the equipment can be sold at auction for $60,000.

(a) Compute the SOYD depreciation schedule for this equipment.

(b) Compute the unit-of-production depreciation schedule.

(c) If interest is 8%, which of the two depreciation methods, SOYD or UOP, is preferred for this profitable firm?

**10-9** A machine is purchased for $1000. It has a useful life and a depreciable life of six years. There is no end-of-useful life salvage value. Assuming a 7% interest rate, what is the preferred depreciation schedule for the machine? Show your computations and the resulting depreciation schedule. Compare your depreciation method to the solution given in Table 10-3.

# Appendix 10-A
Approximate
Economic Analysis Calculations
Using Depreciation

When asked the question: "What uniform annual payment would be required to repay $5000 in five years with interest at 6%?" we have calculated

$$A = P(A/P, 6\%, 5 \text{ years}) = \$5000(0.2374) = \$1187$$

But what does the $1187 represent? A little thought reveals that there are two components.

1. A portion of the $5000 principal sum must be repaid each year by the borrower.
2. The borrower must also pay 6% interest on the amount of the unpaid debt each year.

The details of this repayment plan were described in plan 3 of Table 4-1. We saw that $1187 was the uniform annual payment that was *equivalent* to a present sum of $5000 for the stated situation.

Suppose one wished to estimate the uniform annual payment without using compound interest tables. The only practical thing to do would be to make a reasonable estimate of the annual principal payment and of the annual interest payment. The sum of these two components would be an estimate of the uniform annual payment. One method is described in the next section.

### Annual Cost by Method of Straight Line Depreciation Plus Average Interest

If we assume that a quantity of money is repaid in equal annual payments, then the annual payment is identical to straight line depreciation with zero salvage value. The resulting plot of the remaining debt vs time is shown in

Figure 10A-1. In addition there is an annual interest payment. At the end of the first year it is $Pi$ and in the last year it is $Pi/N$. Since the principal repayment plan is uniform, the annual interest would decline uniformly from the initial $Pi$ to $Pi/N$ in the last year. We may compute the average annual interest as:

$$\text{Average interest} = \frac{\text{first year interest + last year interest}}{2}$$

$$= \frac{Pi + (Pi/N)}{2} = \frac{Pi}{2}\left(1 + \frac{1}{N}\right)$$

$$= \frac{Pi}{2}\left(\frac{N + 1}{N}\right)$$

Combining the components, where $F = 0$:

$$\text{Average annual cost} = \frac{P}{N} + \frac{Pi}{2}\left(\frac{N + 1}{N}\right)$$

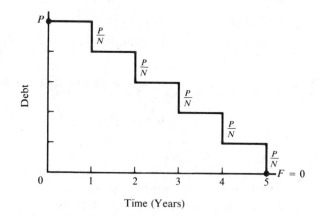

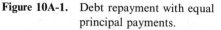

Time (Years)

**Figure 10A-1.**   Debt repayment with equal
                    principal payments.

*EXAMPLE 10A-1*

What will be the average annual payment to repay a debt of $5000 in five years with interest at 6%? Using the method of straight line depreciation plus average interest:

$$\text{Average annual cost} = \frac{P}{N} + \frac{Pi}{2}\left(\frac{N + 1}{N}\right) = \frac{5000}{5} + \frac{5000(0.06)}{2}\left(\frac{6}{5}\right)$$

$$= 1000 + 180 = \$1180$$

What is the difference between the $1180 calculated in Example 10A-1 and the previously calculated $1187? We have seen that the $1187 represents the *equivalent* uniform annual cost while the Example 10A-1 calculation produces the *average* uniform annual cost.

When the straight line depreciation plus average interest method is to be used in computing the *average* annual cost where there is a salvage value, the equation must be adjusted. Where there is a salvage value $F$ as in Figure 10A-2:

$$\text{Straight line depreciation} = \frac{P - F}{N}$$

$$\text{Average interest} = \frac{\text{first year interest} + \text{last year interest}}{2}$$

$$= \frac{(P - F)i + Fi + \dfrac{(P - F)i}{N} + Fi}{2}$$

$$= (P - F)\left(\frac{i}{2}\right)\left(1 + \frac{1}{N}\right) + Fi$$

$$= (P - F)\left(\frac{i}{2}\right)\left(\frac{N + 1}{N}\right) + Fi$$

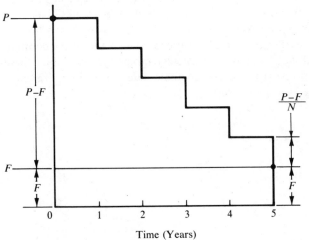

Time (Years)

**Figure 10A-2**

Thus the general equation for the straight line depreciation plus average interest is

$$\text{Average annual cost} = \frac{P - F}{N} + (P - F)\left(\frac{i}{2}\right)\left(\frac{N + 1}{N}\right) + Fi$$

*EXAMPLE 10A-2*

A firm buys equipment with an initial cost of $8000, and an estimated salvage value of $2000 at the end of an eight-year useful life. If interest is 6%, what is (a) the equivalent uniform annual cost? (b) the average uniform annual cost based on straight line depreciation plus average interest?

(a) $\text{EUAC} = (8000 - 2000)(A/P, 6\%, 8 \text{ years}) + 2000(0.06)$

$\qquad = 6000(0.1610) + 2000(0.06)$

$\qquad = 966 + 120 = \$1086$

(b) Average uniform annual cost

$$= \frac{8000 - 2000}{8} + (8000 - 2000)\left(\frac{0.06}{2}\right)\left(\frac{9}{8}\right) + 2000(0.06)$$

$$= 750.00 + 202.50 + 120.00$$

$$= \$1072.50$$

In both of the examples of straight line depreciation plus average interest we find that this approximate method gives values that are smaller than the exact capital recovery calculations. This is always the case. The error becomes larger as $N$ and $i$ increase. Since the exact computation can be readily done if compound interest tables are available, there seems little reason to resort to an approximate calculation like straight line depreciation plus average interest.

# Appendix 10-B
## End-of-Useful Life
## Book Value Coefficients for
## Declining Balance Depreciation

For the declining balance depreciation method the book value curve is a function of only two factors: the depreciation rate ($1.25/N$, $1.5/N$, or $2/N$), and the depreciable life $N$. Since the curve is independent of the salvage value $F$, the declining balance book value curve may result in an end-of-depreciable-life book value $P_N$ that is more than, equal to, or less than the estimated salvage value $F$. This is illustrated in Figure 10B-1. In a given situation it is important to know which of the three graphs in Figure 10B-1 is applicable. In other words, we need to know the relationship between $F$ and $P_N$.

The end of depreciable life book value $P_N$ may be computed in relation to the cost of the asset $P$ using a declining balance book value equation (Eq. 10-11).

Double declining balance

$$\text{Book value} = P\left[1 - \frac{2}{N}\right]^n \tag{10B-1}$$

where $P$ = cost of the asset

$N$ = depreciable life of the asset

$n$ = year at which book value is desired; for book value at the end of the depreciable life, $n = N$

For the end of depreciable life book value, Eq. 10B-1 becomes

$$\text{End of depreciable life book value } P_N = P\left[1 - \frac{2*}{N}\right]^N \tag{10B-2}$$

*Substitute $1.25/N$ or $1.5/N$ as appropriate for the other declining balance depreciation rates.

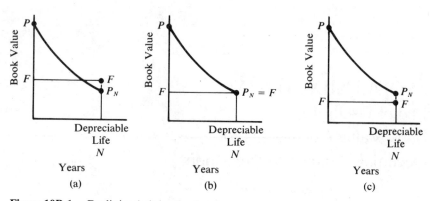

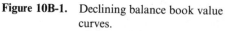

**Figure 10B-1.** Declining balance book value
curves.

Eq. 10B-2 was used to obtain Table 10B-1.

**Table 10B-1.** End of depreciable life book value coefficients for
declining balance depreciation

| Depreciable life N years | Type of declining balance depreciation | | |
|---|---|---|---|
| | $\dfrac{1.25}{N}$ | $\dfrac{1.5}{N}$ | $\dfrac{2}{N}$ |
| | End of depreciable life book value $P_N$ as a fraction of cost P | | |
| 3 | 0.199P | 0.125P | 0.037P |
| 4 | 0.223P | 0.153P | 0.062P |
| 5 | 0.237P | 0.168P | 0.078P |
| 6 | 0.246P | 0.178P | 0.088P |
| 8 | 0.257P | 0.190P | 0.100P |
| 10 | 0.263P | 0.197P | 0.107P |
| 15 | 0.271P | 0.206P | 0.117P |
| 20 | 0.275P | 0.210P | 0.122P |

*EXAMPLE 10B-1*

Some special purpose mining equipment has a cost of $18,000 and a
salvage value of $2500 at the end of its ten-year depreciable life. If
double declining balance depreciation is used, will it be necessary to
convert to straight line depreciation so the book value will equal the
salvage value at the end of ten years?

The question really is, which one of the graphs in Figure 10B-1 is applicable here? From Table 10B-1 we find that $P_N = 0.107P$. Therefore $P_N = 0.107(\$18,000) = \$1926$. Since $P_N$ is less than $F$, we know that the situation here is like Figure 10B-1a. No conversion to straight line is necessary or desirable. This same conclusion could be reached by the use of Eq. 10B-2 directly.

$$\text{End of depreciable life book value } P_N = P\left[1 - \frac{2}{N}\right]^N$$
$$= \$18,000\left[1 - \frac{2}{10}\right]^{10}$$
$$= \$18,000(0.107) = \$1926$$

When $P_N$ and $F$ are known the situation can be identified as falling into one of the following categories and the appropriate depreciation schedule computed.

$P_N < F$    When $P_N$ is less than $F$, the depreciation charges must be halted when the book value has declined to the point where it equals the estimated salvage value. Figure 10B-2a illustrates the resulting adjusted book value curve.

$P_N = F$    In the rare situation where $P_N$ equals $F$, no alteration of the depreciation charges is required. Figure 10B-2b is the same as Figure 10B-1b.

$P_N > F$    When $P_N$ is greater than $F$ the declining balance depreciation method has failed to produce the desired result of allocating the cost of the asset (minus its salvage value) over the depreciable

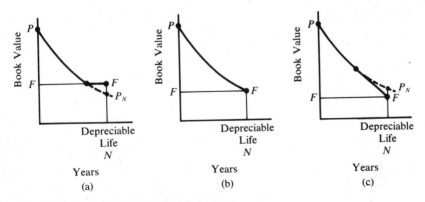

**Figure 10B-2.**   Adjusted declining balance book value curves.

life $N$. As was described in Chapter 10, two approaches may be used to remedy the situation. One solution is to continue the declining balance depreciation if the asset remains in service beyond its depreciable life. A second, and generally preferred method, is to adopt a composite depreciation method of declining balance with conversion to straight line. Figure 10B-2c reflects this latter method.

# 11
# Income
# Taxes

In this chapter we will examine the structure of taxes in the United States. There is, of course, a wide variety of taxes ranging from sales taxes, gasoline taxes, property taxes, state and federal income taxes, and so forth. Here we will concentrate our attention on federal income taxes. Since income taxes are part of all real problems and have a substantial impact on many of them, no realistic analysis can ignore the income tax consequences. As Benjamin Franklin said, two things are inevitable: death and taxes.

First, we must understand the way in which taxes are imposed. The chapter on depreciation is an integral part of this analysis as we shall see. Then, having understood the mechanism, we will see how federal income taxes affect our economic analysis. The various analysis techniques will be used in examples of after-tax calculations.

## A PARTNER IN THE BUSINESS

Probably the most straightforward way to understand the role of federal income taxes is to consider the U.S. Government as a partner in every business activity. As a partner the Government shares in the profits from every successful venture. And in a somewhat more complex way the Government shares in the losses in unprofitable ventures. The tax laws are very complex and hence it is not our purpose here to accurately and fully explain them. Instead we will examine the fundamental concepts of the federal income tax laws, recognizing that there are exceptions and variations to almost every statement we shall make.

## CALCULATION OF TAXABLE
## INCOME

At the mention of income taxes one can visualize all sorts of elaborate and and complex calculations. And there is some truth to that vision, for there can be all sorts of complexities in the computation of income taxes. But some of the difficulty is removed when one recognizes that income taxes are just another disbursement. In economic analysis calculations in prior chapters we have dealt with all sorts of disbursements: operating costs, maintenance, labor and materials, and so forth. Now we simply add one more prospective disbursement—income taxes—to the list.

### Taxable Income of Individuals

The amount of federal income taxes to be paid depends on taxable income and the income tax rates. Therefore, our first concern is the definition of taxable income.

An individual's total earned income is substantially what tax people call *adjusted gross income*. From this amount individuals may subtract two items. One is an *exemption* for each person who depends on the gross income for his living. An individual without dependents would have one exemption ($750) for himself. The second item is allowable deductions. Some of these items of expenses are excessive medical costs, state and local taxes, charitable contributions, interest on borrowed money, and so forth. A *standard deduction* may be taken by people who do not itemize their allowable deductions. The amount of the current *standard deduction* is 16% of adjusted gross income with a minimum deduction of $1700 and a maximum of $2400 for unmarried taxpayers.* Taxpayers should determine their allowable deduction by both methods and then use the method that gives them the larger deduction.

*For individual taxpayers*

> Taxable income = adjusted gross income
> > − deduction for exemptions
> > − itemized deductions or standard deduction

$$(11\text{-}1)$$

### Classification of Business Expenditures

When an individual or a firm operates a business there are three distinct types of business expenditures.

---

*For married couples filing a joint return the minimum is $2100 with a maximum of $2800.

1. Expenditures for depreciable assets
2. Expenditures for nondepreciable assets
3. All other business expenditures

*Expenditures for Depreciable Assets.* When facilities or productive equipment with useful lives in excess of one year are acquired the taxpayer will recover his investment through depreciation charges. In Chapter 10 we examined in great detail the several ways in which the cost of the asset, minus its salvage value, could be allocated over its useful life.

*Expenditures for Nondepreciable Assets.* Land is considered a nondepreciable asset for there is no finite life associated with it. Other nondepreciable assets would be property *not* used in a trade or business or for the production of income. An individual's home and automobile are generally nondepreciable assets. The final category of nondepreciable assets would be those subject to depletion rather than depreciation. Since business firms generally acquire assets for use in the business, their only nondepreciable assets would be land and assets subject to depletion.

*All Other Business Expenditures.* This category is probably the largest of all for it includes all the ordinary and necessary expenditures of operating a business. Labor costs, materials, all direct and indirect costs, facilities and productive equipment with a useful life of one year or less, and so forth are part of the routine expenditures.

Business expenditures in the first two categories, that is for either depreciable or nondepreciable assets, are called *capital expenditures*. In the accounting records of the firm they are "capitalized" while all ordinary and necessary expenditures in the third category are "expensed."

### Taxable Income of Business Firms

The starting point in computing a firm's taxable income is *gross income*. All ordinary and necessary expenses to conduct the business, except capital expenditures, are deducted from gross income. Capital expenditures may not be deducted from gross income. Except for land, business capital expenditures are recovered through depreciation or depletion charges.

*For business firms*

Taxable income = gross income

— all expenditures except capital expenditures

— depreciation and depletion charges     (11-2)

Because of the treatment for tax purposes of capital expenditures the taxable income of a firm may be quite different from the actual cash results.

## EXAMPLE 11-1

During a three year period a firm had the following results.

|  | Year 1 | Year 2 | Year 3 |
|---|---|---|---|
| Gross income from sales | $200 | $200 | $200 |
| Purchase of new production equipment (useful life: 3 years) | − 60 | 0 | 0 |
| All other expenditures | − 140 | − 140 | − 140 |
| Cash results for the year | $0 | $60 | $60 |

Compute the taxable income for each of the three years. The cash results for each year would suggest that year 1 was a poor one while years 2 and 3 were very profitable. A closer look reveals that the firm's cash results were adversely affected in year 1 by the purchase of new production equipment. Since the production equipment has a three year useful life, it is a capital expenditure with its cost allocated over the useful life. For straight line depreciation and no salvage value the annual charge is

$$\text{Annual depreciation charge} = \frac{P - F}{N} = \frac{60 - 0}{3} = 20$$

Applying the equation for taxable income:

Taxable income = gross income
  − all expenditures except capital expenditures
  − depreciation and depletion charges
  = 200 − 140 − 20 = 40

In each of the three years the taxable income is 40.

An examination of the cash results and the taxable income in Example 11-1 indicates that the taxable income is a clearer reflection of the annual performance of the firm.

## INCOME TAX RATES

There are three schedules of federal income tax rates for individuals. Single taxpayers use the Table 11-1 schedule. Married taxpayers filing a joint return use the Table 11-2 schedule. A third schedule not shown here is applicable for unmarried individuals with dependent relatives ("head of a household"). Figure 11-1 shows the incremental tax rates for individuals.

**Table 11-1.** Tax rates—if you are unmarried

| If your taxable income is | | Your tax is | | |
| --- | --- | --- | --- | --- |
| Over | But not over | This | Plus following percentage | Over this |
| $ 0 | $ 500 | $ 0 | 14% | $ 0 |
| 500 | 1,000 | $ 70 | 15% | 500 |
| 1,000 | 1,500 | 145 | 16% | 1,000 |
| 1,500 | 2,000 | 225 | 17% | 1,500 |
| 2,000 | 4,000 | 310 | 19% | 2,000 |
| 4,000 | 6,000 | 690 | 21% | 4,000 |
| 6,000 | 8,000 | 1,110 | 24% | 6,000 |
| 8,000 | 10,000 | 1,590 | 25% | 8,000 |
| 10,000 | 12,000 | 2,090 | 27% | 10,000 |
| 12,000 | 14,000 | 2,630 | 29% | 12,000 |
| 14,000 | 16,000 | 3,210 | 31% | 14,000 |
| 16,000 | 18,000 | 3,830 | 34% | 16,000 |
| 18,000 | 20,000 | 4,510 | 36% | 18,000 |
| 20,000 | 22,000 | 5,230 | 38% | 20,000 |
| 22,000 | 26,000 | 5,990 | 40% | 22,000 |
| 26,000 | 32,000 | 7,590 | 45% | 26,000 |
| 32,000 | 38,000 | 10,290 | 50% | 32,000 |
| 38,000 | 44,000 | 13,290 | 55% | 38,000 |
| 44,000 | 50,000 | 16,590 | 60% | 44,000 |
| 50,000 | 60,000 | 20,190 | 62% | 50,000 |
| 60,000 | 70,000 | 26,390 | 64% | 60,000 |
| 70,000 | 80,000 | 32,790 | 66% | 70,000 |
| 80,000 | 90,000 | 39,390 | 68% | 80,000 |
| 90,000 | 100,000 | 46,190 | 69% | 90,000 |
| 100,000 | | 53,090 | 70% | 100,000 |

## EXAMPLE 11-2

An unmarried student earned $2500 in the summer plus another $500 during the rest of the year. When he files an income tax return he will be allowed one exemption (for himself). What is his taxable income?

Taxable income = adjusted gross income
        − deduction for exemptions
        − itemized deductions or standard deduction

Adjusted gross income = 2500 + 500 = $3000

Deduction for exemptions = 1 × $750 = $750

**Table 11-2.**  Tax rates—if you file a joint return

| If your taxable income is | | Your tax is | | |
| --- | --- | --- | --- | --- |
| Over | But not over | This | Plus following percentage | Over this |
| $     0 | $1,000 | $     0 | 14% | $     0 |
| 1,000 | 2,000 | 140 | 15% | 1,000 |
| 2,000 | 3,000 | 290 | 16% | 2,000 |
| 3,000 | 4,000 | 450 | 17% | 3,000 |
| 4,000 | 8,000 | 620 | 19% | 4,000 |
| 8,000 | 12,000 | 1,380 | 22% | 8,000 |
| 12,000 | 16,000 | 2,260 | 25% | 12,000 |
| 16,000 | 20,000 | 3,260 | 28% | 16,000 |
| 20,000 | 24,000 | 4,380 | 32% | 20,000 |
| 24,000 | 28,000 | 5,660 | 36% | 24,000 |
| 28,000 | 32,000 | 7,100 | 39% | 28,000 |
| 32,000 | 36,000 | 8,660 | 42% | 32,000 |
| 36,000 | 40,000 | 10,340 | 45% | 36,000 |
| 40,000 | 44,000 | 12,140 | 48% | 40,000 |
| 44,000 | 52,000 | 14,060 | 50% | 44,000 |
| 52,000 | 64,000 | 18,060 | 53% | 52,000 |
| 64,000 | 76,000 | 24,420 | 55% | 64,000 |
| 76,000 | 88,000 | 31,020 | 58% | 76,000 |
| 88,000 | 100,000 | 37,980 | 60% | 88,000 |
| 100,000 | 120,000 | 45,180 | 62% | 100,000 |
| 120,000 | 140,000 | 57,580 | 64% | 120,000 |
| 140,000 | 160,000 | 70,380 | 66% | 140,000 |
| 160,000 | 180,000 | 83,580 | 68% | 160,000 |
| 180,000 | 200,000 | 97,180 | 69% | 180,000 |
| 200,000 | | 110,980 | 70% | 200,000 |

Standard deduction = 16% × 3000 or $1700 whichever is larger
                                  = $1700

Taxable income = $3000 − $750 − $1700 = $550

From Table 11-1 the tax is $70 plus 15% on the taxable income in excess of $500.

Federal income tax = $70 + 15%(550 − 500) = $77.50

The tax rates for corporations are much simpler than those for individuals. There are three steps in the 1978 corporate tax rates. The first $25,000 of

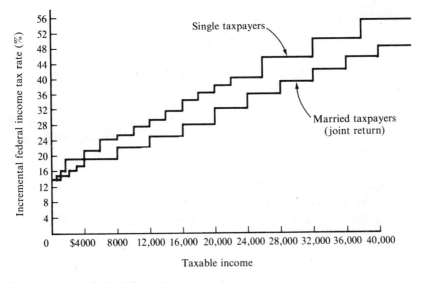

Figure 11-1. Individual federal income tax
rates (based on 1978 rates).

taxable income is taxed at 20%, the next $25,000 at 22%, and all over
$50,000 at 48%. Figure 11-2 graphically illustrates the situation.

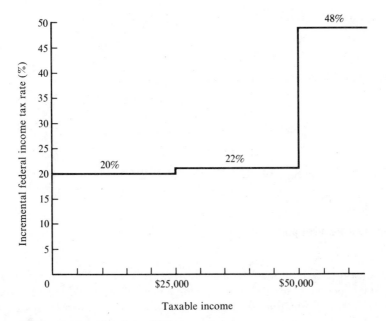

Figure 11-2. Corporation federal income tax
rates (based on 1978 rates).

### EXAMPLE 11-3

The French Chemical Corporation was formed to produce household bleach. In its first year of operation the firm bought land for $22,000, had a $90,000 factory building erected and installed $165,000 worth of chemical equipment. The gross income for the year was $240,000. Supplies and all operating expenses, excluding the capital expenditures, were $100,000. The firm will use double declining balance depreciation and assume there is only a negligible salvage value for the building and the equipment at the end of their useful lives.

(a) What is the first year depreciation charge?
(b) What is the taxable income?
(c) How much will the corporation pay in federal income taxes?

(a) From Table 10-1

|                    | *Useful life* |
|--------------------|---------------|
| Factory buildings  | 45 yrs        |
| Chemical equipment | 11            |

$$\text{First year DDB depreciation} = \frac{2}{45}(90,000) + \frac{2}{11}(165,000)$$
$$= 4000 + 30,000 = \$34,000$$

(b) Taxable income = gross income
— all expenditures except capital expenditures
— depreciation and depletion charges
= 240,000 − 100,000 − 34,000
= $106,000

Federal income tax = 20%(25,000) + 22%(25,000) + 48%(56,000)
= 5000 + 5500 + 26,880 = $37,380

### Combined Federal and State Income Taxes

In addition to federal income taxes most individuals and corporations also pay state income taxes. It would be convenient if we could derive a single tax rate that would represent both the state and federal incremental tax rates. In the computation of taxable income for federal taxes the amount of state taxes paid is one of the allowable itemized deductions. Federal income taxes are not however generally deductible in the computation of state taxable income. Therefore the state income tax is applied to a *larger* taxable income than is the federal income tax rate. As a result the combined incremental tax rate will not be the sum of two tax rates.

For an increment of income (Δincome), state income taxes are (Δstate tax rate)(Δincome). The federal taxable income is (Δincome)(1 − Δstate tax rate), and federal income taxes are (Δfederal tax rate)(Δincome) × (1 − Δstate tax rate).

The total of state and federal income taxes is

[Δstate tax rate + (Δfederal tax rate)(1 − Δstate tax rate)][Δincome]

The terms within the left set of brackets equals the combined incremental tax rate.

Combined incremental tax rate
= Δstate tax rate + (Δfederal tax rate)(1 − Δstate tax rate)        (11-3)

### EXAMPLE 11-4

An engineer has an income that puts him in the 31% federal income tax bracket and at the 10% state incremental tax rate. He has an opportunity to earn an extra $500 by doing a small consulting job. What will be his combined state and federal income tax rate on the additional income? Using Eq. 11-3,

Combined incremental tax rate = 0.10 + 0.31(1 − 0.10)
= 0.379 = 37.9%

### Selecting an Income Tax Rate for Economy Studies

Since income tax rates vary with the level of taxable income for both individuals and corporations, one must decide which tax rate to use in a particular situation. The simple answer is that the tax rate to use is the incremental tax rate that applies to the change in taxable income projected in the economic analysis. If a married man had a taxable income of $22,000 and could increase his income by $2000 what tax rate should be used for the $2000 of incremental income? From Table 11-2 we see the $2000 falls within the 32% tax bracket. All this individual's income between $20,000 and $24,000 is taxed at a 32% incremental tax rate. Now suppose this individual could increase his $22,000 income by $3000. In this situation

Table 11-2 shows that the 32% incremental tax rate should be applied to the first $2000 and a 36% incremental tax rate to the last $1000 of extra income.

For corporations the matter of determining the appropriate incremental tax rate is greatly simplified. If in the economic analysis the increment of taxable income is under $25,000 of total taxable income, the 20% tax rate applies. Between $25,000 and $50,000 the incremental tax rate is 22%. Above $50,000 of total taxable income the incremental tax rate is 48%. Thus while incremental tax rates change for every few thousand dollars of income for individuals, this is not true for corporations. For all but the smallest corporations, the incremental federal income tax rate is 48%. For computational convenience a 50% corporate tax rate is often used.

## ECONOMIC ANALYSIS TAKING INCOME TAXES INTO ACCOUNT

An important step in economic analysis has been to resolve the consequences of an alternative into a cash flow. Because income taxes have been ignored the result has been a before-tax cash flow. This same before-tax cash flow is an essential component in economic analysis that also considers the income tax consequences. The principal elements in an after-tax analysis are:

Before-tax cash flow
Depreciation
Taxable income (before-tax cash flow minus depreciation)
Income taxes (taxable income times incremental tax rate)
After-tax cash flow (before-tax cash flow minus income taxes)

These elements are usually arranged to form a cash flow table. This is illustrated by Example 11-5.

### EXAMPLE 11-5

A medium sized profitable corporation is considering the purchase of a $3000 used 2-ton truck for the use of the shipping and receiving department. During the truck's five year useful life it is estimated the firm will save $800 per year after paying all the costs of owning and operating the truck. Truck salvage value is estimated at $750. (a) What is the before-tax rate of return? (b) What is the after-tax rate of return on this capital expenditure? Assume straight line depreciation.

(a) For a before-tax rate of return, we must first compute the before-tax cash flow.

| Year | Before-tax cash flow |
|------|---------------------|
| 0 | $-\$3000$ |
| 1 | $+800$ |
| 2 | $+800$ |
| 3 | $+800$ |
| 4 | $+800$ |
| 5 | $\begin{cases} +800 \\ +750 \end{cases}$ |

Solve for the rate of return.

$$3000 = 800(P/A,i\%,5) + 750(P/F,i\%,5)$$

Try $i = 15\%$.

$$3000 = 800(3.35) + 750(0.50)$$
$$= 2680 + 375 = 3055 \qquad i \text{ slightly low}$$

Try $i = 20\%$.

$$3000 = 800(2.99) + 750(0.40)$$
$$= 2392 + 300 = 2692 \qquad i \text{ too high}$$

$$i = 15\% + 5\% \left[ \frac{3055 - 3000}{3055 - 2692} \right] = 15\% + 5\%(0.15) = 15.75\%$$

(b) For an after-tax rate of return, we must set up the cash flow table (Table 11-3). The starting point is the before-tax cash flow. Then we will need the depreciation schedule for the truck.

$$\text{Straight line depreciation} = \frac{P - F}{N} = \frac{3000 - 750}{5} = \$450 \text{ per year}$$

Taxable income is the before-tax cash flow minus depreciation. For this medium size profitable corporation the incremental federal income tax rate must be 48%. Therefore income taxes are 48% of taxable income. Finally, the after-tax cash flow equals the before-tax cash flow minus income taxes. These data are used to compute Table 11-3. The after-tax cash flow may be solved to find the after-tax rate of return.

$$3000 = 632(P/A,i\%,5) + 750(P/F,i\%,5)$$

Try $i = 10\%$.

$$3000 = 632(3.79) + 750(0.62)$$
$$= 2395 + 465 = 2860 \qquad i \text{ too high}$$

**Table 11-3.**   Cash flow table

| Year | Before-tax cash flow | Straight line depreciation | Δ Taxable income | 48% income taxes* | After-tax cash flow |
|---|---|---|---|---|---|
| 0 | − $3000 | | | | − $3000 |
| 1 | 800 | $450 | $350 | − $168 | 632 |
| 2 | 800 | 450 | 350 | − 168 | 632 |
| 3 | 800 | 450 | 350 | − 168 | 632 |
| 4 | 800 | 450 | 350 | − 168 | 632 |
| 5 | { 800 / 750 | 450 | 350 | − 168 | { 632 / 750 |

*Sign convention for income taxes: A minus represents a disbursement of money to pay income taxes; a plus represents the receipt of money by a decrease in the tax liability.

Try $i = 8\%$.

$$3000 = 632(3.99) + 750(0.68)$$
$$= 2522 + 510 = 3032 \qquad i \text{ slightly low}$$

$$i = 8\% + 2\% \left[ \frac{3032 - 3000}{3032 - 2860} \right] = 8\% + 2\%(0.19) = 8.38\%$$

The calculations required to compute the after-tax rate of return in Example 11-5 were certainly more elaborate than those for the before-tax rate of return. It must be emphasized, however, that *only* the after-tax rate of return is a meaningful value since income taxes are a major disbursement that cannot be ignored.

## EXAMPLE 11-6

An analysis of a firm's sales activities indicates that a number of profitable sales are lost each year because the firm could not deliver some of its products quickly enough. By investing an additional $20,000 in inventory it is believed that the before-tax profit of the firm will be $1000 higher the first year. The second year before-tax extra profit will be $1500. Subsequent years are expected to continue to increase on a $500 per year gradient. The investment in the additional inventory may be recovered at the end of a four year analysis period simply by selling it and not replenishing the inventory. Compute the (a) before-tax rate of return, and (b) after-tax rate of return assuming an incremental tax rate of 48%. Inventory is not considered a depreciable asset therefore, the investment in additional inventory is not depreciated. The cash flow table for the problem is presented in Table 11-4.

**Table 11-4.** Cash flow table

| Year | Before-tax cash flow | Depreciation | $\Delta$ taxable income | 48% income taxes | After-tax cash flow |
|------|------|------|------|------|------|
| 0 | $-$20,000 | | | | $-$20,000 |
| 1 | 1000 | — | $1000 | $-$480 | 520 |
| 2 | 1500 | — | 1500 | $-720$ | 780 |
| 3 | 2000 | — | 2000 | $-960$ | 1040 |
| 4 | $\begin{cases} 2500 \\ 20,000 \end{cases}$ | — | 2500 | $-1200$ | $\begin{cases} 1300 \\ 20,000 \end{cases}$ |

(a) Before-tax rate of return

$$20,000 = 1000(P/A,i\%,4) + 500(P/G,i\%,4) + 20,000(P/F,i\%,4)$$

Try $i = 8\%$.

$$20,000 = 1000(3.312) + 500(4.650) + 20,000(0.735)$$
$$= 3312 + 2325 + 14,700 = 20,337 \qquad i \text{ too low}$$

Try $i = 10\%$.

$$20,000 = 1000(3.170) + 500(4.378) + 20,000(0.683)$$
$$= 3170 + 2189 + 13,660 = 19,019 \qquad i \text{ too high}$$

$$\text{Before-tax rate of return} = 8\% + 2\%\left[\frac{20,337 - 20,000}{20,337 - 19,019}\right] = 8.5\%$$

(b) After-tax rate of return

The before-tax cash flow gradient is $500.

The resulting after-tax cash flow gradient is $(1 - 0.48)(500) = \$260$

$$20,000 = 520(P/A,i\%4) + 260(P/G,i\%,4) + 20,000(P/F,i\%,4)$$

Try $i = 4\%$.

$$20,000 = 520(3.630) + 260(5.267) + 20,000(0.8548)$$
$$= 1888 + 1369 + 17,096 = 20,353 \qquad i \text{ too low}$$

Try $i = 5\%$.

$$20,000 = 520(3.546) + 260(5.103) + 20,000(0.8227)$$
$$= 1844 + 1327 + 16,454 = 19,625 \qquad i \text{ too high}$$

The after-tax rate of return is 4.4%.

## CAPITAL GAINS AND LOSSES

When a capital asset is sold or exchanged there must be entries in the firm's accounting records to reflect the change. If the selling price of the capital asset exceeds the book value, the excess of selling price over book value is called a capital gain. If the selling price is less than book value the difference is a capital loss.

$$\text{Capital} \begin{Bmatrix} \text{gain} \\ \text{loss} \end{Bmatrix} = \text{selling price} - \text{book value}$$

The rules for the tax treatment of capital gains and losses are quite involved. If the capital asset was held for more than 12 months the result is a "long term" gain or loss; 12 months or less is a "short term" gain or loss. The tax treatment is shown in Table 11-5.

## INVESTMENT TAX CREDIT

One technique that has been used by the U.S. Government to stimulate capital investments is the investment tax credit. The legislation provided that the business could deduct a percentage (usually 7% but it has varied from 4% to 10%) of its new business equipment purchases as a *tax credit*. After computing the income taxes owed by the firm, the tax credit of 7% of the cost of the qualifying equipment is deducted. This means that the net cost of the equipment to the firm is reduced by 7%. At the same time however the basis for computing depreciation remains the full cost of the equipment.

Although taxpayers file an annual income tax return, they pay their income tax liability throughout the year. This is caused by individual tax withholding and the declaration and payment of estimated income taxes for both individuals and corporations. This means that the investment tax credit will be reflected in the firm's cash flow as an almost immediate tax saving. In this book we assume the investment tax credit occurs at the time the business equipment is purchased. All other tax consequences are assumed to be end-of-period cash flows.

Since the investment tax credit may be adopted or removed quickly our treatment here may be different from the situation faced by the reader.

**Table 11-5.** Tax treatment of capital gains and losses

| For individuals | Short term | Long term |
|---|---|---|
| Capital gain | Taxed as ordinary income | 50% of the capital gain taxed as ordinary income, or capital gain taxed at 25%, whichever is smaller |
| Capital loss | Subtract capital losses from any capital gains; balance may be deducted from ordinary income, but not more than $3000 per year | Subtract capital losses from any capital gains; half the balance may be deducted from ordinary income, but not more than $3000 per year |

| For corporations | Short term | Long term |
|---|---|---|
| Capital gain | Taxed as ordinary income | Taxed at ordinary income tax rate (20%, 22% or 48%) or 30% whichever is smaller |
| Capital loss | Corporations may deduct capital losses only to the extent of capital gains. Any capital loss in the current year that exceeds capital gains is carried back three years, and, if not completely absorbed, is then carried forward for up to seven years | |

Unless stated otherwise, we shall assume that the investment tax credit is *not* available.

## ESTIMATING THE AFTER-TAX RATE OF RETURN

There is no shortcut method to compute the after-tax rate of return from the before-tax rate of return. The only exception to this statement is in the situation of nondepreciable assets. In this special case:

After-tax rate of return = (1 − incremental tax rate)
$$\times \text{(before-tax rate of return)}$$

For a 48% incremental tax rate the after-tax rate of return is slightly over half the before-tax rate of return. Example 11-6 represents this situation.

We know therefore that in that problem

After-tax rate of return $= (1 - 0.48)(8.5\%) = 4.4\%$

This relationship may be helpful for selecting a trial after-tax rate of return where the before-tax rate of return is known. It must be emphasized, however, this relationship is only a rough approximation where there is a depreciation schedule, capital gains or losses, or an investment tax credit.


## SUMMARY

Since income taxes are part of all problems, no realistic economic analysis can ignore the income tax consequences. Income taxes make the U.S. Government a partner in every business venture. Thus the Government bene-. fits from all profitable ventures and shares in the losses of unprofitable ventures.

The first step in computing income taxes is to determine taxable income. For individuals the starting point is adjusted gross income. This is substantially the total earned income of the individual. There are two major deductions from adjusted gross income—exemptions for each person dependent on the income, and deductions for certain expenses. Taxable income is the adjusted gross income less the exemptions and itemized or standard deductions.

For corporations taxable income equals gross income minus all ordinary and necessary expenditures (except capital expenditures) and depreciation and depletion charges.

Federal income tax rates for individuals vary from 14% for the lowest bracket of taxable income (first $500 for single taxpayers or $1000 for married taxpayers filing a joint return) to 70% at the top bracket, with 23 different rates in between. For corporations the income tax rates are very simple. For taxable income below $25,000 the tax rate is 20%; between $25,000 and $50,000 the rate is 22%; above $50,000 the tax rate is 48%. In economic analysis the proper rate to use is the incremental tax rate applicable to the increment of taxable income being considered.

Most individuals and corporations pay state income taxes in addition to federal income taxes. Since state income taxes are an allowable deduction in computing federal taxable income, it follows that the taxable income for the federal computation is lower than the state taxable income.

Combined state and federal incremental tax rate

= Δstate tax rate + (Δfederal tax rate)(1 − Δstate tax rate).

To introduce the effect of income taxes in an economic analysis the starting point is a before-tax cash flow. Then the depreciation schedule is deducted from appropriate parts of the before-tax cash flow to obtain taxable income. Income taxes are obtained by multiplying taxable income by the proper tax rate. After-tax cash flow can then be evaluated in any of the ways previously discussed.

One complexity of income taxes is capital gains and losses. When a capital asset is sold, the difference between the net selling price and the book value is a capital gain or loss on disposal. If the asset was held for more than 12 months it is a "long term" gain or loss; less than 12 months is a "short term" gain or loss. The tax treatment depends on whether there is a gain or loss and whether it is classed as short term or long term.

From time to time the U.S. Government has stimulated capital investment by allowing an investment tax credit. A 7% tax credit means that 7% of the cost of qualifying equipment may be deducted from income taxes. Thus, the Government is paying a portion of the cost of the equipment in this manner. Depreciation continues to be based on the full cost of the equipment.

When dealing with nondepreciable assets there is a relationship between before-tax and after-tax rate of return. It is

$$\text{After-tax rate of return} = (1 - \Delta \text{tax rate})(\text{before-tax rate of return})$$

There is no simple relationship between before-tax and after-tax rate of return in the more usual case of investments involving depreciable assets.

## PROBLEMS

**11-1** An unmarried taxpayer with no dependents expects an adjusted gross income of $37,000 in a given year. His nonbusiness deductions are expected to be $3400.

(a) What will his federal income tax be?

(b) He is considering an additional activity expected to increase his adjusted gross income. If this increase should be $6000 and there should be no change in nonbusiness deductions or exemptions, what will be his federal income tax?

**11-2** Bill has an adjusted gross income of $10,800. Ann has an adjusted gross income of $7200. (a) Compute the federal income tax for each of them assuming neither itemize their allowable deductions. (b) If they married and had the same incomes, how would their income tax as a married couple compare with the tax they paid as two unmarried people? (*Answer:* (a) $2518 (b) $2685)

**11-3** Bill Jackson worked during school and the first two months of his summer vacation. After considering his personal exemption (Bill is single) and deductions, he had a total taxable income of $1800. Bill's employer wants him to work an additional month during the summer, but Bill had planned to spend the month hiking. If an additional month's work would increase Bill's taxable income by

$600, how much additional would he have after paying the additional income tax? (*Answer:* $490)

**11-4** A prosperous businessman is considering two alternative investments in bonds. In both cases the first interest payment would be received at the end of the first year. If his personal taxable income is fixed at $18,000 and he is single, which investment produces the greater after-tax rate of return? Compute the after-tax rate of return for each bond to within $\frac{1}{4}$ of 1 percent.

*Ann Arbor Municipal Bonds:* A bond with a face value of $1000 pays $40 per annum. At the end of 15 years the bond becomes due ("matures") at which time the owner of the bond will receive $1000 plus the final $40 annual payment. The bond may be purchased for $800. Since it is a municipal bond, the annual interest is *not* subject to federal income tax. One half of the difference between what the businessman would pay for the bond ($800) and the $1000 face value he would receive at the end of 15 years must be included in taxable income when the $1000 is received.

*Southern Coal Corporation Bonds:* $1000 of these bonds pay $80 per year in annual interest payments. When the bonds mature at the end of 20 years the bondholder will receive $1000 plus the final $80 interest. The bonds may be purchased now for $1000. The income from corporation bonds must be included in federal taxable income.

**11-5** Albert Chan decided to buy a duplex as an investment. After looking for several months, he found a desirable duplex that could be bought for $25,000 cash. He decided that he would rent both sides of the duplex and determined that the total expected income would be $300 per month. The total annual expenses for property taxes, repairs, gardening, and so forth are estimated at $600 per year. For tax purposes, Al plans to depreciate the building by the sum-of-years digits method assuming the building has a 20-year remaining life and no salvage value. Of the total $25,000 cost of the property, $21,000 represents the value of the building and $4000 is the value of the lot. Al is single and has a good income from his job that results in an annual taxable income of $22,000.
In this analysis Al estimates that the income and expenses will remain constant at their present levels. If he buys and holds the property for 20 years, what after-tax rate of return can he expect to receive on his investment using the assumptions below.

(a) Al believes the building and the lot can be sold at the end of 20 years for the $4000 estimated value of the lot.

(b) A more optimistic estimate of the future value of the building and the lot is that the property can be sold for $15,000 at the end of 20 years.

**11-6** Mr. Sam K. Jones, a successful businessman, is considering erecting a building on a commercial lot he owns very close to the center of town. A local furniture company is willing to lease the building for $9000 per year, paid at the end of each year, and in addition, to pay the property taxes, fire insurance, and all other annual costs. The furniture company will require a five-year lease with an option to buy the building and land on which it stands for $125,000 at the end of the five years. Mr. Jones could have the building constructed for $82,000. He could sell the commercial lot now for $30,000, the same price he paid for it.

Mr. Jones is single and has an annual taxable income from other sources of $26,000. He would depreciate the commercial building by the sum-of-years digits method, using a building life, for depreciation purposes, of 40 years, and zero salvage value. Mr. Jones believes that at the end of the five-year lease he could easily sell the property for $125,000. Assuming that the "gain on disposal" is taxed at 25% what after-tax rate of return would Mr. Jones receive from this venture?

**11-7** A store owner, Joe Lang, believes his business has suffered from the lack of adequate automobile parking space for his customers. Thus, when he was offered an opportunity to buy an old building and lot next to his store, he was interested. He would demolish the old building and make off-street parking for 20 customer's cars. Joe estimates that the new parking would increase his business and produce an additional before-income-tax profit of $7000 per year. It would cost $2500 to demolish the old building. Mr. Lang's accountant advised that both the cost of property and the cost to demolish the old building would be considered to be the total value of the land for tax purposes, and it would not be depreciable. Mr. Lang would spend an additional $3000 right away to put a light gravel surface on the lot. This expenditure, he believes, may be charged as an operating expense immediately and need not be capitalized. To compute the tax consequences of adding the parking lot, Joe estimates that his incremental income tax rate will average 40%. If Joe wants a 15% after-tax rate of return from this project, how much could he pay to purchase the adjoining land with the old building? Assume that the analysis period is ten years, and that the parking lot could always be sold to recover the cost of buying the property plus the cost to demolish the old building. (*Answer:* $23,100)

**11-8** Zeon, a large, profitable corporation, is considering adding some automatic equipment to its production facilities. An investment of $120,000 will produce an initial annual benefit of $29,000 but the benefits are expected to decline $3000 per year, making second year benefits $26,000, third year benefits $23,000, and so forth. If the firm uses sum-of-years digits depreciation, an eight-year useful life, and $12,000 salvage value, will it obtain the desired 6% after-tax rate of return? Assume that the equipment can be sold for its $12,000 salvage value at the end of the eight years.

**11-9** A group of businessmen formed a corporation to lease a piece of land for five years at the intersection of two busy streets. The corporation has invested $50,000 in car washing equipment. They will depreciate the equipment by sum-of-years digits depreciation assuming a $5000 salvage value at the end of the five-year useful life. The corporation is expected to have a before-tax cash flow, after meeting all expenses of operation (except depreciation), of $20,000 the first year, and declining $3000 per year in future years (second year = $17,000; third year = $14,000; and so forth). The corporation has no other income, so it is taxed at the lowest corporate tax rate. If the projected income is correct, and the equipment can be sold for $5000 at the end of five years, what after-tax rate of return would the corporation receive from this venture? (*Answer:* 14%)

**11-10** The effective combined tax rate in an owner-managed corporation is 60%. An outlay of $20,000 for certain new assets is under consideration. It is estimated that for the next eight years, these assets will be responsible for annual receipts of $9000 and annual disbursements (other than for income taxes) of $4000. After this time they will be used only for stand-by purposes and no future excess of receipts over disbursements is estimated.

(a) What is the prospective rate of return before income taxes?

(b) What is the prospective rate of return after taxes if these assets can be written off for tax purposes in eight years using straight-line depreciation?

(c) What is the prospective rate of return after taxes if it is assumed that these assets must be written off for tax purposes over the next 20 years using straight-line depreciation?

**11-11** Gerald Adair bought a small house and lot for $15,000. He estimated that $5000 of this amount represented the value of the land. He rented the house for $2000 a year during the four years he owned the house. Expenses for property taxes, maintenance, and so forth were $500 per year. For tax purposes the house was depreciated by the double declining balance method using a 40-year depreciable life and zero salvage value for the house. At the end of four years the property was sold for $13,145. Gerald is married and has a taxable income from his engineering job each year of $13,000. He and his wife file a joint tax return. What after-tax rate of return did Gerald obtain on his investment in the property?

**11-12**  The management of a hospital is considering the installation of an automatic telephone switchboard, which would replace a manual switchboard and eliminate the attendant operator's position. The class of service provided by the new equipment is estimated to be at least equal to the present method of operation. Five operators are needed to provide telephone service three shifts per day, 365 days per year. Each operator earns $8000 per year. Company paid benefits and overhead are 25% of wages. Money costs 8% after income taxes. Combined federal and state income taxes are 55%. Annual property taxes and maintenance are $2\frac{1}{2}$% and 4% of investment, respectively. Depreciation is 15-year straight line. Disregarding inflation, how large an investment in the new equipment can be economically justified by savings obtained by eliminating the present equipment and labor costs? The existing equipment has zero salvage value.

**11-13** A contractor has to choose one of the following alternatives in performing earth moving contracts.

(a) Purchase a heavy duty truck with dump body for $13,000. Salvage value is expected to be $3000 at the end of its seven year depreciable life. Maintenance is $1100 per year. Daily operating expenses are $35.

(b) Hire a similar unit for $83 per day.

Based on a 10% after-tax rate of return, how many days per year must the dump truck be used to justify its purchase? Base your calculations on straight line depreciation and a 50% income tax rate. (*Answer:* $91\frac{1}{2}$ days)

**11-14** The Able Corporation is considering the installation of a small electronic testing device for use in conjunction with a government contract the firm has just won. The testing device will cost $20,000, and have an estimated salvage value of $5000 in five years when the government contract is finished. The firm will depreciate the instrument by the sum-of-years digits method using five years as the useful life and $5000 salvage value. Assume Able Corporation pays 50% corporate income taxes and uses 8% *after tax* in their economic analysis. What minimum equal annual benefit must Able obtain *before taxes* in each of the five years to justify purchasing the electronic testing device? (*Answer:* $5150)

**11-15** A small business corporation is considering whether or not to replace some equipment in the plant. An analysis indicates there are five alternatives in addition to the do-nothing alternative *A*. The alternatives have a five-year useful life with no salvage value. Straight line depreciation would be used.

| Alternatives | Cost | Before-tax uniform annual benefits |
|---|---|---|
| A | $0 thousand | $0 thousand |
| B | 25 | 7.5 |
| C | 10 | 3 |
| D | 5 | 1.7 |
| E | 15 | 5 |
| F | 30 | 8.7 |

If the corporation has taxable income of under $20,000 per year, and expects a 10% after-tax rate of return for any new investments, which alternative should be selected?

# 12
# Replacement Analysis

Until now we have been faced with the problem of selecting which alternative to buy to accomplish a desired task. This has resulted in a rather straightforward analysis of the various alternatives available with the recommendation on which alternative should be adopted. This might be characterized as selecting the equipment for a new plant. But economic analysis is more frequently performed in conjunction with existing facilities. The problem is less frequently one of building a new plant; it is keeping a present plant operating economically.

In replacement analysis we are not choosing between new ways to perform the desired task. Instead, we have equipment performing the task, and the question is, should the existing equipment be retained or replaced. This adversary situation explains the titles *defender* and *challenger*.* The *defender* is the existing equipment and the *challenger* is the best available replacement equipment.

The subject of replacement analysis can be considered as having five fundamental aspects.

1. Understanding what the comparison should be.
2. Remaining life of the defender.
3. Best useful life of the challenger.
4. Replacement analysis techniques.
5. Equipment replacement models.

In this chapter these five topics will be examined.

---

*The use of the titles *defender* and *challenger* was originated by George Terborgh. He is the author of several important books on replacement economy for the Machinery and Allied Products Institute.

## DEFENDER-CHALLENGER
## COMPARISON

In industry, as in government, annual budgets are the usual practice. One important facet of a budget is the allocation of money for new capital expenditures. This may take the form of money for new facilities or for replacing and upgrading existing facilities. Replacement analysis may therefore have as its end product a recommendation that some particular equipment be replaced and that money for the replacement be included in the capital expenditures budget. Of course, if there is not a recommendation to replace the equipment now, this recommendation may be made next year or some subsequent year. This leads us to the first aspect of the comparison. The question is: *shall we replace the defender now, or shall we keep it for one or more additional years*? Thus the question is not whether we are going to remove the defender equipment. At some point the defender equipment will be removed. This may be when the task performed by the equipment is no longer needed or when the task can be better performed by different equipment. The question is not *if* it will be removed, but *when* it will be removed.

Because the defender (we will use this to mean the defender equipment) already is in the plant, there often is more misunderstanding on what value to assign to it in an economic analysis. An example will demonstrate the problem.

### EXAMPLE 12-1

An SK-30 desk calculator was purchased two years ago for $1600 and has been depreciated by straight line depreciation using a four year life and zero salvage value. Because of recent innovations in desk calculators, the current price of the SK-30 calculators has been reduced from $1600 to $995. An office equipment supply firm has offered a trade-in allowance of $350 for the SK-30 in partial payment on a new $1200 electronic EL-40 calculator. Some discussion revealed that without a trade-in, the EL-40 can be purchased for $1050, indicating the true current market value of the SK-30 is only $200. In a replacement analysis what value should be assigned to the SK-30 calculator?

In the example, five different dollar amounts relating to the SK-30 calculator have been outlined.

1. Original cost: the calculator cost $1600.
2. Present cost: the calculator now sells for $995.
3. Book value: the original cost less two years of depreciation is $1600 - (2/4)(1600 - 0) = $800$.

4.   Trade-in value: the offer was $350.

5.   Market value: the estimate was $200.

We have previously seen that an economy study is based on the current situation, not on the past. We referred to past costs as sunk costs to emphasize that since they could not be altered, they were not relevant. The one exception was that past costs may affect present or future income taxes. Here the question is: what value should be used in an economic analysis for the SK-30? The relevant cost is the present market value for the equipment. Neither the original cost, the present cost, nor the book value is relevant.

At first glance the trade-in value would appear to be a suitable present value for the equipment. Often, however, the trade-in price is inflated along with the price for the new item. This is such a common practice in the new automobile field that the phrase "overtrade" is used to describe the excessive portion of the trade-in allowance. (The purchaser, of course, also is quoted a higher price for the new car.) This distortion of the present value of the defender, along with a distorted price for the challenger, can be serious for they do not cancel out in the economic analysis.

From Example 12-1 we see there are several different values that can be assigned to the defender. The appropriate one is the present market value. If a trade-in value is obtained, care should be taken to insure that it actually represents a fair market value.

For the challenger (meaning the challenger equipment) there should be less difficulty on what to use for the installed cost. Of course, there will be a past cost and a book value for the defender, but these have nothing to do with the challenger. There have been times when people added any capital loss on disposal (book value minus net realized salvage value) of the defender to the cost of the challenger. This is incorrect. If there is a capital loss on disposal of the defender it will have a tax consequence, but this certainly will have no bearing on the installed cost of the challenger.

Another aspect of a defender-challenger comparison concerns which equipment is the challenger. If there is to be a replacement of the defender by the challenger, we would want to install the best of the many available alternatives. Prior to this chapter all our attention has been on selecting the best from two or more alternatives, so this aspect is not new. We will however look at the useful life of both the defender and the challenger more critically than previously. This is described in the next two sections.

## REMAINING LIFE OF THE DEFENDER

In replacement analysis the discussion about the defender and the challenger generally amounts to a question of the old vs the new. The old equipment has a relatively short remaining life compared to new equipment. Here we will examine the remaining life of the defender.

How long can the defender be kept operating? Anyone who has seen old machinery in operation, whether it is a 300-year old clock, a 70-year old automobile, or old production equipment has no doubt realized that with proper repair and maintenance almost anything can be kept operating indefinitely. While we might be able to keep a defender going indefinitely, the cost may be excessive. So rather than asking what the remaining life of the defender could be, we really want to know what its economic life is. This we will define as the remaining useful life that results in a minimum equivalent uniform annual cost.

Economic life = life where EUAC is minimum

Example 12-2 illustrates the situation.

### EXAMPLE 12-2

An 11-year old piece of equipment is being considered for replacement. It can be sold for $2000 now and it is believed this same salvage value can also be obtained in future years. The current maintenance cost is $500 per year and is expected to increase $100 per year in future years. If the equipment is retained in service, compute the economic life that results in a minimum EUAC, based on 10% interest.

Here the salvage value is not expected to decline from its present $2000. The annual cost of this invested capital is $Fi = 2000(0.10) = $200$. The maintenance is represented by $500 + $100G$. A year-by-year computation of EUAC is now given.

| Year $n$ | Age of equipment | EUAC of invested capital $= Fi$ | EUAC of maintenance = 500 $+ 100(A/G,10\%,n)$ | Total EUAC |
|---|---|---|---|---|
| 1 | 11 yrs | $200 | $500 | $700 |
| 2 | 12 | 200 | 548 | 748 |
| 3 | 13 | 200 | 594 | 794 |
| 4 | 14 | 200 | 638 | 838 |
| 5 | 15 | 200 | 681 | 881 |

These data are plotted in Figure 12-1. We see that the annual cost of continuing to use the equipment is increasing. It is reasonable to assume that if the equipment is not replaced now, it will be reviewed again next year. Thus the economic life at which EUAC is a minimum is one year.

Example 12-2 represents a common situation. The salvage value is stable but maintenance is increasing. The total EUAC will continue to increase as

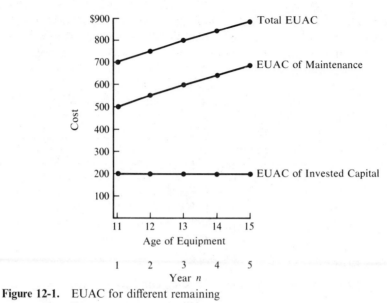

**Figure 12-1.**   EUAC for different remaining
lives.

time passes. This means that an economic analysis to compare the defender at its most favorable remaining life will be based on retaining the
defender one more year. This is not always the case as is shown in Example
12-3.

## EXAMPLE 12-3

A five-year old machine, whose current market value is $5000, is being

| | Estimated salvage value ($F$) end of year $n$ | Estimated maintenance cost for year | If retired at end of year $n$ | | |
|---|---|---|---|---|---|
| Years of remaining life $n$ | | | EUAC of capital recovery $(P-F)$ $(A/P,10\%,n)+Fi$ | EUAC of maintenance $100(A/G,10\%,n)$ | Total EUAC |
| 0 | $P$ = $5000 | | | | |
| 1 | 4000 | $ 0 | $1100 + 400 | $ 0 | $1500 |
| 2 | 3500 | 100 | 864 + 350 | 48 | 1262 |
| 3 | 3000 | 200 | 804 + 300 | 94 | 1198 |
| 4 | 2500 | 300 | 789 + 250 | 138 | 1177 |
| 5 | 2000 | 400 | 791 + 200 | 181 | 1172 |
| 6 | 2000 | 500 | 689 + 200 | 222 | 1111 |
| 7 | 2000 | 600 | 616 + 200 | 262 | 1078 |
| 8 | 2000 | 700 | 562 + 200 | 300 | 1062 |
| 9 | 2000 | 800 | 521 + 200 | 337 | 1058 ← |
| 10 | 2000 | 900 | 488 + 200 | 372 | 1060 |
| 11 | 2000 | 1000 | 462 + 200 | 406 | 1068 |

analysed to determine its economic life in a replacement analysis. Compute its economic life using a 10% interest rate. Salvage value and maintenance estimates are as given. For a minimum EUAC the machine has nine years of remaining life.

From Examples 12-2 and 12-3 we see that the most economical remaining life for an existing machine may be one year or it may be longer. Looking again at the two examples we can see that the examples could represent the same machine being examined at different points in its life. When it was five years old Example 12-3 indicated the economical remaining life would be reached nine years hence making a total of 14 years of service. This would be the point where from age five onward the total EUAC would be a minimum. It is important to recognize that this EUAC is based on the projection of future costs for the five year old machine. We see that the economic life for minimum EUAC depends on future costs, not past or sunk costs. Therefore, the projection for the five year old machine (Example 12-3) is different from the situation when the machine is 11 years old (Example 12-2). Beyond 11 years we found that the EUAC was increasing when computed from age 11 onward.

For older equipment with a negligible or stable salvage value, it is likely that the operating and maintenance costs are increasing. Under these circumstances the useful life at which EUAC is a minimum is one year.

## MOST ECONOMICAL USEFUL LIFE
## OF THE CHALLENGER

In all previous computations to determine which of several alternatives should be selected, useful lives have been assumed. But from the analysis of the most economical remaining life of a defender, we recognize that a similar situation exists for the challenger. If the various costs for the challenger are known along with its year-by-year salvage values, then the most economical useful life may be computed. Example 12-4 illustrates the computation.

*EXAMPLE 12-4*

A piece of machinery costs $10,000 and has no salvage value after it is installed. The manufacturer's warranty will pay all first year maintenance and repairs. In the second year maintenance and repairs will be $600 and will increase on a $600 arithmetic gradient in subsequent years. If interest is 8%, compute the useful life of the equipment that results in minimum EUAC.

| | If retired at end of year n | | |
| Year<br>n | EUAC of<br>capital recovery<br>$10,000(A/P,8\%,n)$ | EUAC of<br>maintenance<br>$600(A/G,8\%,n)$ | Total<br>EUAC |
|---|---|---|---|
| 1 | $10800 | $ 0 | $10800 |
| 2 | 5608 | 289 | 5897 |
| 3 | 3880 | 569 | 4449 |
| 4 | 3019 | 842 | 3861 |
| 5 | 2505 | 1108 | 3613 |
| 6 | 2163 | 1366 | 3529 |
| 7 | 1921 | 1616 | 3537 |
| 8 | 1740 | 1859 | 3599 |
| 9 | 1601 | 2095 | 3696 |

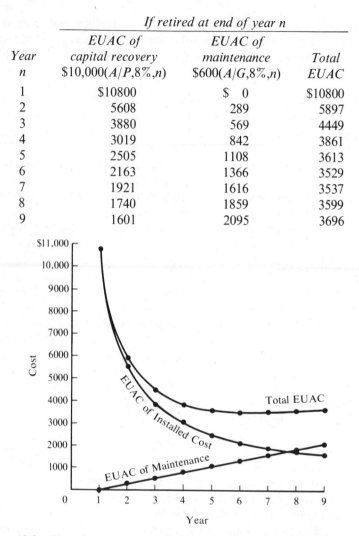

**Figure 12-2.**   Plot of costs for Example 12-4.

The total EUAC data are plotted in Figure 12-2. From either the tabulation or the figure, we see that a useful life of six years results in a minimum EUAC.

## REPLACEMENT ANALYSIS
## TECHNIQUES

From the earlier sections of this chapter we see that determining the most economical remaining life of the defender and the best challenger useful life are substantial problems in themselves. But even when this has been accomplished, we have merely identified the participants in the replacement economic analysis. In this section, we will concentrate on how to proceed with the analysis.

## Defender Remaining Life Equals
## Challenger Useful Life

When we find a situation where the remaining life of the defender equals the useful life of the challenger, we have considerable flexibility in selecting the method of analysis. As a result, these problems may be solved by present worth analysis, annual cash flow analysis, rate of return analysis, benefit-cost ratio analysis, and so forth. Examples 12-5 and 12-6 illustrate before-tax and after-tax replacement analysis for the equal life situation.

### *EXAMPLE 12-5*

Determine whether the SK-30 desk calculator of Example 12-1 should be replaced by the EL-40 electronic calculator. In addition to the data given in Example 12-1, the following estimates have been made. The SK-30 maintenance and service contract costs $80 a year. The EL-40 will require no maintenance. Either calculator is expected to be used for the next five years. At the end of that time the SK-30 will have no value, but the EL-40 probably could be sold for $250. The EL-40 electronic calculator is faster and easier to use than the SK-30 mechanical calculator. This benefit is expected to save about $120 a year by reducing the need for part-time employees. If the MARR is 10% before taxes, should the SK-30 desk calculator be replaced by the EL-40 calculator?

Compute the EUAC for each calculator.

    SK-30

        Present market value $= \$200$
        Future salvage value $= 0$
        Annual maintenance $= \$80$ per year

        $\text{EUAC} = (200 - 0)(A/P,10\%,5) + 80$
               $= 200(0.2638) + 80 = \$132.76$

    EL-40

        Cash price $= \$1050$
        Future salvage value $= \$250$
        Annual benefit of greater usefulness $= \$120$ per year

        $\text{EUAC} = (1050 - 250)(A/P,10\%,5) + 250(0.10) - 120$
               $= 800(0.2638) + 25 - 120 = \$116.04$

The EL-40 calculator has the smaller EUAC and therefore is the preferred alternative in this before-tax analysis.


### EXAMPLE 12-6

Solve Example 12-5 with a MARR equal to 8% after taxes. The EL-40 will be depreciated by straight line depreciation using a four-year depreciable life. The SK-30 already is two years old. The analysis period remains at five years.

*Alternative A:* keep the SK-30 rather than sell it. Compute the after-tax cash flow.

| Year | Before-tax cash flow | Straight line depreciation | Taxable income | 50% income taxes | After-tax cash flow |
|------|------|------|------|------|------|
| 0 | − $200 | | + $600* | − $180* | − $380** |
| 1 | − 80 | $400 | − 480 | + 240 | + 160 |
| 2 | − 80 | 400 | − 480 | + 240 | + 160 |
| 3 | − 80 | 0 | − 80 | + 40 | − 40 |
| 4 | − 80 | 0 | − 80 | + 40 | − 40 |
| 5 | $\left\{\begin{array}{l}-80\\ 0 \text{ salvage}\end{array}\right.$ | 0 | − 80 | + 40 | − 40 |

*If sold for $200 there would be a $600 long term capital loss on disposal. If $600 long term capital gains were offset by the loss during the year, there would be no tax on the capital gain, saving 30% × $600 = $180. If the SK-30 is not sold, this loss is not realized and the resulting income taxes will be $180 higher than if it had been sold.

**This is the sum of the $200 selling price foregone plus the $180 income tax saving foregone.

Compute the EUAC. Note that in computing a cost the signs appear opposite to those shown in the after-tax cash flow.

$$\text{EUAC} = [380 - 160(P/A,8\%,2) + 40(P/A,8\%,3)(P/F,8\%,2)](A/P,8\%,5)$$
$$= \$45.86$$

*Alternative B:* purchase an EL-40. Compute the after tax cash flow.

| Year | Before-tax cash flow | Straight line depreciation | Taxable income | 50% income taxes | After-tax cash flow |
|------|------|------|------|------|------|
| 0 | −$1050 | | | | −$1050 |
| 1 | +120 | $200 | −$80 | +$40 | +160 |
| 2 | +120 | 200 | −80 | +40 | +160 |
| 3 | +120 | 200 | −80 | +40 | +160 |
| 4 | +120 | 200 | −80 | +40 | +160 |
| 5 | { +120<br>{ +250 salvage | 0 | +120 | −60 | +310 |

Compute the EUAC.

$$\text{EUAC} = [1050 - 160(P/A,8\%,4) - 310(P/F,8\%,5)](A/P,8\%,5)$$
$$= \$77.43$$

Based on this after-tax analysis the SK-30 calculator is the preferred alternative.

*EXAMPLE 12-7*

Solve Example 12-6 by computing the rate of return on the difference between the alternatives. In Example 12-6 the two alternatives were keep the SK-30, or buy an EL-40. The difference between the alternatives would be:

| Buy an EL-40 | rather than | Keep the SK-30 |
|------|------|------|
| or alternative *B* | minus | alternative *A*. |

The after-tax cash flow for the difference between the alternatives may be computed as follows.

| Year | A | B | B-A |
|------|------|------|------|
| 0 | −$380 | −$1050 | −$670 |
| 1 | +160 | +160 | 0 |
| 2 | +160 | +160 | 0 |
| 3 | −40 | +160 | +200 |
| 4 | −40 | +160 | +200 |
| 5 | −40 | +310 | +350 |

The rate of return on the difference between the alternatives may be computed as follows:

PW of cost = PW of benefit

$$670 = 200(P/A,i\%,2)(P/F,i\%,2) + 350(P/F,i\%5)$$

Try $i = 8\%$.

670 = 200(1.783)(0.8573) + 350(0.6806)
670 = 543.9

The PW of benefit is too low indicating that the interest rate is too high. Try $i = 3\%$.

670 = 200(1.913)(0.9426) + 350(0.8626)
670 = 662.6

The rate of return is less than 3%.

This example shows how a rate of return may be computed for an equipment replacement problem. The cash flows for both alternatives are computed and then the rate of return is computed on the difference between the alternatives. The rate of return is far less than the 8% MARR. The investment in buying an EL-40 rather than keeping the SK-30 does not meet the MARR and therefore it should not be made.

## Defender Remaining Life Different From Challenger Useful Life

When the alternative lives are equal to the analysis period, we can generally solve problems in a variety of ways. But when the alternatives (defender and challenger) have different lives there may be difficulties. For unequal lived alternatives annual cash flow analysis is generally the most suitable method of analysis. In Chapter 6 it was stated that a comparison of equivalent uniform annual costs for unequal lived alternatives is suitable only if the following assumptions are valid.

1. When an alternative has reached the end of its useful life it is assumed to be replaced by an identical replacement (with the same costs, performance, and so forth).
2. The analysis period is a common multiple of the useful lives of the alternatives, or there is a continuing or perpetual requirement for the selected alternative.

But there are defender-challenger situations where these conditions cannot be met. It is reasonable to assume that the challenger can be replaced by an identical replacement. This is not, however, a reasonable assumption for the defender when it reaches the end of its economic life. The defender is typically an older piece of equipment with a modest current selling price.

An identical replacement, even if it could be found, probably would have an installed cost far in excess of the current selling price of the defender.

Thus the assumptions we made for using an annual cash flow analysis on unequal lived alternatives in Chapter 6 frequently cannot be met in the usual defender-challenger situation. But a careful examination of the equipment replacement problem indicates that those assumptions are not essential here. There are two alternatives in equipment replacement.

1. Replace the defender now.
2. Retain the defender for the present.

The question is not really one of selecting the defender or the challenger, but rather that of deciding if *now* is the time to replace the defender. When the defender is replaced, it will be by the challenger—the best available replacement. The challenger could be a piece of new equipment, or it could be used or reconditioned equipment. Thus in equipment replacement, an annual cash flow analysis does not assume that the defender has an identical replacement at the end of its useful life. The replacement is always by the challenger.

If the defender-challenger problem assumes a continuing requirement for the equipment, an annual cash flow analysis of the unequal lived alternatives is proper. When there is a definite analysis period, after which the equipment will not be needed, then a careful analysis is needed to see how the analysis period affects the alternatives.

For unequal lived alternatives neither present worth nor rate of return are practical methods of analysis. Both methods require that the consequences of the alternatives be evaluated over the analysis period. For this reason, these methods are attempted only when there is a well defined analysis period.

### EXAMPLE 12-8

An economic analysis is to be made to determine if some existing (defender) equipment in the plant should be replaced. A $4000 overhaul must be done now if the equipment is to be retained in service. Maintenance is estimated at $1800 in each of the next two years, after which it is expected to increase on a $1000 linear gradient. The defender has no present or future salvage value. The equipment described in Example 12-4 is the challenger. Make a replacement analysis to determine whether to retain the defender or replace it by the challenger if 8% interest is used.

The first step is to determine the economic life of the defender. The pattern of overhaul and maintenance costs (Figure 12-3) suggests that if the overhaul is done, the remaining economic life of the equipment will be several years. The computation is now given.

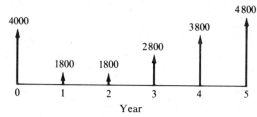

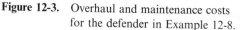

**Figure 12-3.** Overhaul and maintenance costs
for the defender in Example 12-8.

| | | *If retired at end of year n* | |
|---|---|---|---|
| *Year* | *EUAC of overhaul* | *EUAC of maintenance* $1800 + $1000 | *Total* |
| *n* | $4000(A/P,8\%,n)$ | *gradient from year 3 on* | *EUAC* |
| 1 | $4320 | $1800 | $6120 |
| 2 | 2243 | 1800 | 4043 |
| 3 | 1552 | 1800 + 308* | 3660 |
| 4 | 1208 | 1800 + 683† | 3691 |
| 5 | 1002 | 1800 + 1079 | 3881 |

*For the first 3 years the maintenance is $1800, $1800, and $2800. Thus EUAC = 1800
+ 1000(A/F,8\%,3) = 1800 + 308.
†EUAC = 1800 + 1000(P/G,8\%,3)(P/F,8\%,1)(A/P,8\%,4) = 1800 + 683.

For minimum EUAC the remaining economic life of the defender is
three years. In Example 12-4 we determined that the best useful life of
the challenger is six years and that the resulting EUAC is $3529. Since
the EUAC of the challenger ($3529) is less than the EUAC of the de-
fender ($3660), the challenger should be installed now to replace the
defender.

## A Closer Look At The Challenger

We defined the challenger as the best available alternative to replace the
defender. But as time passes the best available alternative can change. And
given the trend in our technological society, it seems likely that future challengers will be better than the present challenger. If this is so, the prospect
of improved future challengers may affect the present decision between the
defender and the challenger. Figure 12-4 illustrates two possible estimates
of future challengers. In many technological areas it seems likely that the
equivalent uniform annual costs associated with future challengers will
decrease by a constant amount each year. There are other fields, however,
where a rapidly changing technology will produce a sudden and substantially improved challenger (with decreased costs or increased benefits). The
uniform decline curve of Figure 12-4 assumes that each future challenger

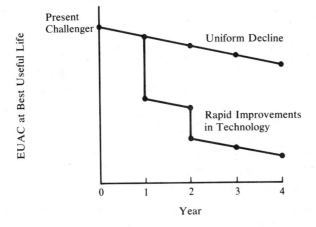

Figure 12-4. Two possible ways the EUAC of future challengers may decline.

has a minimum EUAC that is a fixed amount less than the previous year's challenger. This, of course, is only one of many possible assumptions that could be made regarding future challengers.

If future challengers will be better than the present challenger, what impact will this have on an analysis now? The prospect of better future challengers makes it more desirable to retain the defender and to reject the present challenger. By keeping the defender for now we will be able to replace it later by a better future challenger. Or to state it another way, the present challenger is made less desirable by the prospect of improved future challengers. Example 12-9 illustrates the situation.

*EXAMPLE 12-9*

Recompute Example 12-8. This time it is estimated that future challengers will be improved so that each year the EUAC for their six-year best useful life will decline by $100. This new situation means that there are a series of challengers—the present challenger and the improved future challengers. The available alternatives are listed.

 A. Keep the defender.
 B. Replace the defender with the present challenger.
 C. Keep the defender one year and then replace it with a better challenger next year.
 D. Keep the defender two years and then replace it with a still better challenger two years hence.
 E. Keep the defender three years and then replace it with a still better challenger three years hence.

    *F.*   Keep the defender four years and then replace it with a still better challenger four years hence.

And so on. At this point we must emphasize that the above alternatives are designed to provide an answer to the fundamental replacement analysis question: *shall we replace the defender now or shall we keep it one or more additional years*? If alternative *B* above is the most economical, the decision is to replace the defender now. If any of the other alternatives is more economical the decision will be to retain the defender for now and to review the situation again next year.

    We already have done the computations to determine the EUAC of the defender and the present challenger at their best useful lives.

    *A.*   Keep the defender. From Example 12-8 the minimum EUAC is $3660.

    *B.*   Replace the defender with the present challenger.

        The computations were done in Example 12-4.

        Minimum EUAC is $3529.

A diagram of alternatives *B* through *F* is given in Figure 12-5. As described in the problem statement, the EUAC for each subsequent year challenger is $100 less than the prior challenger. From Example 12-8 the EUAC for retaining the defender was computed as follows.

| Years defender retained | EUAC for period |
|:---:|:---:|
| 1 yr | $6120 |
| 2 | 4043 |
| 3 | 3660 |
| 4 | 3692 |

With these data we can compute the EUAC for alternatives *C* through *F* which are the combination of retaining the defender one or more years plus six years of a future challenger. For alternatives *C* through *F*:

$$\text{EUAC} = [(\text{defender EUAC})(P/A,8\%,n) + (\text{challenger EUAC})$$
$$\times (P/A,8\%,6)(P/F,8\%,n)][(A/P,8\%,n+6)]$$

Alternative *C*:
$$\text{EUAC} = [6120(P/A,8\%,1) + 3429(P/A,8\%,6)(P/F,8\%,1)][(A/P,8\%,7)]$$
$$= [6120(0.926) + 3429(4.623)(0.962)][0.1921] = \$3909$$

Alternative *D*:
$$\text{EUAC} = [4043(P/A,8\%,2) + 3329(P/A,8\%,6)(P/F,8\%,2)][(A/P,8\%,8)]$$
$$= [4043(1.783) + 3329(4.623)(0.8573)][0.1740] = \$3550$$

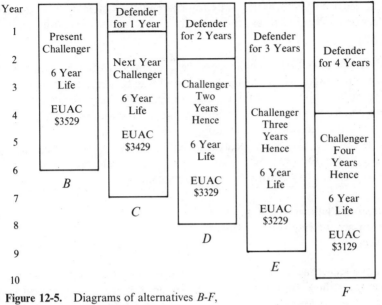

**Figure 12-5.** Diagrams of alternatives *B-F*,
Example 12-9.

Alternative *E*:

EUAC = $[3660(P/A,8\%,3) + 3229(P/A,8\%,6)(P/F,8\%,3)][(A/P,8\%,9)]$
    = $[3660(2.577) + 3229(4.623)(0.7938)][0.1601] = \$3407$

Alternative *F*:

EUAC = $[3692(P/A,8\%,4) + 3129(P/A,8\%,6)(P/F,8\%,4)][(A/P,8\%,10)]$
    = $[3692(3.312) + 3129(4.623)(0.7350)][0.1490] = \$3406$

The alternatives may be summarized as follows.

| Alternative | Description | Equivalent uniform annual cost |
|---|---|---|
| A | Keep the defender | $3660 |
| B | Replace with the challenger | 3529 |
| C | Defender 1 year; better challenger 6 years | 3909 |
| D | Defender 2 years; better challenger 6 years | 3550 |
| E | Defender 3 years; better challenger 6 years | 3407 |
| F | Defender 4 years; better challenger 6 years | 3406 |

The least cost alternative calls for keeping the defender for now. Our decision will be to retain the defender and to restudy the situation next year.

In Example 12-8 future challengers were assumed to be the same as the present challenger. As a result, the decision was to select the present challenger rather than the defender. But in Example 12-9 we assumed that future challengers would be an improvement over the present challenger. This time the decision was to retain the defender. The contradictory decision in the two examples illustrates that while the present challenger is preferred over the defender, a still more attractive alternative is to delay the replacement to obtain a better future challenger.

### Shortening Challenger Life to Compensate for Improved Future Challengers

We have seen that if better future challengers can be expected to become available in the future, this decreases the desirability of the present challenger. The assumption of a series of improved future challengers makes the two alternative defender-challenger problem into a multiple alternative problem with many more calculations. To reduce the amount of calculations, one frequently used technique is to assume that the challenger has a shortened life.

If the challenger life in Example 12-8 were assumed to be four years, instead of six years, the challenger EUAC would have been $3861 (as computed in Example 12-4). This would have changed the decision in Example 12-8 to keep the defender for the time being. Thus the shortening of the challenger life by one-third produced the same conclusion as Example 12-9. While the general approach throughout this book has been to advocate accurate calculations, shortening the challenger life to reflect the prospect of better future challengers seems like a practical technique to avoid longer calculations.

### EQUIPMENT REPLACEMENT MODELS

In the equipment replacement problems so far discussed, the assumptions have been carefully made to keep the computations to a manageable level. But more complex replacement problems exist and must be solved. A variety of equipment replacement mathematical models have been developed for this purpose. While these models are generally beyond the scope of this book, an introduction to the work of George Terborgh of the Machinery and Allied Products Institute (MAPI) will be given.

The several versions of the MAPI replacement analysis system are described in a series of publications.* The present MAPI system, described in *Business Investment Management*, has these features.

1. Worksheets help to organize the problem by providing a format for the analysis.
2. The analysis considers both the changes in costs and the changes in benefits.
3. Future challengers are recognized as being better than the present challenger.
4. Standard projections are provided for certain estimates that are difficult to make.
5. The computations result in an incremental after-tax rate of return based on the difference between the proposed project and a stated alternative (like continuing with the existing situation).

The MAPI system assumes that when a piece of capital equipment is purchased, one is in effect buying "service values" that are the flow of future benefits to the owner. The "service values" or benefits represent the earnings of the equipment over the years compared to the then available challengers. Since future challengers are expected to be a steady improvement over the present challenger, the earnings are assumed to decline uniformly over the life of the equipment. When there is no salvage value, the projected before-tax earnings are assumed to decline to zero at the end of the service life. Figure 12-6 illustrates the earnings assumption. When there is a salvage value, the before-tax earnings are assumed to decline to an amount equal to the cost of continuing the asset in service.

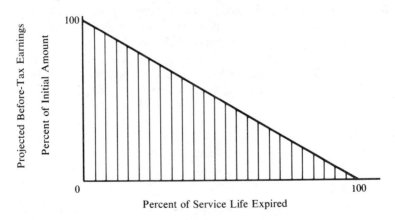

**Figure 12-6.** MAPI earnings projection for an asset without salvage value.

*Dynamic Equipment Policy* (1949); *MAPI Replacement Manual* (1950); *Business Investment Policy* (1958); *Business Investment Management* (1967). Washington, D.C.: Machinery and Allied Products Institute.

Using the service value concept, we see that the purchase of a piece of capital equipment yields the owner the present worth of all future service values plus the present worth of the salvage value. We therefore assume that

Purchase price = PW of service values + PW of future salvage value.

A key feature in the MAPI system is determining the retention value of the asset at the end of the comparison period. This is done by determining the year-by-year capital consumption based on the following allocation of before-tax earnings.

1.  To pay income taxes (assuming a 50% tax rate) on the excess of the earnings over the deduction for depreciation of the asset and interest on the debt portion of the investment.
2.  To pay 3% interest on the one-quarter of the unrecovered investment that is assumed to be borrowed.
3.  To provide a 10% after-tax return on the three-quarters of the unrecovered investment that is equity capital (not borrowed).
4.  To provide for annual capital consumption, which is the difference between the unrecovered investment at the beginning of the year and the unrecovered investment at the end of the year. At any time the unrecovered investment is to be equal to the present worth of the after-tax earnings plus the present worth of the salvage value.

The MAPI capital consumption model uses the four conditions above to devise a relationship between before-tax earnings and the initial investment. For any initial investment there is a single before-tax earnings value that will yield the MAPI capital consumption model.

For a $10,000 investment, five-year service life, straight line depreciation, and no salvage value, the initial before-tax earnings have been computed from *Business Investment Management* (page 328, Eq. 8) as $4748. Using this value the capital consumption may be computed.

Table 12-1 has been constructed to satisfy the MAPI conditions we have previously stated. A column-by-column examination of the table will aid in understanding the capital consumption model.

*Column.*

(1)  Before-tax earnings. The before-tax earnings follow the pattern of Figure 12-6, that is, a declining gradient series. From a MAPI formula we computed the value of the gradient series that would provide the exact amount of money needed to pay income taxes, interest on the borrowed money, a 10% after-tax return on the equity capital, and provide for the capital consumption. We must empha-

size that the purpose of doing all this is to compute the capital consumption in (9) on the MAPI assumptions.

**Table 12-1.** Computation of year-by-year capital consumption using the MAPI assumptions

| | (1) | (2) | (3) | (4) | (5) | (6) | (7) | (8) 10% return on equity capital | (9) |
|---|---|---|---|---|---|---|---|---|---|
| Year | Before-tax earnings | SL deprec | Interest payment | Taxable earnings | 50% Income taxes | After tax earnings | Unrecovered investment | | Capital consumption |
| 1 | $4748 | $2000 | $75 | $2673 | −$1336 | $3412 | $10000 | $750 | $2587 |
| 2 | 3799 | 2000 | 56 | 1743 | −872 | 2927 | 7413 | 556 | 2315 |
| 3 | 2849 | 2000 | 38 | 811 | −406 | 2443 | 5098 | 382 | 2023 |
| 4 | 1899 | 2000 | 23 | −124 | +62 | 1961 | 3075 | 231 | 1707 |
| 5 | 950 | 2000 | 10 | −1060 | +530 | 1480 | 1368 | 102 | 1368 |
| | | | | | | | | 0 | $10000 |

(2) Straight line depreciation. For an initial cost of $10,000 and no salvage value, the annual straight line depreciation is $2000.

(3) Interest payment. One-quarter of the unrecovered investment (7) is assumed to be borrowed. Column (3) is 3% interest on the debt. For year 1: interest $= 0.03 \times 0.25 \times 10,000 = \$75$.

(4) Taxable earnings. Taxable earnings $=$ before-tax earnings $-$ interest $-$ depreciation.

(5) 50% income taxes. Taxes $= 50\%$ of taxable earnings.

(6) After-tax earnings. Before-tax earnings $-$ income taxes.

(7) Unrecovered investment. Initially unrecovered investment is the $10,000 initial cost. The second year, unrecovered investment is the first year unrecovered investment ($10,000) minus the first year capital consumption ($2587) and so on.

(8) 10% return on equity capital. 75% of the unrecovered investment is assumed to be equity capital. For the first year, 10% return on the equity capital $= 0.10 \times 0.75 \times 10,000 = \$750$.

(9) Capital consumption. Capital consumption $=$ after-tax earnings $-$ interest payment $-$ return on equity capital. For the first year, capital consumption $= 3412 - 75 - 750 = \$2587$.

An important assumption of the MAPI capital consumption model is that the unrecovered investment at any time equals the present worth of the future after-tax earnings of the asset plus the present worth of any salvage value, computed at 8.25% interest. For the data tabulated above, we can compute the present worth of the after-tax earnings for each year as follows.

| Year | (P/F,8.25%,1) | (P/F,8.25%,2) | (P/F,8.25%,3) | (P/F,8.25%,4) | (P/F,8.25%,5) | |
|---|---|---|---|---|---|---|
| 1 | 3412(0.9238) + | 2927(0.8534) + | 2443(0.7883) + | 1961(0.7283) + | 1480(0.6728) = | $10,000 |
| 2 | 2927(0.9238) + | 2443(0.8534) + | 1961(0.7883) + | 1480(0.7283) = | $7413 | |
| 3 | 2443(0.9238) + | 1961(0.8534) + | 1480(0.7883) = | $5098 | | |
| 4 | 1961(0.9238) + | 1480(0.8534) = | $3075 | | | |
| 5 | 1480(0.9238) = | $1368. | | | | |

The computed present worths of the after-tax earnings are equal to the tabulation of unrecovered investment (7) in Table 12-1. The capital consumption model therefore meets all the MAPI assumptions. From Table 12-1 we can compute the percent retention value each year for the case of five-year service life, straight line depreciation and no salvage value. The first year $2587 of the $10,000 cost is consumed. The retention value is $10,000 − 2587 or $7413. This is 74.1% of the initial cost. By similar computations the percent retention values may be computed for the various comparison periods.

| Comparison period | Value of % retention |
|---|---|
| 1 yr | 74.1% |
| 2 | 50.9 |
| 3 | 30.7 |
| 4 | 13.6 |
| 5 | 0 |

Figure 12-7 is the MAPI Chart 3A for Percent Retention Values for a one-year comparison period and straight line depreciation. Our 74.1% retention value is one point on this chart. You should be able to locate the point and see that it agrees with the value on the chart. The other percent retention values may be located in Figure 12-8 on the MAPI Chart 3B for longer comparison periods.

The MAPI model charts were computed using a 50% tax rate, 25% of the investment borrowed, 3% interest on the borrowed money, and a 10% after-tax return on equity capital, as has been described above. Different values could be inserted in the model formulas and a new set of MAPI charts computed to more accurately reflect a specific situation.

## MAPI Worksheets

A two-page worksheet is needed to solve a problem by the MAPI system. These are shown in Figures 12-9 and 12-10. On Sheet 1 the annual oper-

# MAPI CHART No. 3A

### (ONE-YEAR COMPARISON PERIOD AND STRAIGHT-LINE TAX DEPRECIATION)

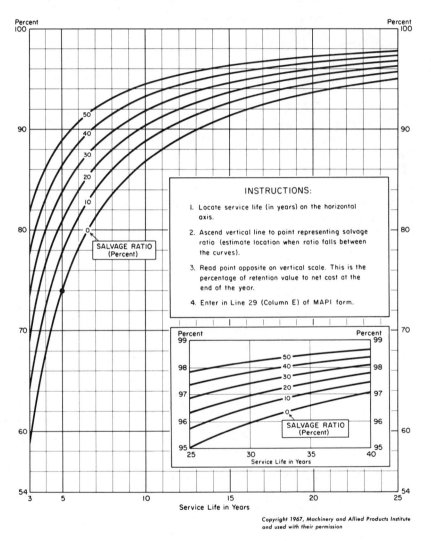

INSTRUCTIONS:

1. Locate service life (in years) on the horizontal axis.

2. Ascend vertical line to point representing salvage ratio (estimate location when ratio falls between the curves).

3. Read point opposite on vertical scale. This is the percentage of retention value to net cost at the end of the year.

4. Enter in Line 29 (Column E) of MAPI form.

**Figure 12-7.** MAPI Chart 3A: percent retention values for a one-year comparison period and straight line depreciation.

ating advantage is computed, assuming the project is implemented rather than some alternative (like continuing with the present situation). The comparison period should be selected as the shorter of the economic lives of

# MAPI CHART No. 3B

### (LONGER THAN ONE-YEAR COMPARISON PERIODS AND STRAIGHT-LINE TAX DEPRECIATION)

INSTRUCTIONS:

1. Locate on horizontal axis percentage which comparison period is of service life.

2. Ascend vertical line to point representing salvage ratio (estimate location when ratio falls between the curves).

3. Read point opposite on vertical scale. This is the percentage of retention value to net cost at end of comparison period.

4. Enter in line 29 (Column E) of MAPI form.

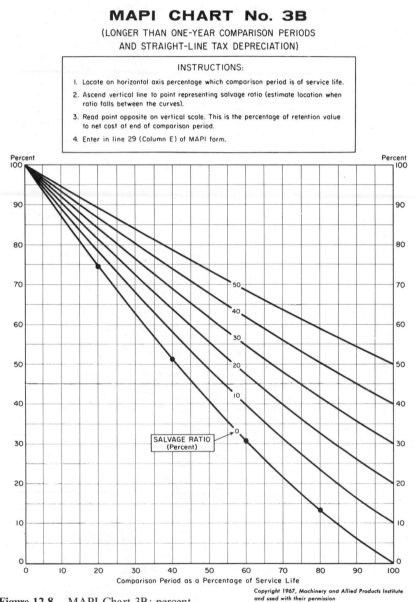

**Figure 12-8.** MAPI Chart 3B: percent retention values for longer than one-year comparison periods and straight line depreciation.

the project or its alternative. As previously described, this often is one year, but it may be longer. On Sheet 2 the form is divided into three sections.

PROJECT NO._____

## MAPI SUMMARY FORM
### (AVERAGING SHORTCUT)

PROJECT_____

ALTERNATIVE_____

COMPARISON PERIOD (YEARS)　　　　　　　　　　　　(P)_____

ASSUMED OPERATING RATE OF PROJECT (HOURS PER YEAR)　　_____

### I. OPERATING ADVANTAGE
(NEXT-YEAR FOR A 1-YEAR COMPARISON PERIOD,* ANNUAL AVERAGES FOR LONGER PERIODS)

#### A. EFFECT OF PROJECT ON REVENUE

|  |  | INCREASE | DECREASE |  |
|---|---|---|---|---|
| I | FROM CHANGE IN QUALITY OF PRODUCTS | $ | $ | I |
| 2 | FROM CHANGE IN VOLUME OF OUTPUT |  |  | 2 |
| 3 | TOTAL | $　　　　X | $　　　　Y | 3 |

#### B. EFFECT ON OPERATING COSTS

|  |  | INCREASE | DECREASE |  |
|---|---|---|---|---|
| 4 | DIRECT LABOR | $ | $ | 4 |
| 5 | INDIRECT LABOR |  |  | 5 |
| 6 | FRINGE BENEFITS |  |  | 6 |
| 7 | MAINTENANCE |  |  | 7 |
| 8 | TOOLING |  |  | 8 |
| 9 | MATERIALS AND SUPPLIES |  |  | 9 |
| 10 | INSPECTION |  |  | 10 |
| 11 | ASSEMBLY |  |  | 11 |
| 12 | SCRAP AND REWORK |  |  | 12 |
| 13 | DOWN TIME |  |  | 13 |
| 14 | POWER |  |  | 14 |
| 15 | FLOOR SPACE |  |  | 15 |
| 16 | PROPERTY TAXES AND INSURANCE |  |  | 16 |
| 17 | SUBCONTRACTING |  |  | 17 |
| 18 | INVENTORY |  |  | 18 |
| 19 | SAFETY |  |  | 19 |
| 20 | FLEXIBILITY |  |  | 20 |
| 21 | OTHER |  |  | 21 |
| 22 | TOTAL | $　　　　Y | $　　　　X | 22 |

#### C. COMBINED EFFECT

| 23 | NET INCREASE IN REVENUE (3X−3Y) | $ | 23 |
|---|---|---|---|
| 24 | NET DECREASE IN OPERATING COSTS (22X−22Y) | $ | 24 |
| 25 | ANNUAL OPERATING ADVANTAGE (23+24) | $ | 25 |

---

\* Next year means the first year of project operation. For projects with a significant break-in period, use performance after break-in.

**Figure 12-9.** MAPI summary form: Sheet 1.

## II. INVESTMENT AND RETURN

### A. INITIAL INVESTMENT

26  INSTALLED COST OF PROJECT                         $ _____
    MINUS INITIAL TAX BENEFIT OF                  $ _____        (Net Cost)    $ _____        26
27  INVESTMENT IN ALTERNATIVE
    CAPITAL ADDITIONS MINUS INITIAL TAX BENEFIT              $ _____
    PLUS: DISPOSAL VALUE OF ASSETS RETIRED
        BY PROJECT *                              $ _____        $ _____        27
28  INITIAL NET INVESTMENT (26−27)                                              $ _____        28

### B.  TERMINAL INVESTMENT

29  RETENTION VALUE OF PROJECT AT END OF COMPARISON PERIOD
    (ESTIMATE FOR ASSETS, IF ANY, THAT CANNOT BE DEPRECIATED OR EXPENSED. FOR OTHERS, ESTIMATE
    OR USE MAPI CHARTS.)

| Item or Group | Installed Cost, Minus Initial Tax Benefit (Net Cost) A | Service Life (Years) B | Disposal Value, End of Life (Percent of Net Cost) C | MAPI Chart Number D | Chart Percentage E | Retention Value $\left(\dfrac{A \times E}{100}\right)$ F |
|---|---|---|---|---|---|---|
|  | $ |  |  |  |  | $ |

    ESTIMATED FROM CHARTS  (TOTAL OF COL. F)        $ _____
    PLUS: OTHERWISE ESTIMATED                       $ _____        $ _____        29
30  DISPOSAL VALUE OF ALTERNATIVE AT END OF PERIOD *                          $ _____        30
31  TERMINAL NET INVESTMENT (29−30)                                          $ _____        31

### C.  RETURN

32  AVERAGE NET CAPITAL CONSUMPTION $\left(\dfrac{28-31}{P}\right)$        $ _____        32

33  AVERAGE NET INVESTMENT $\left(\dfrac{28+31}{2}\right)$        $ _____        33

34  BEFORE-TAX RETURN $\left(\dfrac{25-32}{33} \times 100\right)$        % _____        34

35  INCREASE IN DEPRECIATION AND INTEREST DEDUCTIONS        $ _____        35
36  TAXABLE OPERATING ADVANTAGE (25−35)                    $ _____        36
37  INCREASE IN INCOME TAX (36×TAX RATE)                   $ _____        37
38  AFTER-TAX OPERATING ADVANTAGE (25−37)                  $ _____        38
39  AVAILABLE FOR RETURN ON INVESTMENT (38−32)             $ _____        39

40  AFTER-TAX RETURN $\left(\dfrac{39}{33} \times 100\right)$        % _____        40

* After terminal tax adjustments.

**Figure 12-10.**   MAPI summary form: Sheet 2.

A. Initial Investment

Line 26. The net cost is the installed cost of the project minus any investment tax credit.

Line 27. It may be that if the project is not adopted it would be necessary to make a capital expenditure to continue with the present situation or whatever alternative is used. If so, it is added here. If, however, the project is selected there may be alternative assets that would be retired and they would be added here.

Line 28. The initial net investment is simply the net cost of the project minus any net capital additions that would be needed by the alternative, minus any assets retired by the project. In short, it is the net incremental cost of proceeding with the project rather than the alternative.

B. Terminal Investment

Line 29. Our elaborate capital consumption computations were to show how the retention value is computed for use on this line.

    A. The net cost is taken from line 26.

    B. Service life is estimated for the project. The model assumes better future challengers so the service life should not be reduced here.

    E. The percent retention value from the chart is entered here.

Line 30. This is the salvage value of the alternative if it were retained.

C. Return

Line 32. The *P* in the denominator is the comparison period on Sheet 1.

Line 35. To solve for the after-tax rate of return it is necessary to take into account the depreciation and interest deductions that the project produces in excess of the deductions for the alternative. For a one-year comparison it is the first year increase; for longer comparison periods it is the average annual increase.

The use of the MAPI system and the associated worksheets can be illustrated by an example problem.

### EXAMPLE 12-10

An existing milling machine is being considered for replacement by a new milling machine. If the old milling machine is continued in service,

**MAPI SUMMARY FORM**
(AVERAGING SHORTCUT)

PROJECT __New Milling Machine__

ALTERNATIVE __Continue with existing milling machine__

COMPARISON PERIOD (YEARS)

ASSUMED OPERATING RATE OF PROJECT (HOURS PER YEAR)

(P) __1 year__

**I. OPERATING ADVANTAGE**
(NEXT-YEAR FOR A 1-YEAR COMPARISON PERIOD,* ANNUAL AVERAGES FOR LONGER PERIODS)

**A. EFFECT OF PROJECT ON REVENUE**

| | | INCREASE | DECREASE | |
|---|---|---|---|---|
| I | FROM CHANGE IN QUALITY OF PRODUCTS | $ 500 | $ | I |
| 2 | FROM CHANGE IN VOLUME OF OUTPUT | 1000 | | 2 |
| 3 | TOTAL | $ 1500 X | $ Y | 3 |

**B. EFFECT ON OPERATING COSTS**

| | | INCREASE | DECREASE | |
|---|---|---|---|---|
| 4 | DIRECT LABOR | $ | $ 500 | 4 |
| 5 | INDIRECT LABOR | 200 | | 5 |
| 6 | FRINGE BENEFITS | 50 | | 6 |
| 7 | MAINTENANCE | | 400 | 7 |
| 8 | TOOLING | | | 8 |
| 9 | MATERIALS AND SUPPLIES | | 2100 | 9 |
| 10 | INSPECTION | | 100 | 10 |
| 11 | ASSEMBLY | | | 11 |
| 12 | SCRAP AND REWORK | | 1600 | 12 |
| 13 | DOWN TIME | | | 13 |
| 14 | POWER | 50 | | 14 |
| 15 | FLOOR SPACE | | | 15 |
| 16 | PROPERTY TAXES AND INSURANCE | 100 | | 16 |
| 17 | SUBCONTRACTING | | | 17 |
| 18 | INVENTORY | | | 18 |
| 19 | SAFETY | | | 19 |
| 20 | FLEXIBILITY | | | 20 |
| 21 | OTHER | | | 21 |
| 22 | TOTAL | $ 400 Y | $ 4700 X | 22 |

**C. COMBINED EFFECT**

| | | | |
|---|---|---|---|
| 23 | NET INCREASE IN REVENUE (3X−3Y) | $ 1500 | 23 |
| 24 | NET DECREASE IN OPERATING COSTS (22X−22Y) | $ 4300 | 24 |
| 25 | ANNUAL OPERATING ADVANTAGE (23+24) | $ 5800 | 25 |

* Next year means the first year of project operation. For projects with a significant break-in period, use performance after break-in.

its remaining economic life would be one year. The operating advantage of the new machine over the existing machine is tabulated on the MAPI worksheet 1 in Figure 12-11.

The new milling machine, which would cost $42,000, qualifies for a 7% investment tax credit. It has an estimated 20-year useful life and could be sold for 20% of its net cost at the end of that time. It will be depreciated by the straight line method. The next year increase in depreciation and interest deductions is estimated to be $1700.

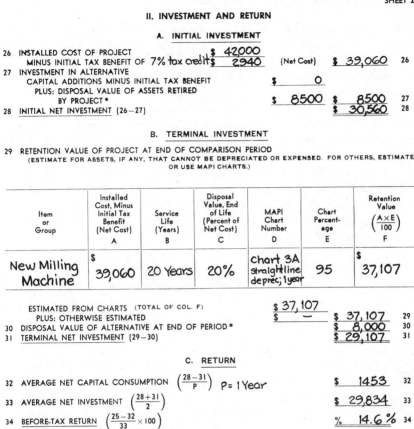

## II. INVESTMENT AND RETURN

### A. INITIAL INVESTMENT

26 INSTALLED COST OF PROJECT     $ 42000
    MINUS INITIAL TAX BENEFIT OF 7% tax credit $ 2940    (Net Cost)   $ 39,060   26
27 INVESTMENT IN ALTERNATIVE
    CAPITAL ADDITIONS MINUS INITIAL TAX BENEFIT     $    O
    PLUS: DISPOSAL VALUE OF ASSETS RETIRED
      BY PROJECT *         $ 8500   $ 8500   27
28 INITIAL NET INVESTMENT (26−27)             $ 30,560   28

### B. TERMINAL INVESTMENT

29 RETENTION VALUE OF PROJECT AT END OF COMPARISON PERIOD
    (ESTIMATE FOR ASSETS, IF ANY, THAT CANNOT BE DEPRECIATED OR EXPENSED. FOR OTHERS, ESTIMATE
                      OR USE MAPI CHARTS.)

| Item or Group | Installed Cost, Minus Initial Tax Benefit (Net Cost) A | Service Life (Years) B | Disposal Value, End of Life (Percent of Net Cost) C | MAPI Chart Number D | Chart Percentage E | Retention Value $\left(\frac{A \times E}{100}\right)$ F |
|---|---|---|---|---|---|---|
| New Milling Machine | $ 39,060 | 20 Years | 20% | Chart 3A straightline deprec; 1 year | 95 | $ 37,107 |

    ESTIMATED FROM CHARTS (TOTAL OF COL. F)       $ 37,107
    PLUS: OTHERWISE ESTIMATED           $  —    $ 37,107   29
30 DISPOSAL VALUE OF ALTERNATIVE AT END OF PERIOD *        $ 8,000   30
31 TERMINAL NET INVESTMENT (29−30)           $ 29,107   31

### C. RETURN

32 AVERAGE NET CAPITAL CONSUMPTION $\left(\frac{28-31}{P}\right)$ P= 1 Year     $ 1453   32

33 AVERAGE NET INVESTMENT $\left(\frac{28+31}{2}\right)$            $ 29,834   33

34 BEFORE-TAX RETURN $\left(\frac{25-32}{33} \times 100\right)$          % 14.6%   34
35 INCREASE IN DEPRECIATION AND INTEREST DEDUCTIONS      $ 1700   35
36 TAXABLE OPERATING ADVANTAGE (25−35)          $ 4100   36
37 INCREASE IN INCOME TAX (36×TAX RATE) 50%       $ 2050   37
38 AFTER-TAX OPERATING ADVANTAGE (25−37)          $ 3750   38
39 AVAILABLE FOR RETURN ON INVESTMENT (38−32)       $ 2297   39
40 AFTER-TAX RETURN $\left(\frac{39}{33} \times 100\right)$          % 7.7%   40

---

* After terminal tax adjustments.

**Figure 12-11.** MAPI worksheets 1 and 2 for Example 12-10.

The existing machine could be sold for $8500 now or about $8000 a year hence. Other data for both milling machines are shown on the MAPI worksheets. Compute the incremental after-tax rate of return if the existing milling machine (the alternative) is replaced by the new milling machine (the project).

From the Figure 12-11 MAPI worksheets we see that the incremental after-tax rate of return is 7.7%.

**Applying Equipment Replacement Models**

From the discussion of the MAPI system it is apparent that the incremental after-tax rate of return computed by the MAPI model is based on somewhat different assumptions than our previous direct computations. Thus our direct computations and the MAPI computations will produce slightly different results. Which is correct, if either one is, depends on how well the actual situation is portrayed in the computational models.

While we usually make simplifying assumptions in direct computations, these assumptions are readily apparent and their probable impact on the answer may be estimated. In replacement models, like the MAPI system, the complexity obscures the underlying assumptions and makes it harder to judge the sensitivity of the answer to the assumptions. A reasonable conclusion is that mathematical models may be very valuable in equipment replacement analysis, but care must be taken to insure that the model fits the situation to which it is to be applied.

## SUMMARY

In selecting equipment for a new plant the question is, which of the machines available on the market will be more economical. But when one has a piece of equipment that is now performing the desired task, the analysis is made more complicated. The existing equipment (called the defender) is already in place so the question is, shall we replace it now, or shall we keep it for one or more years? When a replacement is indicated, it will be by the best available replacement equipment (called the challenger). When we already have equipment there may be a tendency to use past costs in the replacement analysis. But it is only present and future costs that are relevant.

For the defender, the economic life is the remaining useful life where the EUAC is a minimum. In situations where costs are increasing the remaining useful life is considered to be one year. For the challenger a similar calculation is made to determine its most economical useful life. The replacement analysis should be based on the defender at its economical remaining useful life compared to the challenger at its economical useful life. When the useful lives are the same, a variety of analysis techniques (present worth, annual cash flow, rate of return, and so forth) may be used. But for different useful lives the annual cash flow analysis is generally the most suitable method of analysis.

Until this chapter we have assumed that future replacements will be the same as the present replacement. But it may be more realistic to assume that future challengers will be an improvement over the present challenger. The prospect of improved future challengers makes the present challenger a less desirable alternative. By continuing with the defender we may later acquire an improved challenger. One method of adjusting a replacement analysis to account for improved future challengers is to reduce the economic life of the present challenger.

Equipment replacement models have been developed for more complex replacement problems. An example is the MAPI replacement model. The model is carefully organized to provide an accurate incremental after-tax rate of return based on the difference between the proposed project and a stated alternative. The difficult problem in using any replacement model is determining whether the assumptions of the model fit the situation to which the model is to be applied.

## PROBLEMS

**12-1** The Quick Manufacturing Co., a large profitable corporation, is considering the replacement of a production machine tool. A new machine would cost $3700, have a four year useful and depreciable life, and no salvage value. If purchased, the firm would be allowed a 7% investment tax credit. For tax purposes sum-of-years digits depreciation would be used. The existing machine tool was purchased four years ago at a cost of $4000, and has been depreciated by straight line depreciation assuming an eight year life and no salvage value. It could be sold now to a used equipment dealer for $1000 or be kept in service for another four years. It would then have no salvage value. An analysis indicates the new machine would save about $900 per year in operating costs compared to the existing machine.

(a) Compute the before-tax rate of return on the replacement proposal of installing the new machine rather than keeping the existing machine.

(b) Compute the after-tax rate of return on this replacement proposal.

(*Answer:* (a) 12.6% (b) 11.5%)

**12-2** The Plant Department of the local telephone company purchased four special pole hole diggers eight years ago for $14,000 each. They have been in constant use to the present time. Due to an increased workload, it is considered that additional machines will soon be required. Recently it was announced that an improved model of the digger has been put on the market. The new machines have a higher production rate and lower maintenance expense than the old machines, but their cost will be $32,000 each. The service life of the new machines is estimated to be eight years with a salvage estimated at $750 each. The four original diggers have an immediate salvage of $2000 each and an estimated salvage of $500 each eight years hence. The estimated average annual maintenance expense associated with the old machines is approximately $1500 each compared to $600 each for the new machines. A field study and trial indicates that the workload would require three additional new type machines if the old machines are continued in service. However, if the old machines are all retired from service, the present workload plus the estimated increased load could be carried by six new machines with an annual savings of $12,000 in operator costs. Because the new machines employ a new principle of operation, it is contemplated that a special personnel training program will be necessary before the machines can be placed in operation. It is estimated that this training program will cost about $700 per new machine. If the MARR is 9% before taxes, what should the company do?

**12-3** Fifteen years ago the Acme Manufacturing Company bought a propane powered forklift truck for $4800. They depreciated it using straight line depreciation, a 12-year life and zero salvage value. Over the years the forklift has been a good piece of equipment, but lately the maintenance cost has risen sharply. Estimated end-of-year maintenance costs for the next ten years are as follows:

| Year | Maintenance cost |
|------|------------------|
| 1    | $400             |
| 2    | 600              |
| 3    | 800              |
| 4    | 1000             |
| 5–10 | 1400/year        |

The old forklift has no present or future net salvage value as its scrap metal value just equals the cost to haul it away. A replacement is now being considered for the old forklift. A modern unit can be purchased for $6500. It has an economic life equal to its ten-year depreciable life. Straight line depreciation will be employed with zero salvage value at the end of the ten-year depreciable life. At any time the new forklift can be sold for its book value. Maintenance on the new forklift is estimated to be a constant $50 per year for the next ten years. After that maintenance is expected to increase sharply. Should Acme Manufacturing keep its old forklift truck for the present, or replace it now with a new forklift truck? The firm expects an 8% after-tax rate of return on its investments. Income to this large profitable corporation is taxed at 48%.
(*Answer:* Keep the old forklift truck)

**12-4** An industrial establishment secures its water supply from a system of wells. They are using a six-inch, single-stage centrifugal pump that is in good condition. The pump was purchased three years ago for $1350 and has a present book value of $1005, based on straight line depreciation and an expected life of 10 years. However, owing to improvements of design, the demand for a pump of this type is so small that its present sale value is only $500. It is anticipated that the pump will have a salvage value of $200 seven years from now. An improved pump of the same type can now be purchased for $1700 and will have an estimated life of ten years with a salvage value of $200 at the end of that time. The pumping demand is 225 cubic feet per minute against an average head of 200 feet. The old pump has an efficiency of 75% when furnishing this demand. The new pump has an efficiency of 81% when furnishing the same demand. Power costs 4¢ per kw-hr (equal to 3¢ per horsepower-hour), and either pump must operate 2400 hours per year. (1 horsepower = 33,000 foot-pounds per minute.)

(a) If the firm expects a 12% before-tax rate of return, should the new pump be purchased?

(b) Assuming straight line depreciation, a 50% income tax rate, and a required 8% after-tax rate of return, should the new pump be purchased?

**12-5** A firm is concerned about the condition of some of its plant machinery. Bill

James, a newly hired engineer, was assigned the task of looking into the situation and determining what alternatives are available. After a careful analysis, Bill reports that there are five feasible mutually exclusive alternatives.

*Alternative A.* Spend $44,000 now repairing various items. The $44,000 can be charged as a current operating expense (rather than capitalized), and deducted from other taxable income immediately. These repairs are anticipated to keep the plant functioning for the next seven years with operating costs remaining at present levels.

*Alternative B.* Purchase $49,000 of general purpose equipment. Depreciation would be straight line, with the depreciable life equal to the seven-year useful life of the equipment. The equipment will have no end-of-useful-life salvage value. The new equipment will reduce operating costs $6000 per year below their present levels.

*Alternative C.* Purchase $56,000 of new specialized equipment. This equipment would be depreciated by sum-of-years digits depreciation over its seven-year useful life. This equipment would reduce operating costs $12,000 per year below their present levels. It will have no end-of-useful-life salvage value.

*Alternative D.* This alternative is the same as alternative *B*, except that this particular equipment would reduce operating costs $7000 per year below their present levels.

*Alternative E.* This is the "do nothing" alternative. If nothing is done, future annual operating costs are expected to be $8000 above the present level.

This profitable firm pays 50% corporate income taxes. In their economic analysis, they require a 10% after-tax rate of return. Which of the five alternatives should the firm adopt?

**12-6** In a replacement analysis problem the following facts are known:

Initial cost: $12,000

Annual Maintenance: None for the first three years.
$2000 at the end of the fourth year.
$2000 at the end of the fifth year.
Increasing $2500 per year after the fifth year. ($4500 at the end of the sixth year; $7000 at the end of the seventh year, and so forth.)

Actual salvage value in any year is zero. Assume a 10% interest rate and ignore income taxes. Compute the best useful life for this challenger. (*Answer:* five years)

**12-7** A new machine is being purchased for $5500. It is estimated that the machine could be sold at the end of any year for an amount equal to its book value, based on sum-of-years digits depreciation, a ten-year depreciable life and zero salvage value at the end of ten years. During the first year there will be no maintenance costs as the manufacturer's warranty will pay for all maintenance. End-of-year maintenance costs are expected to be $170 the second year, $340 the third year, $510 the fourth year, and so forth, continuing on a uniform gradient. If a 10% interest rate is used, what is the most economical life of the machine?

# 13
## Inflation
## and
## Deflation

Until now we have assumed a stable economic situation, that is, where prices are relatively unchanged over substantial periods of time. Unfortunately, this is not always a realistic assumption. This chapter will describe how either inflation or deflation may be incorporated into an economic analysis. We will see that if all costs and benefits are changing at equal rates, then inflation has no net effect on before-tax economic analyses. But even in this idealized situation, inflation or deflation has an impact on the after-tax results. Because it has been the more frequent situation, the emphasis of the chapter will be on inflation.

### INFLATION AND DEFLATION
### DEFINED

Inflation, as all readers will recognize, is the situation where prices of goods and services are increasing. Apartment rents, textbook prices, haircuts, groceries and almost everything else increase as time passes. This general upward movement of prices in an inflationary period, does not necessarily produce a single rate of inflation. The prices of different items change at different rates, due to the complexities of a free competitive economy (plus monopolistic and other not-so-competitive influences).

If inflation means prices are rising, what about the value of money? Ten years ago the wholesale price of a box of oranges was $4.75, while recently it was $8.75. In other words $4.75 will not buy today what it would buy ten years ago. The purchasing power (or value) of money has declined. Inflation makes future dollars less valuable than present dollars.

## EXAMPLE 13-1

Ten years ago the owner of an orange grove went to the bank and borrowed \$4750. This represented an amount of money equal to 1000 boxes of oranges. The loan was for ten years at 6% interest. The orchardist promised to pay

$$4750(F/P,6\%,10) = 4750(1.791) = \$8507$$

at the end of ten years. He computed that he would have to sell 8507/4.75 = 1791 boxes of oranges to repay the debt.

Now oranges sell for \$8.75 per box. To repay the debt the orchardist has to sell 8507/8.75 = 972 boxes of oranges. Although the orchardist borrowed money equivalent to 1000 boxes of oranges, the increase in their selling price means that he is repaying the equivalent of fewer oranges than he borrowed.

Inflation benefits long term borrowers of money—this is because they will repay their debt in dollars with reduced purchasing power. If inflation helps long term borrowers, it just as clearly is unfavorable for long term lenders of money.

Deflation is just the reverse of inflation. Prices tend to decline with the result that future dollars have more purchasing power than present dollars. Here the lenders are benefited as they lend present dollars and must be subsequently repaid in future dollars with greater purchasing power.

## PRICE CHANGES

Price changes may or may not need to be considered in an economic analysis. We know that costs and benefits must be computed in comparable units. It would make no sense to compute costs in year 1977 dollars and then measure benefits in year 1985 dollars, if 1985 dollars are a different unit of measure than 1977 dollars. It would be like measuring costs in apples and benefits in oranges. On the other hand, one might have a situation where future benefits fluctuate along with the inflation or deflation of the period. This could result in a situation where the future benefits are constant when measured in year 0 dollars. If this had been the original assumption (constant benefits in year 0 dollars), the subsequent inflation or deflation would have no effect on a before-tax economic analysis. We would need to be concerned only with any differential inflation or deflation effects which distort the cash flow when it is measured in comparable units.

In after-tax economy studies the income tax consequences are computed on the taxable portion of the cash flow in a given year. To compute the 1985 taxable income one must use 1985 dollars and 1985 data. At the same time the 1985 depreciation deduction would be based on the original cost of the asset. Therefore, in after-tax economic analysis the impact of inflation or deflation cannot be eliminated by simply adjusting future before-tax benefits. Initially the effect of inflation or deflation will be presented without considering income taxes. Then the more difficult after-tax situation will be considered in the last part of the chapter.

### Measuring the Rate of Price Change

Price indexes are based on the cost of an assortment of goods at one period compared to some prior period. For example, the U.S. Government computes a Wholesale Industrial Price Index that compares the price of an assortment of industrial products now, to their average price in 1967. The average price during 1967 is taken as 100 on the Wholesale Price Index. In 1978 the Wholesale Price Index was 215, indicating prices were 215% of the 1967 prices for comparable items.

A convenient way of stating the rate of price change would be to compute a compound price change (CPC) assuming annual compounding. For the Wholesale Price Index over the 11 year period 1967–1978:

$$F = P(F/P, \text{compound price change}, n \text{ years})$$

$$215 = 100(F/P, \text{compound price change}, 11 \text{ yrs})$$

$$(F/P, \text{CPC}, 11) = \frac{215}{100} = 2.15$$

Interpolating from compound interest tables, compound price change (CPC) = 7.2% per year.

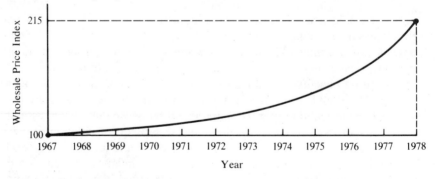

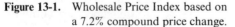

**Figure 13-1.**   Wholesale Price Index based on
a 7.2% compound price change.

It is important to note that a compound price change (CPC) of 7.2% implies how the change took place. Figure 13-1 is a graph of a compound price change of 7.2%. Figure 13-2, on the other hand, is a graph of the actual way the Wholesale Price Index changed. In projecting future changes in prices a frequent assumption (that may or may not be valid) is that the past experience will continue into the future.

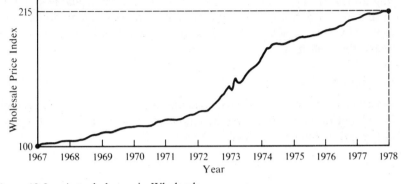

**Figure 13-2.** Actual change in Wholesale Price Index.

### Are Price Indexes Suitable for Estimating Future Prices?

Often one must project the future price for some specific item, like a piece of machinery or a raw material item. Rather than use some general index, like the Consumer Price Index or the Wholesale Price Index, it would be better to consider the specific cost components of the item whose cost is being projected. It may be, for example, that although wholesale prices are rising, a specific item may more reasonably be expected to remain unchanged, or even decline. There could be many reasons for such a situation: improved technology, expanded production facilities, increased price competition, and so forth. Thus general price indexes may—or may not—be suitable for predicting future prices of specific items.

## CONSIDERING PRICE CHANGES IN BEFORE-TAX CALCULATIONS

Estimates of price changes in the future reflecting increased or decreased purchasing power of money certainly should be included in economy studies. In this situation a disbursement at one point in time no longer yields the same amount of an item as the same disbursement at another point in time. Thus dollars at one point in time represent more or less buying power than dollars at another point in time. This variation in buying power must be adjusted. The obvious way is to convert all components of a cash flow to equivalent dollars with equal buying power. The standard dollar could be dollars with 1967 buying power, or 1976 buying power, or what have you. The essential item is that a standard be adopted. Economic analyses typically are analyses "now" concerning alternative future actions. Since the buying power of "now" dollars is both known and a reasonable standard, the usual practice is to adjust the cash flow to equivalent dollars now (year 0). This may be accomplished by making a year by year computation to adjust the cash flow.

*EXAMPLE 13-2*

For the cash flow it is expected that the future benefits will be received in inflated dollars. What are the equivalent year 0 dollars if the compound price change (CPC) is +4% per year?

| Year | Cash flow |
|------|-----------|
| 0    | $-$50    |
| 1    | $+10$     |
| 2    | $+25$     |
| 3    | $+20$     |
| 4    | $+20$     |

| Year | Unadjusted cash flow | Multiplied by | | Cash flow in year 0 dollars |
|------|----------------------|---------------|---|-----------------------------|
| 0 | $-$50 | 1 | $=$ | $-$50 |
| 1 | $+10$ | $(1 + 0.04)^{-1}$ | $=$ | $+9.6$ |
| 2 | $+25$ | $(1 + 0.04)^{-2}$ | $=$ | $+23.1$ |
| 3 | $+20$ | $(1 + 0.04)^{-3}$ | $=$ | $+17.8$ |
| 4 | $+20$ | $(1 + 0.04)^{-4}$ | $=$ | $+17.1$ |

The general form of the multiplier is $(1 + \text{CPC})^{-n}$. When CPC is positive (indicating increasing prices and inflation), the multiplier may be read from compound interest tables as $(P/F,\text{CPC},n)$. A negative CPC indicates prices decreasing (deflation).

If the cash flow fluctuates as in Example 13-2, each year must be individually adjusted. Then to compute the present worth of the future benefits, for example, a series of single payment present worth factors are used.

**Impact of Price Changes on Constant
Future Receipts or Disbursements**

In situations of loans, annuities, and so forth, there is a commitment to a series of uniform end-of-period payments or receipts. These payments or receipts are unaffected by price changes. Yet the value of these future cash flow payments, computed in year 0 dollars, is greatly affected by the price change rate. For equal future payments or receipts, the actual cash flow is represented by Figure 13-3. If there were a positive compound price change (for example, CPC = +5%), the actual cash flow, measured in year 0 dollars would look like Figure 13-4. The present worth of these future payments or receipts may be computed in year 0 dollars by the method demonstrated in Example 13-3.

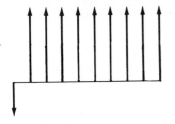

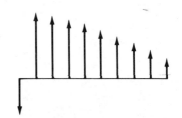

**Figure 13-3.** Actual cash flow for a loan or annuity.

**Figure 13-4.** Cash flow measured in year 0 dollars for a loan or annuity with a positive price change.

*EXAMPLE 13-3*

Mr. Dale Hart is considering the purchase of an annuity that will pay $1000 per year for ten years, beginning at the end of the first year. A friend tells him he believes there will be a compound price change of +6% per year for the next ten years. If Mr. Hart decides that he should obtain a 5% rate of return, after considering the effect of the 6% inflation, how much would he be willing to pay now for the annuity?

The individual $1000 annuity payments may be adjusted by the compound price change rate (CPC) to obtain their equivalent values in year 0 dollars. The CPC multiplier may be used to convert an actual amount of money $F_{actual}$ in any year to the equivalent sum at the same point in time but in year 0 dollars, $F_{0 \text{ dollars}}$.

$$F_{0 \text{ dollars}} = F_{actual} (1 + CPC)^{-n}$$

If this computation were carried out, the changed form of the cash flow diagram would be like Figure 13-5. But the computation need not be

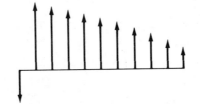

**Figure 13-5.** Cash flow for Example 13-3 in year 0 dollars.

made. Instead, write the equation for the present worth of $F$ in year 0 dollars:

$$\text{PW of } F \text{ in year 0 dollars} = F_{act.} \frac{(1 + CPC)^{-n}}{(1 + i)^n} = F_{act.} \frac{1}{(1 + i)^n(1 + CPC)^n}$$

$$= F_{act.} \frac{1}{[1 + i + CPC + (i)(CPC)]^n}$$

This is similar to the single payment present worth factor: $1/(1 + i)^n$
The present worth factor may be applied here if we substitute for $i$:

Equivalent $i = [i + CPC + (i)(CPC)]$

This equivalent value of $i$ may be used with either single payment or uniform series present worth factors. In this problem: CPC = $+6\%$ $i = 5\%$.

Equivalent $i = [0.05 + 0.06 + (0.05)(0.06)] = 0.113 = 11.3\%$

The amount Mr. Hart would be willing to pay:

PW $= 1000(P/A,11.3\%,10) = 1000(5.816) = 5816$

## Impact of Price Changes on a Fixed Quantity of Future Goods or Services

Future payments or receipts may be associated with a fixed quantity of goods. In the situation of positive (increasing) price change, for example, more dollars would be required to obtain a fixed quantity of goods. Of course it takes more or less actual dollars depending on whether there are price increases or price decreases. Figure 13-6 is the actual cash flow for a positive compound price change like CPC = $+5\%$. Figure 13-7 shows that since a fixed quantity of goods or services are being priced, the cash flow measured in year 0 dollars is constant. Thus present worth of cost of these goods or services in year 0 dollars would simply be

PW of cost $= A(P/A,i\%,n)$.

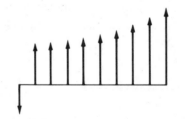

**Figure 13-6.**   Actual cash flow for a fixed
quantity of goods or services
with positive compound price
change (CPC).

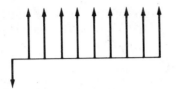

**Figure 13-7.** Cash flow measured in year 0 dollars for
a fixed quantity of goods or services.

## INFLATION AFFECT ON
## AFTER-TAX CALCULATIONS

In the previous sections we have noted the impact of inflation on before-tax calculations. We found that if the subsequent benefits brought constant quantities of dollars, then inflation will diminish the true value of the future benefits and hence the real rate of return. If, however, the future benefits keep up with the rate of inflation the rate of return will not be adversely affected due to the inflation. Unfortunately, we are not so lucky when we consider a real situation with income taxes. This is illustrated by Example 13-4.

*EXAMPLE 13-4*

A $12,000 investment will return annual benefits for six years with no salvage value at the end of six years. Assume straight line depreciation and a 48% corporate income tax rate. The problem is to be solved for before and after-tax rates of return for two situations.

1.  No inflation: the annual benefits are constant at $2918 per year.
2.  Inflation equal to a 5% compound price change: the benefits from the investment increase at this same rate, so that they continue to be the equivalent of $2918 in year 0 dollars.

Compute the benefit schedule for the two situations.

| Year | Annual benefit for both situations Year 0 dollars | No inflation Actual dollars received | 5% CPC inflation factor* | 5% CPC inflation Actual dollars received |
|---|---|---|---|---|
| 1 | $2918 | $2918 | $(1.05)^1$ | $3064 |
| 2 | 2918 | 2918 | $(1.05)^2$ | 3217 |
| 3 | 2918 | 2918 | $(1.05)^3$ | 3378 |
| 4 | 2918 | 2918 | $(1.05)^4$ | 3547 |
| 5 | 2918 | 2918 | $(1.05)^5$ | 3724 |
| 6 | 2918 | 2918 | $(1.05)^6$ | 3910 |

*May be read from the 5% compound interest table as $(F/P,5\%,n)$.

Since both situations (no inflation and 5% inflation) have an annual benefit, stated in year 0 dollars of $2918, they have the same before-tax rate of return.

PW of cost = PW of benefit

$$12,000 = 2918(P/A,i\%,6)$$

$$(P/A,i\%,6) = \frac{12,000}{2,918} = 4.11$$

From interest tables: before-tax rate of return equals 12%.

*No inflation: After-tax rate of return*

| Year | Before-tax cash flow | Straight line depreciation | Taxable income | 48% income taxes | *Actual dollars and year 0 dollars* after-tax cash flow |
|------|------|------|------|------|------|
| 0    | −$12,000 |        |       |        | −$12,000 |
| 1–6  | +2,918   | $2000  | $918  | −$441  | +2,477   |

PW of cost = PW of benefit

$$12,000 = 2477(P/A,i\%,6)$$

$$(P/A,i\%,6) = \frac{12,000}{2,477} = 4.84$$

From interest tables: after-tax rate of return equals 6.5%.

*5% CPC: After-tax rate of return*

| Year | Before-tax cash flow | Straight line depreciation | Taxable income | 48% income taxes | *Actual dollars* After-tax cash flow |
|------|------|------|------|------|------|
| 0 | −$12000 |       |        |        | −$12000 |
| 1 | +3064   | $2000 | $1064  | −$511  | +2553   |
| 2 | +3217   | 2000  | 1217   | −584   | +2633   |
| 3 | +3378   | 2000  | 1378   | −661   | +2717   |
| 4 | +3547   | 2000  | 1547   | −743   | +2804   |
| 5 | +3724   | 2000  | 1724   | −828   | +2896   |
| 6 | +3910   | 2000  | 1910   | −917   | +2993   |

Converting to year 0 dollars and solving for the rate of return:

| Year | Actual dollars After-tax cash flow | Conversion factor | | Year 0 dollars After-tax cash flow | Present worth at 5% | Present worth at 4% |
|---|---|---|---|---|---|---|
| 0 | −$12000 | | | −$12000 | −$12000 | −$12000 |
| 1 | +2553 | × (1.05)$^{-1}$ | = | +2431 | +2315 | +2338 |
| 2 | +2633 | × (1.05)$^{-2}$ | = | +2388 | +2166 | +2208 |
| 3 | +2717 | × (1.05)$^{-3}$ | = | +2347 | +2027 | +2086 |
| 4 | +2804 | × (1.05)$^{-4}$ | = | +2307 | +1898 | +1972 |
| 5 | +2896 | × (1.05)$^{-5}$ | = | +2269 | +1778 | +1865 |
| 6 | +2993 | × (1.05)$^{-6}$ | = | +2233 | +1666 | +1765 |
| | | | | | −150 | +234 |

Linear interpolation between 4% and 5%:

$$\text{After-tax rate of return} = 4\% + 1\%[234/(234 + 150)] = 4.6\%$$

From Example 13-4 we see that the before-tax rate of return for both situations (no inflation and 5% inflation) is the same. This was expected because the benefits in the inflation situation increased in proportion to the inflation. This shows that where future benefits fluctuate with changes in inflation or deflation, the effects do not alter the year 0 dollar estimates. Thus no special calculations are needed in before-tax calculations when future benefits are expected to respond to inflation or deflation rates.

The after-tax calculations illustrate a different result. The two situations, with equal before-tax rates of return, do not produce equal after-tax rates of return.

| Situation | Before-tax rate of return | After-tax rate of return |
|---|---|---|
| No inflation | 12% | 6.5% |
| 5% inflation | 12% | 4.6% |

Thus 5% inflation resulted in a smaller after-tax rate of return even though the benefits increased at the same rate as the inflation. A review of the cash flow table will reveal that while benefits increased, the depreciation schedule did not. Thus the inflation resulted in increased taxable income and hence larger income tax payments, but there were not sufficient increases in benefits to offset these additional disbursements. The result was that while the after-tax cash flow in actual dollars was increased, it was not large enough to offset both inflation and increased income taxes. This is readily apparent when the equivalent year 0 dollar after-tax cash flow is examined. With inflation, the year 0 dollar after-tax cash flow is smaller than the year 0 dollar after-tax cash flow without it. Inflation might cause equipment to have a salvage value that was not forecast, or a larger one than had been projected. This would tend to reduce the unfavorable effect of inflation on the after-tax rate of return.


## SUMMARY

Inflation is characterized by rising prices for goods and services, while deflation produces a fall in prices. An inflationary trend makes future dollars have less purchasing power than present dollars. This helps long term borrowers of money for they may repay a loan of present dollars in the future with dollars of reduced buying power. The help to borrowers is at the expense of lenders.

Deflation has the opposite effect from inflation. If money is borrowed at one point in time, and then a deflationary period occurs, the borrower has to repay his loan with dollars of greater purchasing power than he borrowed. This would be advantageous to lenders at the expense of borrowers.

While price changes occur in a variety of ways, one method of stating a price change is as a uniform rate of price change per year, or a compound price change. For this method the proper single payment multiplier to adjust a future cash flow to year 0 buying power is

Year 0 dollars $= [(1 + \text{CPC})^{-n}]$ year $n$ dollars

When there is a uniform future cash flow series, the year 0 buying power has the same general form as the compound interest factor. An equivalent interest rate

$$i_{\text{equiv}} = [i + \text{CPC} + i(\text{CPC})]$$

may be computed and then used in conjunction with the present worth factor to compute the present worth of the future cash flow in year 0 dollars.

The effect of inflation on the computed rate of return for an investment depends on how future benefits respond to the inflation. If benefits produce constant dollars, which are not increased by inflation, the effect of inflation is to reduce the before-tax rate of return of the investment. If, on the other hand, the dollar benefits increase to keep up with the inflation, the before-tax rate of return will not be adversely affected by the inflation. This is not true when an after-tax analysis is made. Even if the future benefits increase to match the inflation rate, the allowable depreciation schedule does not increase. The result will be increased taxable income and income tax payments. This reduces the available after-tax benefits and therefore the after-tax rate of return. The important conclusion is that estimates of future inflation or deflation may be important in evaluating capital expenditure proposals.

## PROBLEMS

**13-1** One economist has predicted that there will be a 7% per year inflation of prices during the next ten years. If this compound price change proves to be correct, an item that presently sells for $10 would sell for what price ten years hence? (*Answer:* $19.67)

**13-2** A man bought a 5% tax free municipal bond. It cost $1000 and will pay $50 interest each year for 20 years. The bond will mature at the end of the 20 years and return the original $1000. If there is a 2% CPC annual inflation during this period, what rate of return will the investor receive after considering the effect of inflation?

**13-3** A firm is having a large piece of equipment overhauled. It anticipates the machine will be needed for the next 12 years. The firm has an 8% minimum attractive rate of return. The contractor has suggested three alternatives.

(a) A complete overhaul for $6000 that should permit 12 years of operation.

(b) A major overhaul for $4500 that can be expected to provide 8 years of service. At the end of 8 years a minor overhaul would be needed.

(c) A minor overhaul now. At the end of 4 and 8 years additional minor overhauls would be needed.

If minor overhauls cost $2500, which alternative should the firm select? If minor overhauls, which now cost $2500, increase in cost at +5% per year, which alternative should the firm select? (*Answer:* Alt. c, Alt. a)

**13-4** A man wishes to set aside money for his daughter's college education. His goal is to have a bank savings account containing an amount equivalent to $20,000 with today's purchasing power of the dollar, at the time of the girl's 18th birthday. The estimated inflation rate is 8% compound price change. If the bank pays 5% compounded annually, what lump sum of money should he deposit in the bank savings account on his daughter's fourth birthday? (*Answer:* $29,670)

**13-5** One economist has predicted that for the next five years the United States will have an 8% annual inflation rate (compound price change), followed by five years at a 6% inflation rate, compounded annually. This is equivalent to what average price change per year for the entire ten-year period?

**13-6** A homebuilder's advertising has the caption, "Inflation to Continue for Many Years." The advertisement continues with the explanation that if one bought a home now for $37,000, and inflation continued at a 7% annual rate, compounded annually, the home would be worth $102,000 in 15 years. This means, the advertisement says, a profit of $65,000 from the purchase of a home now. Do you agree with the homebuilder's logic? Explain.

# 14
## Estimation
## of
## Future Events

Economic analysis, if it is to be of any value, must concern itself with present and future consequences. We know that a post audit is desirable to see how well prior estimates were actually achieved, but that is not the central thrust of economic analysis. Our task is to take the present situation and our appraisal of the future and make sound decisions based upon them. That is probably a lot easier said than done. It may be relatively easy to determine the present situation, but it certainly is not easy to look into the future.

One ironic aspect of an engineering education is that engineers are well trained to analytically evaluate a situation and yet there is very little attention given to predicting the future. This seems strange when we recognize that engineering inevitably deals with both the present and the future. Engineering—including economic analysis—requires the estimation of future events.

In this chapter we will consider the problem of evaluating the future. The easiest way to begin will be by making an exact estimate. Then we will examine the possibility of predicting a range of possible outcomes. Finally, we will consider the situation where the probabilities of the various outcomes are known or may be estimated.

## PRECISE ESTIMATES

In an economic analysis we need to evaluate the future consequences of an alternative. While that cannot be easy, it must be done. In practically every chapter of this book there are cash flow tables where the costs and benefits for future years are precisely described. Do we really believe that we can foretell the future of, for example, an exact maintenance cost? No one believes he can predict the future with certainty. Instead the goal is to select a single value which represents the *best* estimate that can be made of the future.

Once our best estimates are made of the various future consequences, we must put them into the economic analysis. The way they are used is really quite different from the way they were determined. In estimating the future consequences we recognize that they are not precise and that the actual values likely will be somewhat different from our estimates. But once the estimates are in the economic analysis we have proceeded on the assumption that these estimates are correct. We know the estimates will not always be correct, and yet we assume that they will be correct in the economic analysis. This can lead to trouble. If the actual costs and benefits are somewhat different from the estimates, it could be that an undesirable alternative will be selected simply because the variability of the future consequences is concealed by assuming that the best estimates will actually occur. This is illustrated by Example 14-1.

## EXAMPLE 14-1

Two alternatives are being considered. The best estimates for the various consequences are as follows.

|  | *A* | *B* |
|---|---|---|
| Cost | $1000 | $2000 |
| Net annual benefit | 150 | 250 |
| Useful life | 10 yrs | 10 yrs |
| End of useful life salvage value | 100 | 400 |

If interest is $3\frac{1}{2}\%$, which alternative should be selected?

*Alternative A*

$$NPW = -1000 + 150(P/A,3\tfrac{1}{2}\%,10) + 100(P/F,3\tfrac{1}{2}\%,10)$$
$$= -1000 + 150(8.317) + 100(0.7089)$$
$$= -1000 + 1248 + 71 = +\$319$$

*Alternative B*

$$NPW = -2000 + 250(P/A,3\tfrac{1}{2}\%,10) + 400(P/F,3\tfrac{1}{2}\%,10)$$
$$= -2000 + 250(8.317) + 400(0.7089)$$
$$= -2000 + 2079 + 284 = +\$363$$

Alternative *B*, with its larger NPW, would be selected.

Suppose that at the end of ten years the actual salvage value turned out to be $300 instead of the $400 best estimate. If all the other estimates were correct, was *B* still the preferred alternative?

*Corrected B*

$$\begin{aligned}
\text{NPW} &= -2000 + 250(P/A,3\tfrac{1}{2}\%,10) + 300(P/F,3\tfrac{1}{2}\%,10) \\
&= -2000 + 250(8.317) + 300(0.7089) \\
&= -2000 + 2079 + 213 = +\$292
\end{aligned}$$

Under these circumstances $A$ is now the preferred alternative.

Example 14-1 shows that the change in the salvage value of $B$ actually resulted in a change of preferred alternative.

## EXAMPLE 14-2

Using Example 14-1 data, compute the sensitivity of the decision to alternative $B$ salvage value. For alternative $A$ NPW = +319. For breakeven between the alternatives,

$$\begin{aligned}
\text{NPW}_A &= \text{NPW}_B \\
+319 &= -2000 + 250(P/A,3\tfrac{1}{2}\%,10) + \text{salvage value}_B(P/F,3\tfrac{1}{2}\%,10) \\
&= -2000 + 250(8.317) + \text{salvage value}_B(0.7089)
\end{aligned}$$

At the breakeven point

$$\text{Salvage value}_B = \frac{319 + 2000 - 2079}{0.7089} = \frac{240}{0.7089} = 339$$

For alternative $B$ salvage value > \$339, $B$ is preferred; for alternative $B$ salvage value < \$339, $A$ is preferred.

Breakeven and sensitivity analysis provide one means of examining the impact of the variability of some estimate on the outcome. It helps by answering the question, how much variability can a parameter have before the decision will be affected? This approach does not, however, answer the basic problem of how to take the inherent variability of parameters into account in an economic analysis. This will be considered next.

## A RANGE OF ESTIMATES

Realistically, the *true* situation is that there is a range of possible values for the parameter. One could, for example, estimate values for the optimistic estimate, the most likely estimate, or the pessimistic estimate. Then the economic analysis could be performed on each set of data to determine if the decision is sensitive to the range of projected values.

## EXAMPLE 14-3

A firm is considering an investment. Three estimates for the various parameters are as follows.

|  | Optimistic value | Most likely value | Pessimistic value |
|---|---|---|---|
| Cost | $1000 | $1000 | $1052 |
| Net annual benefit | 200 | 198 | 190 |
| Useful life | 12 yrs | 12 yrs | 9 yrs |
| End of useful life salvage value | 100 | 0 | 0 |

If a 10% before-tax minimum attractive rate of return is required, is the investment justified under all three estimates? Compute the rate of return for each estimate.

*Optimistic estimate*

PW of cost = PW of benefit

$1000 = 200(P/A,i\%,12) + 100(P/F,i\%,12)$

Try $i = 18\%$

$1000 = 200(4.793) + 100(0.1372) = 973$

18% is too high. $i \approx 17\%$

*Most likely estimate*

$1000 = 198(P/A,i\%,12)$

$(P/A,i\%,12) = \dfrac{1000}{198} = 5.05$

From compound interest tables, $i = 16.7\%$.

*Pessimistic estimate*

$1052 = 190(P/A,i\%,9)$

$(P/A,i\%,9) = \dfrac{1052}{190} = 5.54$

From compound interest tables, $i = 11\%$.

From the calculations we conclude that the rate of return for this investment is most likely to be 16.7%, but might range from 11% to 17%. The investment meets the 10% MARR criterion for all estimates.

Example 14-3 required that three separate calculations be made for the investment, one each for the optimistic values, the most likely values, and the pessimistic values. This approach emphasizes the unlikely situations where all parameters prove to be very favorable or very unfavorable. Neither is likely to happen. Rather, there is likely to be a blend of results with the parameters assuming values near to the most likely estimate, but with due consideration for the possible range of values. One way to accomplish this is to estimate an average or mean* value for each parameter, based on the following weighting factors.†

$$\text{Mean value} = \frac{\text{optimistic value} + 4(\text{most likely value}) + \text{pessimistic value}}{6}$$

This approach is illustrated in Example 14-4.

## EXAMPLE 14-4

Solve the Example 14-3 problem by applying the weighting factors given in the text. Compute the resulting mean rate of return.

|  | Optimistic value | Most likely value | Pessimistic value |
|---|---|---|---|
| Cost | $1000 | $1000 | $1052 |
| Net annual benefit | 200 | 198 | 190 |
| Useful life | 12 yrs | 12 yrs | 9 yrs |
| End of useful life salvage value | 100 | 0 | 0 |

$$\text{Mean cost} = \frac{1000 + 4(1000) + 1052}{6} = 1009$$

$$\text{Mean net annual benefit} = \frac{200 + 4(198) + 190}{6} = 197$$

$$\text{Mean useful life} = \frac{12 + 4(12) + 9}{6} = 11.5 \text{ years}$$

$$\text{End of useful life salvage value} = \frac{100}{6} = 17$$

Compute the mean rate of return.

PW of cost = PW of benefit

$1009 = 197(P/A,i\%,11.5) + 17(P/F,i\%,11.5)$

From compound interest tables, rate of return is approximately 16%.

---

*The mean value is defined as the sum of all the values divided by the number of values.

†If you are interested, these weighting factors represent an approximation of the beta distribution.

Example 14-4 gave a mean rate of return (16%) that was different from the most likely rate of return (16.7%) computed in Example 14-3. The immediate question is, why are these values different? The reason for the difference can be seen from the way the two values were calculated. One rate of return was based exclusively on the most likely values. The mean rate of return, on the other hand, took into account not only the most likely values, but also the variability of the parameters. In examining the data we see that the pessimistic values are further away from the most likely values than are the optimistic values. This causes the resulting weighted mean values to be less favorable than the most likely values. As a result the mean rate of return, in this example, is less than the rate of return based on the most likely values.

## PROBABILITY AND RISK

Probability can be considered to be the long run relative frequency of occurrence of an outcome. There are just two possible outcomes from flipping a coin (a head or a tail). If, for example, a coin is flipped over and over we can expect in the long run that half the time heads will appear and half the time tails. We would say the probability of flipping a head is 0.50 and of flipping a tail is 0.50. Since probabilities are defined so that the sum of probabilities for all possible outcomes is one, the situation is

Probability of flipping a head = 0.50
Probability of flipping a tail  = 0.50
Sum of all possible outcomes = 1.00

A more complex situation is given in the following example.

### EXAMPLE 14-5

If one were to roll one die (that is, one half of a pair of dice), what is the probability that either a 1 or a 6 would result?
Since a die is a perfect 6-sided cube, the probability of any side appearing is 1/6.

$$
\begin{aligned}
\text{Probability of rolling a } 1 &= P(1) = 1/6 \\
2 &= P(2) = 1/6 \\
3 &= P(3) = 1/6 \\
4 &= P(4) = 1/6 \\
5 &= P(5) = 1/6 \\
6 &= P(6) = 1/6 \\
\text{Sum of all possible outcomes} &= 6/6 = 1
\end{aligned}
$$

The probability of rolling either a 1 or a 6 = $1/6 + 1/6 = 1/3$.

In the two examples the probability of each outcome was the same. This need not be the case.

### EXAMPLE 14-6

In the game of Blackjack a perfect hand is a ten or a facecard plus an ace. What is the probability of being dealt a ten or a facecard from a newly shuffled deck of 52 cards? What is the probability of being dealt an ace in this same situation?

    The three outcomes being examined are to be dealt a ten or a facecard, an ace, or some other card. Every card in the deck represents one of these three possible outcomes. There are 4 aces, 16 tens, jacks, queens and kings, and 32 other cards.

| | |
|---|---|
| The probability of being dealt a ten or a facecard | $= 16/52 = 0.31$ |
| The probability of being dealt an ace | $= 4/52 = 0.08$ |
| The probability of being dealt some other card | $= 32/52 = \dfrac{0.61}{1.00}$ |

The term *risk* has a special meaning when it is used in statistics. It is defined as a situation where there are two or more possible outcomes and the probability associated with each outcome is known. In the two previous examples there is a risk situation. We could not know in advance what playing card would be dealt or what number would be rolled by the die. However, since the various probabilities could be computed, our definition of risk has been satisfied.

    Probability and risk are not restricted to gambling games. In a particular engineering course, for example, a student has computed the probability for each of the letter grades he might receive as follows.

| Outcome | | Probability |
|:---:|:---:|:---:|
| *Grade* | *Grade point* | *P(grade)* |
| A | 4.0 | 0.10 |
| B | 3.0 | 0.30 |
| C | 2.0 | 0.25 |
| D | 1.0 | 0.20 |
| F | 0 | 0.15 |
| | | 1.00 |

From the table we see that the grade with the highest probability is B. This therefore, is the most likely grade. We see from the table that there is a substantial probability that some grade other than B will be received. And the probabilities indicate that if a B is not received, the grade will probably be something less than a B. But in saying the most likely grade is a B, these other outcomes are ignored. In the next section we will show that a composite statistic may be computed using all the data.

## EXPECTED VALUE

In the example in the previous section we saw that the most likely grade of B in an engineering class had a probability of 0.30. That is not a very high probability. In some other course, like a math class, we might estimate a probability of 0.65 of obtaining a B, making the B the most likely grade. While a B is most likely in both the classes, it is more certain in the math class. In the early part of this chapter we computed a weighted mean to give a better understanding of the total situation as represented by various possible outcomes. We can do the same thing here. An obvious selection of the weighting factors is to use the probabilities for the various outcomes. Then since the sum of the probabilities equals one, the computation is:

$$\text{Weighted mean} = \frac{\text{outcome}_A \times P(A) + \text{outcome}_B \times P(B) + \cdots}{1}$$

When the probabilities are used as the weighting factors we call the result the *expected value* and write the equation as:

$$\text{Expected value} = \text{outcome}_A \times P(A) + \text{outcome}_B \times P(B) + \cdots$$

*EXAMPLE 14-7*

Compute the student's expected grade in the engineering course using the probabilities given in the text.

| Grade | Grade point | P(grade) | Grade point × P(grade) |
|-------|-------------|----------|------------------------|
| A | 4.0 | 0.10 | 0.40 |
| B | 3.0 | 0.30 | 0.90 |
| C | 2.0 | 0.25 | 0.50 |
| D | 1.0 | 0.20 | 0.20 |
| F | 0 | 0.15 | 0 |
| | | Expected(GP) = | 2.00 |

The expected grade point (GP) of 2.00 indicates a grade of C.

From the calculations in Example 14-7 we find that for a given set of probabilities the most likely grade is B and the expected grade is C. How can we resolve these conflicting results? First, the results are correct. The most likely grade *is* B and the expected grade *is* C.

The two values tell us different things about the probabilities of the outcomes. Suppose 1000 students took courses in which they each believed the Example 14-7 distribution of probabilities were correct. Each person could correctly state that their most likely grade would be B with an expected grade of C. If the projected probabilities proved to be correct, suppose the 1000 students might receive grades as follows.

| Grade | Number of students |
|-------|--------------------|
| A | 100 |
| B | 300 |
| C | 250 |
| D | 200 |
| F | 150 |
| | 1000 |

We would note immediately that only 300 students would have received B grades. Most students would have received some other grade. If the average grade were computed, what would it be? To average A, B, C, D, and F we would assign the numerical values, 4, 3, 2, 1, and 0. The computation is

| Grade | Number of students | Grade × students |
|-------|--------------------|-----------------|
| A = 4.0 | 100 | 400 |
| B = 3.0 | 300 | 900 |
| C = 2.0 | 250 | 500 |
| D = 1.0 | 200 | 200 |
| F = 0 | 150 | 0 |
| | Sum = | 2000 |

$$\text{Average grade} = \frac{\text{sum}}{\text{number of students}} = \frac{2000}{1000} = 2.0$$

The average grade is C.

We recognize that the average grade is exactly the expected grade. This helps us to understand expected value. If a situation were to occur over and over again then the accumulated results will approach the expected value. Nevertheless, B remains the most likely grade to be received.

*EXAMPLE 14-8*

Just before a horserace is about to begin a spectator decides that the situation on the four horse race is as follows.

| Horse | Probability of winning | Outcome of a $10 bet if horse wins* |
|:---:|:---:|:---:|
| 1 | 0.15 | $48.00 |
| 2 | 0.15 | 58.00 |
| 3 | 0.50 | 16.50 |
| 4 | 0.20 | 42.00 |
| | 1.00 | |

*In horserace betting a ticket is purchased for $10. The outcome of a winning ticket represents the refund of the $10 bet plus the amount won.

(a) What would be the expected value of the ticket if he bet $10 on horse 3 to win?

$$\text{Expected value} = \text{Outcome if 3 wins} \times P(\text{win}_3)$$
$$+ \text{Outcome if 3 loses} \times P(\text{loss}_3)$$
$$= \$16.50(0.50) + \$0(0.50) = \$8.25$$

Thus a ticket purchased for $10 has an expected value of $8.25. (Is this the way to get rich?) One also notes that there is no way for the bettor to actually win the expected value. He must either win $16.50 or nothing. There are no other possibilities. Figure 14-1 illustrates the situation. In this example we find that betting $10 on the favorite horse (the one with the highest probability of winning) could be expected to result in a net loss to the bettor.

(b) An alternate strategy would be to bet $10 on each of the horses to win. In this way one is certain to have a winning ticket. What will be the expected value of the four tickets for this betting scheme?

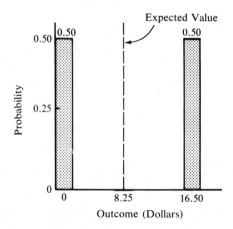

**Figure 14-1.** Example 14-8a situation.

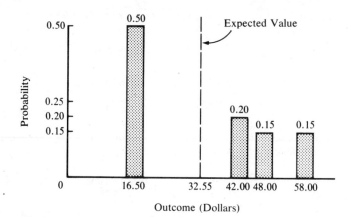

Outcome (Dollars)

**Figure 14-2.** Example 14-8b situation.

Expected value = $48.00(0.15) + $58.00(0.15) + $16.50(0.50)
+ $42.00(0.20)
= $7.20 + 8.70 + 8.25 + 8.40
= $32.55

Figure 14-2 is a plot of this betting plan.

(c) Which betting scheme is better? Expected result from a single $10 bet = $8.25 − $10.00 = $1.75 loss. Expected result from four $10 bets = $32.55 − $40.00 = $7.45 loss. On this basis the single $10 bet is preferred. Of course, it is clear that the best decision would be to make no bet at all. (Or would it?)

Expected value may be a useful analysis tool in certain kinds of situations as it represents the long term outcome if the particular situation occurs over and over again. If one were to put money into a slot machine and play it over and over for an hour or two, the expected value may be a rather accurate estimate of the results from playing the slot machine a couple of hundred times. But would it be useful in estimating the results from playing the machine once? Obviously not. The most likely result would be the loss of the coin. Much less frequently one might receive three or more coins back. Thus the expected value (at possibly 0.75 coin returned per play) is not very useful when we are trying to evaluate a situation that is not repeated. Example 14-8 represented a one-time event. A particular horse race will be run once. Thus the expected value cannot tell us much about the outcome from a particular race; it does say that people who continue to bet on horseraces (or play slot machines) must expect to lose money over the long term.

## DISTRIBUTION OF OUTCOMES

In the three previous sections we have considered ways of treating situations where the outcomes vary. There was no discussion of the distributions of the outcomes as in Figures 14-1 and 14-2. In this section we will examine two specific distributions (uniform and normal) and describe how to randomly sample from these or any other distributions. The ability to randomly sample from a distribution is prerequisite for the discussion of simulation in the next section.

### Uniform Distribution

In Example 14-5 we saw that a die, being a perfect 6-sided cube, can be rolled to provide each of six numbers with an equal probability of 1/6. The distribution is discrete since only integers appear and values in between are not possible. When the outcomes have an equal probability, this is called a uniform distribution and is illustrated by Figure 14-3. When we would like to know what the outcome is from the roll of a die, the easiest way is simply to roll the die. On the other hand we could simulate rolling the die in a variety of ways. One way would be to make up six slips of paper and assign the numbers 1 through 6 to them. Then shake them in a hat and choose one of them at random. In this manner we would have simulated the rolling of a die. Since there are six papers in the hat and they are shaken up so each piece of paper is equally likely to be selected, we really have created another uniform distribution (Figure 14-4). It is easy to use six pieces of paper and create a uniform distribution with outcomes of 1 through 6. A more general approach to simulating the uniform distribution can now be devised.

Suppose someone had put each of the possible two-digit numbers (00 through 99) on slips of paper and put the 100 slips in a hat. Then after careful mixing one is chosen at random and the two-digit value written

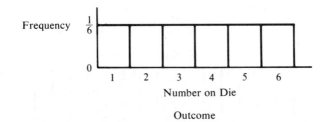

**Figure 14-3.** Uniform distribution—outcomes of a die.

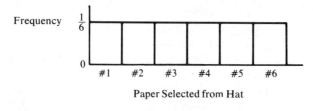

Paper Selected from Hat

Outcome

**Figure 14-4.**   Uniform distribution—papers
in a hat.

down. The slip of paper is returned to the hat and the process repeated.
Table 14-1 is a sample of random two-digit numbers. A careful examination of Table 14-1 shows that some two-digit numbers appear several times
and some do not appear at all. This is, of course, possible for the slips of
paper were replaced in the hat after they were drawn.

The table of random numbers represents random sampling from a uniform distribution of values from 00 through 99 as indicated by Figure 14-5.

**Table 14-1.**   150 two-digit random numbers

| 77 | 41 | 71 | 94 | 66 | 14 | 16 | 77 | 50 | 17 | 65 | 61 | 85 | 23 | 15 |
|----|----|----|----|----|----|----|----|----|----|----|----|----|----|----|
| 06 | 40 | 75 | 90 | 34 | 75 | 45 | 34 | 96 | 74 | 34 | 92 | 24 | 52 | 99 |
| 46 | 19 | 39 | 60 | 20 | 50 | 05 | 80 | 16 | 33 | 79 | 03 | 27 | 22 | 37 |
| 02 | 66 | 67 | 29 | 92 | 03 | 24 | 58 | 08 | 05 | 56 | 57 | 47 | 72 | 02 |
| 68 | 40 | 65 | 20 | 78 | 69 | 76 | 30 | 39 | 21 | 00 | 22 | 40 | 61 | 19 |
| 43 | 10 | 23 | 08 | 48 | 55 | 14 | 45 | 52 | 22 | 90 | 71 | 15 | 89 | 04 |
| 80 | 87 | 71 | 81 | 45 | 19 | 30 | 72 | 88 | 08 | 44 | 24 | 18 | 41 | 97 |
| 35 | 17 | 82 | 18 | 84 | 00 | 77 | 87 | 38 | 83 | 42 | 38 | 55 | 17 | 31 |
| 87 | 21 | 94 | 49 | 66 | 74 | 96 | 10 | 70 | 09 | 76 | 34 | 21 | 06 | 55 |
| 15 | 69 | 32 | 47 | 88 | 87 | 14 | 99 | 19 | 27 | 41 | 61 | 40 | 53 | 03 |

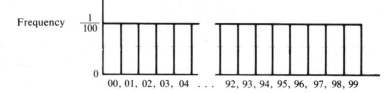

**Figure 14-5.**   Uniform distribution—two digit
random numbers.

We can use the table of random numbers to simulate the outcome of a die by assigning 1/6 of the random numbers to a 1 on the die, another 1/6 of the random numbers to a 2 on the die, and so forth. Since 100/6 equals 16.6, it will not come out even. We will let 16 random numbers represent each face of the die. The assignment of numbers is as follows.

| Random numbers | Outcome on the die |
|---|---|
| 00–15 | 1 |
| 16–31 | 2 |
| 32–47 | 3 |
| 48–63 | 4 |
| 64–79 | 5 |
| 80–95 | 6 |
| 96–99 | random numbers not used |

Graphically, the situation would look like Figure 14-6.

*EXAMPLE 14-9*

Using the table of two-digit random numbers (Table 14-1) and Figure 14-6, simulate the rolling of a die three times.

The first task is to obtain two-digit random numbers from Table 14-1. The proper way to use a table of random numbers is to enter the table at some random point and read in a consistent manner. Following this

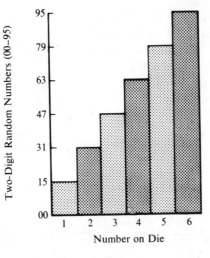

**Figure 14-6.** Plot of outcome on die vs random numbers.

method we might read the numbers 50, 96, and 16 from the ninth column in the table. Enter the Figure 14-6 graph with 50 as the random
number. Read from the y-axis across to the stair-step curve and read
down to 4 on the x-axis. This indicates that the first roll of the die is a 4.
The second random number 96 is not being used in the simulation. One
should discard the number and go on to the next random number selected. A random number of 16 corresponds to a 2 on the die. Another
random number must be selected from Table 14-1. The selected 08 corresponds to 1 on the die. Using a table of random numbers we have
simulated the rolling of a die with the numbers 4, 2, and 1 having been
selected.

With the technique illustrated in Example 14-9, any uniform distribution
may be simulated with a table of random numbers.

**Normal Distribution**

Possibly the best known of all distributions is the normal distribution. It is
defined by the equation:

$$y(x) = \frac{1}{\sigma\sqrt{2\pi}} \exp -\tfrac{1}{2}\left(\frac{x - \mu}{\sigma}\right)^2$$

where $\mu$ = mean, $\sigma$ = standard deviation.

Since all values are possible the normal distribution is a continuous distribution. We see that a particular normal distribution is defined by its mean
$\mu$ and standard deviation $\sigma$. Different values of $\mu$ and $\sigma$ give normal distributions that look different from one another. Two normal distributions
are shown in Figure 14-7.

Because the equation defining the normal distribution is cumbersome
to use, data are usually obtained from a table of the distribution with mean

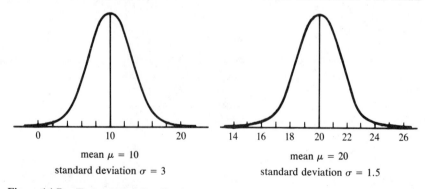

mean $\mu$ = 10
standard deviation $\sigma$ = 3

mean $\mu$ = 20
standard deviation $\sigma$ = 1.5

**Figure 14-7.**   Two normal distributions with
different means and standard
deviations.

$\mu$ equal to 0 and standard deviation $\sigma$ equal to 1. For continuous distributions the area under the curve equals the probability. As the sum of the probabilities for all possible outcomes is one, the total area under the normal curve is one. We can find the area under portions of the distribution from Figure 14-8. The figure shows that 68.3% of the area under the nor-

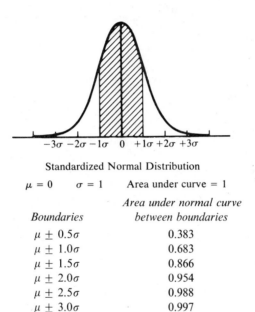

$-3\sigma \;\; -2\sigma \;\; -1\sigma \quad 0 \quad +1\sigma \; +2\sigma \; +3\sigma$

Standardized Normal Distribution

| $\mu = 0$ | $\sigma = 1$ | Area under curve = 1 |

| Boundaries | Area under normal curve between boundaries |
|---|---|
| $\mu \pm 0.5\sigma$ | 0.383 |
| $\mu \pm 1.0\sigma$ | 0.683 |
| $\mu \pm 1.5\sigma$ | 0.866 |
| $\mu \pm 2.0\sigma$ | 0.954 |
| $\mu \pm 2.5\sigma$ | 0.988 |
| $\mu \pm 3.0\sigma$ | 0.997 |
| $\mu \pm 3.5\sigma$ | 0.999 |

**Figure 14-8.** Area under the normal distribution curve.

mal curve lies between $\mu - 1.0\sigma$ and $\mu + 1.0\sigma$. Thus in any situation we could expect a normally distributed variable to be within this range about 68.3% of the time.

A comment about the notation is in order. When one is referring to the normal distribution of some population (like the ages of *all* college students) we say the mean is $\mu$ and the standard deviation $\sigma$. But when we are referring to a *sample* of ages of college students, we say the mean is $\bar{x}$ and the standard deviation $s$. Thus

|  | *Sample of college students* | *All college students* |
|---|---|---|
| Mean | $\bar{x}$ | $\mu$ |
| Standard deviation | $s$ | $\sigma$ |

EXAMPLE 14-10

The ages of 20 college students are as follows:

$$21 \quad 22 \quad 24 \quad 23 \quad 25 \quad 23 \quad 22 \quad 22 \quad 26 \quad 22$$
$$23 \quad 21 \quad 23 \quad 24 \quad 25 \quad 23 \quad 24 \quad 23 \quad 22 \quad 24$$

For this group of people it is believed their ages are normally distributed. Compute the mean and standard deviation for the 20 students. If the 20 ages are added, their sum $\Sigma x = 462$. The mean of this sample of 20 students is

$$\text{mean } \bar{x} = \frac{\Sigma x}{n} = \frac{462}{20} = 23.1 \text{ years}$$

The standard deviation was once called the root-mean-square deviation for this is one of the methods of its calculation. The standard deviation is the square root of the sum of the square of the deviations about the mean, divided by the sample size minus 1.

$$\text{Standard deviation } s = \sqrt{\frac{\Sigma(x - \bar{x})^2}{n - 1}}$$

The 20 ages may be grouped and this equation used to find the standard deviation.

| Number in age group $n$ | Age group $x$ | $x - \bar{x}$ | $(x - \bar{x})^2$ | $n(x - \bar{x})^2$ |
|---|---|---|---|---|
| 2 | 21 | −2.1 | 4.4 | 8.8 |
| 5 | 22 | −1.1 | 1.2 | 6.0 |
| 6 | 23 | −0.1 | .0 | .0 |
| 4 | 24 | +0.9 | 0.8 | 3.2 |
| 2 | 25 | +1.9 | 3.6 | 7.2 |
| 1 | 26 | +2.9 | 8.4 | 8.4 |
| 20 | | | $\Sigma(x - \bar{x})^2 =$ | 33.6 |

$$s = \sqrt{\frac{\Sigma(x - \bar{x})^2}{n - 1}} = \sqrt{\frac{33.6}{19}} = 1.33$$

It is frequently easier to solve for the standard deviation when the equation is rewritten as follows:

$$\text{Standard deviation } s = \sqrt{\frac{\Sigma x^2}{n - 1} - \frac{(\Sigma x)^2}{n(n - 1)}}$$

In this problem

$$\Sigma x^2 = 21^2 + 23^2 + 22^2 + 21^2 + 24^2 + \cdots = 10{,}706$$
$$\Sigma x = 462$$

$$s = \sqrt{\frac{10706}{19} - \frac{462^2}{20(19)}} = \sqrt{563.47 - 561.69} = \sqrt{1.78} = 1.33$$

We have computed for the 20 college students $\bar{x} = 23.1$, $s = 1.33$. The sample mean and standard deviation are our best estimates of the population. We therefore estimate that $\mu = 23.1$ and $\sigma = 1.33$.

To define a particular normal distribution it is necessary to specify the mean $\mu$ and standard deviation $\sigma$. Suppose, for example, you believed that the useful life of a particular type of equipment was normally distributed with a mean life of 15 years and a standard deviation of 2.4 years. How could you obtain random samples from this distribution?

A convenient method is based on the standardized normal distribution. Figure 14-9 shows both the standardized normal distribution and the particular useful life normal distribution we want. The deviation of any value of $x$ from the mean of the distribution may be expressed in number of standard deviations.

$$z = \frac{x - \mu}{\sigma} \qquad \text{or} \qquad x = z\sigma + \mu$$

From this equation we see that any point $x$ on a specific distribution (with $\mu$ and $\sigma$) has an equivalent point $z$ on the standardized normal distribution. This relationship allows us to relate the standardized normal distribution to any other normal distribution.

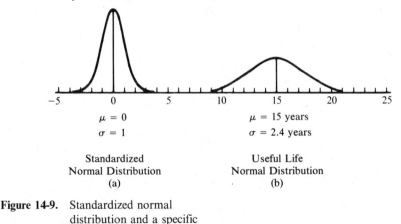

|  |  |
|---|---|
| $\mu = 0$ | $\mu = 15$ years |
| $\sigma = 1$ | $\sigma = 2.4$ years |
| Standardized Normal Distribution (a) | Useful Life Normal Distribution (b) |

**Figure 14-9.** Standardized normal distribution and a specific useful life normal distribution.

In our useful life example, two standard deviations above the mean would be at $x = 15 + 2(2.4) = 19.8$ years on the useful life distribution, Figure 14-9b. The equivalent point on the standardized normal distribution is

$$z = \frac{x - \mu}{\sigma} = \frac{19.8 - 15}{2.4} = +2.0$$

This interrelationship means that if we randomly sample from the standardized normal distribution, we can relate this to an equivalent random sample for any normal distribution.

A point on the standardized normal distribution ($\mu = 0$, $\sigma = 1$) is fully defined by specifying the number of standard deviations the point is to the left (negative) or to the right (positive) of the mean. Thus the value $+1.02$ would indicate the point is 1.02 standard deviations to the right of the mean. Table 14-2 represents random samples from the standardized normal distribution. The use of random normal numbers is illustrated in Example 14-11.

### EXAMPLE 14-11

On a particular portion of a highway, observations indicate that the speed of automobiles is normally distributed with mean $\mu = 104$ kilometers per hour and standard deviation $\sigma = 11$ kilometers per hour. Obtain a random sample of five automobiles on the highway.

From Table 14-2 randomly select a value of random normal number ($z$). The value of $z$ selected is $-0.94$. This value of $z$ represents a vehicle on the highway travelling at a speed of

$$x = z\sigma + \mu = -0.94(11) + 104 = 93.7 \text{ kilometers per hour}$$

**Table 14-2.** 100 random normal numbers ($z$)

| | | | | | | | | | |
|---|---|---|---|---|---|---|---|---|---|
| −0.22 | −0.87 | 2.32 | −0.94 | 0.63 | 0.81 | 0.74 | 1.08 | −1.82 | 0.07 |
| −0.89 | −0.39 | 0.29 | −0.27 | 1.06 | −0.42 | 2.26 | −0.35 | 1.09 | −2.55 |
| 0.06 | 1.28 | −1.74 | 2.47 | 0.58 | 0.69 | 1.41 | −1.19 | 2.37 | −0.06 |
| −0.01 | −0.40 | 0.64 | −2.22 | 1.10 | 0.47 | −0.09 | −0.35 | −0.72 | 0.30 |
| −0.87 | −1.34 | 0.85 | 0.27 | −1.35 | 0.58 | −1.72 | 1.88 | −0.45 | 0.82 |
| 0.24 | 0.40 | 0.50 | 1.41 | −1.95 | −0.02 | −1.00 | −0.20 | −1.08 | −0.78 |
| −1.05 | −0.06 | 0.27 | −0.04 | 0.99 | −0.78 | 0.46 | −1.18 | 0.37 | 1.07 |
| −0.57 | 0.24 | −1.02 | 0.86 | 0.78 | −1.69 | −0.17 | −0.23 | −0.87 | −0.45 |
| −1.24 | −0.63 | 0.03 | −0.83 | 0.25 | −0.89 | −0.77 | 0.90 | −0.27 | 0.94 |
| 2.11 | 0.78 | −1.69 | −0.17 | 0.21 | 0.48 | −2.82 | −0.86 | 1.40 | −1.20 |

Four other random normal numbers, 1.06, 0.69, $-0.09$, and 1.88 were selected. The indicated automobile speeds are

$x = 1.06(11) + 104 = 115.7$
$x = 0.69(11) + 104 = 111.6$
$x = -0.09(11) + 104 = 103.0$
$x = 1.88(11) + 104 = 124.7$

Through the use of random normal numbers the speed of five random automobiles has been computed.

## Sampling From Any Distribution
## Using Random Numbers

In the two previous sections we saw how to obtain a random sample from the uniform distribution and from the normal distribution. While these are two important situations, there are times when we want to obtain a random sample from some other distribution. We might, for example, wish to obtain a random sample from the grade distribution of Figure 14-10. The procedure is to replot the data to represent a cumulative distribution of grades. This has been done in Figure 14-11. Where the

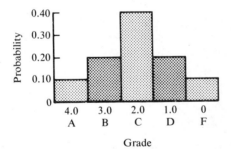

| Grade | Grade Point Average | Probability |
|-------|:-------------------:|:-----------:|
| A | 4.0 | 0.10 |
| B | 3.0 | 0.20 |
| C | 2.0 | 0.40 |
| D | 1.0 | 0.20 |
| F | 0 | 0.10 |
|   |   | 1.00 |

**Figure 14-10.** Distribution of grades.

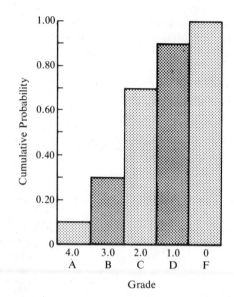

| Grade | Probability | Cumulative probability |
|-------|-------------|------------------------|
| A | 0.10 | 0.10 |
| B | 0.20 | 0.30 |
| C | 0.40 | 0.70 |
| D | 0.20 | 0.90 |
| F | 0.10 | 1.00 |

**Figure 14-11.** Cumulative distribution
  of grades.

*x*-axis represents all possible outcomes, as in this case, the cumulative probability on the *y*-axis must vary from 0 to 1.00. To facilitate random sampling, random numbers may be assigned to each segment of the ordinate in proportion to its probability. Figure 14-12 shows the data of Figure 14-11 with two-digit random numbers assigned to each segment of the *y*-axis.

To obtain a random sample from the distribution of grades (Figure 14-10), the first step is to randomly select a two-digit number from the table of random numbers, Table 14-1. This number is then entered as the ordinate in Figure 14-12. Read across from the ordinate to the curve and then read the corresponding value on the *x*-axis. The random number 35, for example, corresponds to a grade of C on the *x*-axis.

The procedure described for sampling from the grade distribution can be used for any discrete or continuous distribution. In fact, looking back

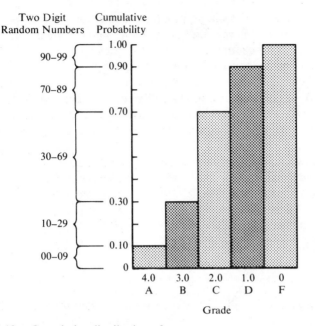

**Figure 14-12.** Cumulative distribution of grades with two digit random numbers assigned to the *y*-axis.

at the discussion of the uniform distribution reveals that we were actually using a cumulative distribution (Figure 14-6) to randomly sample from the uniform distribution.

## SIMULATION

Simulation may be described as the repetitive analysis of a mathematical model. When we can accurately predict all the various consequences of an investment project with precise estimates, a single computation will give us the results. Sometimes future events are stated in terms of a range of values like optimistic, most likely, and pessimistic estimates. In this circumstance one approach is to compute the results (like rate of return) for each of the three estimates. Here for the first time multiple computations are used to describe the results.

When values for an economic analysis are stated in terms of one or more probability distributions, the analysis is further complicated. The number of different combinations of values quickly becomes so large (often infinite) that one can no longer consider evaluating each possibility. The only practical solution is to randomly sample from each of the variables and compute the results. Is a single random sample adequate to portray the situation? The answer is yes and no. Surely we know more about the situation than if we had no random sample, but not much more.

We could compare this with a random sample of a piece of chicken from a bucket of chicken. If the one piece is a leg, does that mean the bucket consists only of chicken legs, or that all pieces are as good as the chicken leg? We simply cannot say. The random sample that produced the chicken leg proves at least that there was one in the bucket, but it says very little about the rest of the contents.

To obtain greater information about an economic analysis the procedure is to continue to take additional random samples for evaluation. This technique is simulation. In Example 14-9 we simulated the rolling of a die three times. We now see that this was a very simple simulation. Example 14-12 illustrates an economic analysis simulation.

## EXAMPLE 14-12

If a more accurate scales is installed on a production line it will reduce the error in computing postage charges and save $250 a year. The useful life of the scales is believed to be uniformly distributed and range from 12 to 16 years. The initial cost of the scales is estimated to be normally distributed with a mean of $1500 and a standard deviation of $150.

Simulate 25 random samples of the problem and compute the rate of return for each sample. Construct a graph of rate of return vs frequency of occurrence.

The useful life is uniformly distributed between 12 and 16 years. This is illustrated in Figure 14-13. The data are replotted as a cumulative distribution function in Figure 14-14.

We can assign 20 two-digit random numbers to each of the five possible useful lives as follows:

| Random numbers | Useful life |
|:---:|:---:|
| 00–19 | 12 yrs |
| 20–39 | 13 |
| 40–59 | 14 |
| 60–79 | 15 |
| 80–99 | 16 |

For the first random sample we randomly enter the random number table (Table 14-1) and read the number 82. We previously assigned this number to represent a useful life of 16 years.

Next, to sample from the normal distribution we read a random

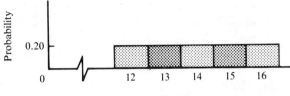

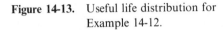

**Figure 14-13.** Useful life distribution for Example 14-12.

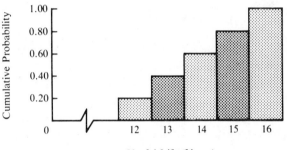

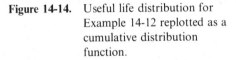

**Figure 14-14.** Useful life distribution for Example 14-12 replotted as a cumulative distribution function.

normal number of $-1.82$ from Table 14-2. To obtain the cost of the scales, multiply this value by the standard deviation and add the mean.

Cost of scales $= -1.82(\$150) + \$1500 = \$1227$.

With these values computed, the cash flow is:

| Year | Cash flow |
|------|-----------|
| 0    | $-\$1227$ |
| 1–16 | 250       |

For the cash flow the rate of return may be computed as follows:

PW of cost = PW of benefit
$$1227 = 250(P/A,i\%,16) \qquad i = 19\%$$

Table 14-3 shows the results when this process is repeated to obtain 25 values of rate of return. The rates of return shown in Table 14-3 are plotted in Figure 14-15. An additional 75 random samples have been computed (but not detailed here) and the data for all 100 samples are shown in Figure 14-16.

**Table 14-3.**   Results of 25 random samples for Example 14-12

| Random number | Useful life | Random normal number | Cost | Computed rate of return |
|---|---|---|---|---|
| 82 | 16 years | −1.82 | $1227 | 19% |
| 17 | 12 | 1.08 | 1662 | 11% |
| 35 | 13 | 0.74 | 1611 | 12% |
| 87 | 16 | 0.81 | 1622 | 13% |
| 21 | 13 | −0.42 | 1437 | 14% |
| 94 | 16 | 0.69 | 1604 | 14% |
| 49 | 14 | 0.47 | 1570 | 13% |
| 66 | 15 | 1.10 | 1665 | 12% |
| 74 | 15 | −2.22 | 1167 | 20% |
| 96 | 16 | 0.64 | 1596 | 14% |
| 10 | 12 | −0.40 | 1440 | 14% |
| 70 | 15 | −0.01 | 1498 | 14% |
| 09 | 12 | 0.06 | 1509 | 13% |
| 76 | 15 | 1.28 | 1692 | 12% |
| 34 | 13 | −0.39 | 1442 | 14% |
| 21 | 13 | −0.87 | 1370 | 15% |
| 06 | 12 | 2.32 | 1848 | 8% |
| 55 | 14 | 0.29 | 1544 | 13% |
| 03 | 12 | −0.27 | 1460 | 13% |
| 53 | 14 | 1.06 | 1659 | 12% |
| 40 | 14 | −0.42 | 1437 | 15% |
| 61 | 15 | 2.26 | 1839 | 11% |
| 41 | 14 | 1.41 | 1712 | 11% |
| 27 | 13 | −0.09 | 1486 | 14% |
| 19 | 12 | −1.72 | 1242 | 17% |

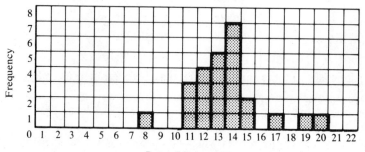

Rate of Return (%)

**Figure 14-15.**   Graph of rate of return vs
frequency for 25 random
samples in Example 14-12.

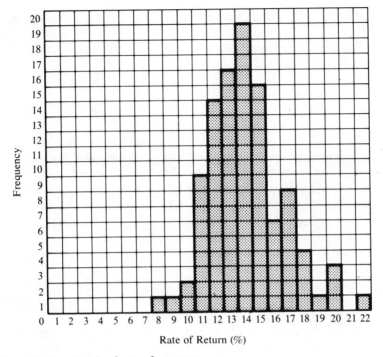

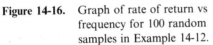

Rate of Return (%)

**Figure 14-16.** Graph of rate of return vs frequency for 100 random samples in Example 14-12.

Instead of recognizing the variability of the cost of the scales in Example 14-12, or its useful life, we might have used the single best estimates for them. The problem would have become:

| | |
|---|---|
| Cost of scales | $1500 |
| Annual benefit | $250 |
| Useful life | 14 yrs |

These values yield a 14% rate of return. Thus all our elaborate computations in this case did not change the results very much. The computations, however, give a picture of the possible variability of the results. Without recognizing the variability of the data and doing the simulation computations we would have no way of knowing the prospective variation in the rate of return. Simulation may add a dimension to our knowledge of a problem. We can learn about a projected result and also the likelihood that the result or some other one will occur.

From Example 14-12 it is apparent that if one goes from a situation of best estimates to one of probability distributions, the computations are multiplied about a hundred times. Fortunately, most of these calculations may be performed on a digital computer. In Chapter 18, a computer program is presented to solve simulation problems. This makes simulation a practical tool.

## SUMMARY

Estimation of the future is an important element of an engineer's work. In economic analysis we have several ways of describing the future. Precise estimates will not be exactly correct, but they are considered to be the best single values to represent what we think will happen.

A simple way to represent variability is through a range of estimates for future events. Frequently this is done by making three estimates: optimistic, most likely, and pessimistic. If the problem is solved three times, using first the optimistic values, followed by the most likely and pessimistic values, the full range of prospective results may be examined. The disadvantage of this approach is that it is extremely unlikely that either the optimistic results (based on optimistic estimates for all the components of the problem) or the pessimistic results will occur.

A variation to the solution of the range of estimates is to assign relative weights to the various estimates and solve the problem using the weighted values. The weights suggested are:

| Estimate | Relative weight |
|---|---|
| Optimistic | 1 |
| Most likely | 4 |
| Pessimistic | 1 |

When the probability of a future event is known or may be reasonably predicted, the technique of *expected value* may be used. Here the probabilities are applied as the relative weights.

Expected value = outcome$_A$ × probability$_A$ + outcome$_B$ × probability$_B$ + · · ·

Expected value is a useful technique in projecting the long term results when a situation occurs over and over again. For situations where the event will only occur once or a few times, expected value gives little insight into the infrequent event.

Two probability distributions are described. The uniform distribution is where each outcome has an equal probability of occurrence. The 52 different cards in a deck of cards, when shuffled, represent a uniform distribution. For computations, two-digit random numbers are a practical representation of the uniform distribution.

The normal distribution is possibly the best known of all distributions. A normal distribution is defined by its two parameters, the mean (a measure of central tendency), and standard deviation (a measure of the variability of the distribution). A random sample from any normal distribution may be obtained through the use of random normal numbers. A random sample may also be obtained from other distributions if the cumulative distribution function can be plotted.

Where the elements of an economic analysis are stated in terms of probability distributions, a repetitive analysis of a random sample is often done. This simulation is based on the premise that a random sampling of increasing size becomes a better and better estimate of the possible outcomes. The large number of computations limit the usefulness of the technique when hand methods are used.

## PROBLEMS

**14-1** Two instructors announced that they "grade on the curve," that is, give a fixed percentage of each of the various letter grades to each of their classes. Their curves are:

| Grade | Instructor A | Instructor B |
|-------|------|------|
| A | 10% | 15% |
| B | 15% | 15% |
| C | 45% | 30% |
| D | 15% | 20% |
| F | 15% | 20% |

If a student came to you and said that his objective was to enroll in the class where his expected grade point average would be greatest, which instructor would you recommend? (*Answer:* Instructor *A*)

**14-2** A man wants to determine whether or not to invest $1000 in a friend's speculative venture. He will do so if he thinks he can get his money back. He believes the probabilities of the various outcomes at the end of one year are:

| Result | Probability |
|--------|-------------|
| $2000 (double his money) | 0.3 |
| 1500 | 0.1 |
| 1000 | 0.2 |
| 500 | 0.3 |
| 0 (lose everything) | 0.1 |

What would be his expected outcome if he invests the $1000?

**14-3** The University football team has ten games scheduled for next season. The business manager wishes to estimate how much money the team can be expected to gross during the season, including any post-season "bowl game" to which they might be invited. From records for the past season and estimates by informed people. the business manager has assembled the following data.

| Situation | Probability | Situation | Gross income |
|---|---|---|---|
| Regular season | | Regular season | |
| Win 3 games | 0.10 | Win 5 or less | |
| Win 4 games | 0.15 | games | $250,000 |
| Win 5 games | 0.20 | | |
| Win 6 games | 0.15 | Win 6 to 8 | |
| Win 7 games | 0.15 | games | 400,000 |
| Win 8 games | 0.10 | | |
| Win 9 games | 0.07 | Win 9 or 10 | |
| Win 10 games | 0.03 | games | 600,000 |
| Play in post-season bowl game | 0.10 | Play in post-season bowl game | Additional income of $100,000 |

Based on the business manager's data, what is the expected gross income for the team next season? (*Answer:* $355,000)

**14-4** Telephone poles are an example of items that have varying useful lives. Telephone poles, once installed in a location, remain in useful service until one of a variety of events occur.

(a) Name three reasons why a telephone pole might be removed from useful service at a particular location.

(b) You are to estimate the total useful life of telephone poles. If the pole is removed from an original location while it is still serviceable, it will be installed elsewhere. Estimate the optimistic life, most likely life, and pessimistic life for telephone poles. What percentage of all telephone poles would you expect to have a total useful life greater than your estimated optimistic life?

**14-5** In the Nevada gaming casinos, the crap table is a popular gambling game. One of the many bets available is the "hard way 8" or "Big 8." A dollar bet in this fashion will win the player $4 if in the game the pair of dice come up 4 and 4 prior to one of the other ways of totaling 8. For a $1 bet, what is the expected result? (*Answer:* 80¢)

**14-6** When a pair of dice are tossed the results may be any number from 2 through 12. In the game of craps one can win by tossing either a 7 or an 11 on the first roll. What is the probability of doing this? Hint: There are 36 ways that a pair of 6 sided dice can be tossed. What portion of them result in either a 7 or an 11? (*Answer:* 8/36)

**14-7** Your grade point average for the current school term probably cannot yet be determined with certainty.

(a) Estimate the probability of obtaining a grade point average in each of the five categories below.

| Grade point average (A = 4.00) | Probability of obtaining the GPA |
|---|---|
| 0.00–0.80 | |
| 0.81–1.60 | |
| 1.61–2.40 | |
| 2.41–3.20 | |
| 3.21–4.00 | |

(b) Plot the data from part (a) with the grade point average as the *x*-axis and cumulative probability (of obtaining the GPA category or some lower one) as the *y*-axis.

(c) Assign the values between 00 and 99 to represent the cumulative probability 0.00–1.00 on the *y*-axis. Obtain a random number from Table 14-1 and determine the matching cumulative probability. Read and record the corresponding value of the GPA on the *x*-axis. In this manner, obtain 25 values of GPA. Graph the results as a bar graph with the *x*-axis as GPA and the *y*-axis as frequency. How does this bar graph compare with the table in part (a)?

**14-8** A decision has been made to perform certain repairs on the outlet works of a small dam. For a particular 36-inch gate valve, there are three available alternatives: (a) leave the valve as it is, (b) repair the valve, or (c) replace the valve.

If the valve is left as it is, the probability of a failure of the valve seats, over the life of the project, is 60%; the probability of failure of the valve stem is 50%; and the probability of failure of the valve body is 40%.

If the valve is repaired, the probability of a failure of the seats, over the life of the project, is 40%; of failure of the stem is 30%; and of failure of the body is 20%.

If the valve is replaced, the probability of a failure of the seats, over the life of the project, is 30%; of failure of the stem is 20%; and of failure of the body is 10%.

The present worth of cost of future repairs and service distruption of a failure of the seats is $10,000; the present worth of cost of a failure of the stem is $20,000; and the present worth of cost of a failure of the body is $30,000. The cost of repairing the valve now is $10,000; and of replacing it is $20,000. If the criterion is to minimize expected costs, which alternative is best?

**14-9** A man went to Las Vegas with $500 and placed 100 bets of $5 each on a number on the roulette wheel. There are 38 numbers on the wheel and the gaming casino pays 35 times the amount bet if the ball drops into the bettor's numbered slot in the roulette wheel. In addition, the bettor receives back the original $5 bet. Estimate how much money the man is expected to win or lose in Las Vegas.

328

**14-10** A heat exchanger is being installed as part of a plant modernization program. It cost $10,000, including installation, and is expected to reduce the overall plant fuel cost by $2500 per year. Estimates of the useful life of the heat exchanger range from an optimistic 12 years to a pessimistic 4 years. The most likely value is 5 years. Using the range of estimates, what is the estimated before-tax rate of return   (*Answer:* 13%)

**14-11** A factory building is located in an area subject to occasional flooding by a nearby river. You have been brought in as a consultant to determine whether or not floodproofing of the building is economically justified. The alternatives are as follows.

(a) Do nothing. Damage in a moderate flood is $10,000 and in a severe flood, $25,000.

(b) Alter the factory building at a cost of $15,000 to withstand moderate flooding without damage and to withstand severe flooding with $10,000 damages.

(c) Alter the factory building at a cost of $20,000 to withstand a severe flood without damage.

In any year the probability of flooding is as follows: 0.70, no flooding of the river; 0.20, moderate flooding; and 0.10, severe flooding. If interest is 15% and a 15-year analysis period is used, what do you recommend?

# 15
# Selection of a Minimum Attractive Rate of Return

In the chapters that precede this one we have said very little about what interest rate or minimum attractive rate of return is suitable for use in a particular situation. Since the problem is quite complex, there is no single answer that is always appropriate. A discussion of a suitable interest rate to use must inevitably begin with an examination of the sources of capital, followed by a look at the prospective investment opportunities. Only in this way can an intelligent decision be made on the choice of an interest rate or minimum attractive rate of return.

## SOURCES OF CAPITAL

In broad terms there are four sources of capital available to the firm. They are money generated from the operation of the firm, borrowed money, sale of mortgage bonds, and sale of capital stock.

### Money Generated from the Operation of the Firm

A major source of capital investment money is through the retention of profits resulting from the operation of the firm. Since only about half of the profits of industrial firms are paid out to stockholders, the half that is retained is an important source of funds for all purposes, including capital investments. In addition to profit, there is money generated in the business equal to the annual depreciation charges on existing capital assets. In other words, a profitable firm will generate money equal to its depreciation charges plus its retained profits. Even a firm that earns zero profit will still generate money from operations equal to its depreciation charges. (A firm with a loss, of course, will have still less funds.)

**External Sources of Money**

When a firm requires money for a few weeks or months it typically borrows the money from banks. Longer term unsecured loans of from one to four years may also be arranged through banks. While banks undoubtedly finance a lot of capital expenditures, regular bank loans cannot be considered a source of permanent financing.

Longer term secured loans may be obtained from banks, insurance companies, pension funds or even from the public. The security for the loan is frequently a mortgage on specific property of the firm. When sold to the public, this financing is by mortgage bonds. The sale of stock in the firm is still another source of money. While bank loans and bonds represent debt that has a maturity date, stock is a permanent addition to the ownership of the firm.

**Choice of Source of Funds**

The choice of the source of funds for capital expenditures is a decision for the board of directors and sometimes requires approval of the stockholders. In situations where internal operations generate adequate funds for the desired capital expenditures, external sources of money probably are seldom used. But when the internal sources are inadequate, then either external sources must be employed or the capital expenditures deferred or cancelled.

## COST OF BORROWED MONEY

A first step in deciding on a minimum attractive rate of return might be to determine the interest rate at which money can be borrowed. Longer term secured loans may be obtained from banks, insurance companies, or the variety of places where substantial amounts of money accumulates (like the oil producing nations).

A large, profitable corporation might be able to borrow money at the prime rate, that is, the interest rate that banks charge their best and most sought after customers. All other firms are charged a higher interest rate that might be from one-half to several percent higher. In addition to the financial strength of the borrower and his ability to repay the loan, the interest rate will vary depending on the duration of the loan.

## COST OF CAPITAL

Another interest rate that is frequently thought to be relevant is the cost of capital. The general assumption concerning the cost of capital is that all the money the firm uses for investments is drawn from all the components of the overall capitalization of the firm. The mechanics of the computation is given in Example 15-1.

## EXAMPLE 15-1

For a particular firm the purchasers of common stock require an 11% rate of return, mortgage bonds are sold at a 7% interest rate and bank loans are available at 9%. Compute the cost of capital for the following capital structure:

|  |  | Rate of return | Annual amount |
|---|---|---|---|
| $20 million | Bank loan | 9% | $1.8 million |
| 20 million | Mortgage bonds | 7% | 1.4 million |
| 60 million | Common stock and retained earnings | 11% | 6.6 million |
| $100 million |  |  | $9.8 million |

The annual amount is the amount of money required to satisfy the loan and bond requirement and to meet the stockholders' expectations. In this situation the calculated cost of $100 million of capital is $9.8 million after income taxes, or 9.8%. Thus the after-tax cost of capital is 9.8%. If the firm pays 50% income taxes, it must earn $13.2 million before taxes to equal the $6.6 million after taxes for its stockholders. The loan and bond interest payments are deductible business expenses, hence they are paid from before-tax funds of the firm. The cost of the $100 million of capital before taxes is $16.4 million (1.8 + 1.4 + 13.2) or 16.4%.

In an actual situation the cost of capital is difficult to compute. The fluctuation in the price of common stock, for example, makes it difficult to pick a cost, and the fluctuating prospects of the firm makes it even more difficult to estimate the future benefits the purchasers of the stock expect to receive. Given the fluctuating costs and prospects of future benefits, what rate of return do stockholders require? There is no precise answer, but we can obtain an approximate answer. Similar assumptions must be made for the other components of a firm's capitalization.

## INVESTMENT OPPORTUNITIES

An industrial firm has many more places that it can invest its money than does an individual. A firm has larger amounts of money and this alone makes certain kinds of investment possible that are unavailable to individual investors, with their more limited investment funds. The U.S. Government, for example, borrows money for short terms of 90 or 180 days by issuing certificates called treasury bills that frequently yield a greater interest rate than savings accounts. The customary minimum purchase is $25,000. More important, however, is the fact that a firm conducts a business and this business offers many investment opportunities. While exceptions can be found, a good generalization is that the opportunities for investment of money within the firm are superior to the investment opportunities outside the firm. Consider the following situation. A tabulation of the available investment opportunities for a particular firm is outlined in Table 15.1. A plot of these projects by rate of return vs investment is shown in Figure 15-1. The cumulative investment required for all projects at or above a given rate of return is given in Figure 15-2.

The two figures illustrate that a firm may have a broad range of investment opportunities available at varying rates of return. It may take some study and searching to identify the better investment projects available to a firm. If this is done, the available projects will almost certainly exceed the money the firm budgets for capital investment projects.

**Table 15-1**

| Project number | Project | Cost ($\times 10^3$) | Estimated rate of return |
|---|---|---|---|
| Investment Related to Current Operations | | | |
| 1 | New equipment to reduce labor costs | $150 | 30% |
| 2 | New equipment to reduce labor costs | 50 | 45% |
| 3 | Overhaul particular machine to reduce material costs | 50 | 38% |
| 4 | New test equipment to reduce defective products produced | 100 | 40% |
| New Operations | | | |
| 5 | Manufacture parts that previously had been purchased | 200 | 35% |
| 6 | Further processing of products previously sold in semifinished form | 100 | 28% |
| 7 | Further processing of other products | 200 | 18% |
| New Production Facilities | | | |
| 8 | Relocate production to new plant | 250 | 25% |
| External Investments | | | |
| 9 | Investment in a different industry | 300 | 20% |
| 10 | Investment in a different industry | 300 | 10% |
| 11 | Overseas investment | 400 | 15% |
| 12 | Buy Treasury bills | Unlimited | 8% |

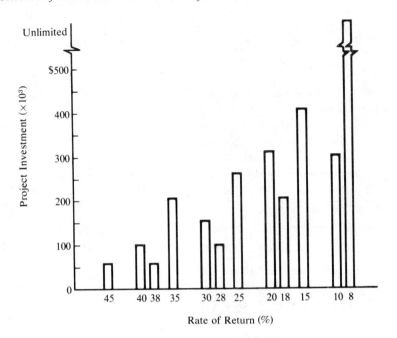

**Figure 15-1.** Rate of return vs project investment.

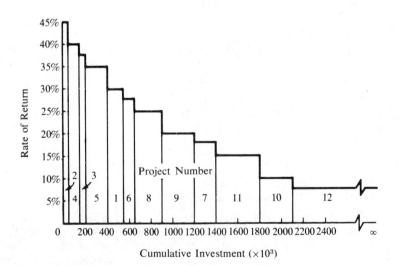

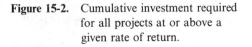

**Figure 15-2.** Cumulative investment required for all projects at or above a given rate of return.

## OPPORTUNITY COST

We have seen that there are two basically independent situations. One is the source and quantity of money available for capital investment projects. The other is the investment opportunities available to the firm. These are typically out-of-balance with investment opportunities exceeding the available money supply. Thus some of the investment opportunities can be selected and the rest must be rejected. Obviously, we want to insure that all the selected projects are better than the best rejected project. To do this we must know something about the rate of return on the best rejected project. The best rejected project is the best opportunity foregone and this is defined as the opportunity cost.

Opportunity cost = cost of the best opportunity foregone
= rate of return on the best rejected project

If one could predict in advance what the opportunity cost would be for some future period (like the next 12 months), this rate of return could be one way to judge whether to accept or reject any proposed capital expenditure.

### *EXAMPLE 15-2*

Consider the situation represented by Figures 15-1 and 15-2. For a capital expenditure budget of $1200 ($\times 10^3$), what is the opportunity cost? From Figure 15-2 we see that the eight projects with a rate of return of 20% or more require a cumulative investment of $1200 ($\times 10^3$). We would take on these projects and reject the other four (7, 11, 10, and 12) with rates of return of 18% or less. The best rejected project is number 7 and it has an 18% rate of return. This indicates the opportunity cost is 18%.

## SELECTING A MINIMUM
## ATTRACTIVE RATE OF RETURN

Using the three concepts on the cost of money (the cost of borrowed money, the cost of capital, and opportunity cost), which, if any, of these values should be used as the minimum attractive rate of return (MARR) in economic analyses?

Fundamentally, we know that unless the benefits of a project exceed the cost of the project, we cannot add to the profitability of the firm. A lower boundary for the minimum attractive rate of return must be the cost of the money invested in the project. It would be unwise, for example, to borrow money at 8% and invest it in a project yielding a 6% rate of return.

Further, we know that no firm has an unlimited ability to borrow money. Bankers, and others who evaluate the limits of a firm's ability to borrow money, look at both the profitability of the firm and the relationship between the components in the firm's capital structure. This means that continued borrowing of money will require that additional stock must be sold to maintain an acceptable ratio between ownership and debt. In other words, borrowing for a particular investment project is only a block of money from the overall capital structure of the firm. This suggests that the minimum attractive rate of return should not be less than the cost of capital. Finally, we know that the MARR should not be less than the rate of return on the best opportunity foregone. Stated simply,

Minimum attractive rate of return (MARR) should be equal to the highest one of the following: cost of borrowed money; cost of capital; or opportunity cost.

## ADJUSTING MARR TO ACCOUNT FOR RISK AND UNCERTAINTY

We know that in estimating the future, what actually occurs is often different from the estimate. When we are fortunate enough to be able to assign probabilities to the possible future outcomes, we call this a risk situation. We saw in Chapter 14 that techniques like expected value and simulation may be used when the probabilities are known.

Uncertainty is the term used to describe a different condition. When the probabilities are *not* known, this is called uncertainty. Thus if the probabilities of future outcomes are known we have risk, and if they are not known we have uncertainty.

One way to reduce the likelihood of undertaking projects that do not produce satisfactory results is to pass up marginal projects. In other words, no matter what projects are undertaken, some will turn out better than anticipated and some worse. Some undesirable results can be prevented by selecting only the best projects and avoiding those where the expected results are relatively close to a minimum standard. In theory, at least, the selected projects will provide results above the minimum standard even if they do considerably worse than anticipated.

In projects where there is normal business risk and uncertainty, the MARR is used without adjustment. For projects with greater risk or uncertainty, the MARR is increased. This is certainly not the best way to handle conditions of risk. A preferable way is to deal explicitly with the probabilities using one of the techniques of Chapter 14. This may be more acceptable as an adjustment for uncertainty. When the interest rate (MARR) used in economic analysis calculations is raised to adjust for risk or uncertainty, greater emphasis is placed on immediate or short term results and less emphasis on longer term results. This may not be an ideal method of risk or uncertainty adjustment and is illustrated by Example 15-3.

*EXAMPLE 15-3*

Consider the two following alternatives. The MARR has been raised from 10% to 15% to take into account the greater risk and uncertainty that the alternative B results may not be as favorable as indicated. What is the impact of this change of MARR on the decision?

| Year | Alternative A | Alternative B |
|------|---------------|---------------|
| 0 | −80 | −80 |
| 1–10 | 10 | 13.86 |
| 11–20 | 20 | 10 |

| Year | Alternative A | NPW at 14.05% | NPW at 10% | NPW at 15% |
|------|---------------|---------------|------------|------------|
| 0 | −80 | −80.00 | −80.00 | −80.00 |
| 1–10 | 10 | 52.05 | 61.45 | 50.19 |
| 11–20 | 20 | 27.95 | 47.38 | 24.81 |
|  |  | 0 | +28.83 | −5.00 |

| Year | Alternative B | NPW at 15.48% | NPW at 10% | NPW at 15% |
|------|---------------|---------------|------------|------------|
| 0 | −80 | −80.00 | −80.00 | −80.00 |
| 1–10 | 13.86 | 68.31 | 85.14 | 69.56 |
| 11–20 | 10 | 11.99 | 23.69 | 12.41 |
|  |  | 0 | +28.83 | +1.97 |

*Computations at MARR of 10% ignoring risk and uncertainty.* Both alternatives have the same positive NPW (+28.83) at a MARR of 10%. Also, the differences in the benefits schedules (A − B) produce a 10% incremental rate of return. (The calculations are not shown here.) This must be true if NPW for the two alternatives is to remain constant at a MARR of 10%.

*Considering risk and uncertainty with MARR of 10%.* At 10% both alternatives are equally desirable. Since B is believed to have greater risk and uncertainty, a logical conclusion is to select alternative A rather than B.

*Increase MARR to 15%.* At a MARR of 15% alternative A has a negative NPW and B has a positive NPW. Alternative B is preferred under these circumstances.

*Conclusion.* Based on a business-risk MARR of 10%, the two alternatives are equivalent. Recognizing some greater risk of failure for *B* makes *A* the preferred alternative. If the MARR is increased to 15%, to add a margin of safety against risk and uncertainty, the computed decision is to select *B*. Since *B* has been shown to be less desirable than *A*, the decision, based on a MARR of 15%, may be an unfortunate one. The difficulty is that the same risk adjustment (increase the MARR by 5%) is applied to both alternatives even though they have different amounts of risk.

The conclusion to be drawn from Example 15-3 is that increasing the MARR to compensate for risk and uncertainty is only an approximate technique and may not always achieve the desired result. Nevertheless, adjusting the MARR upward for increased risk and uncertainty is commonly done in industry.

## REPRESENTATIVE VALUES OF MARR USED IN INDUSTRY

We have seen that the minimum attractive rate of return (MARR) should be established at the highest one of the following: cost of borrowed money, cost of capital, or the opportunity cost.

The cost of borrowed money will vary from enterprise to enterprise with the lowest rate being the prime interest rate. The prime rate may change several times in a year and is widely reported in newspapers and business publications. The interest rate for firms that do not qualify for the prime interest rate may be $\frac{1}{2}\%$ to several percent higher.

The cost of capital of a firm is an elusive value. There is no widely accepted way to compute it, but we know that as a composite value for the capital structure of the firm, it conventionally is higher than the cost of borrowed money. The cost of capital must consider the market valuation of the shares (common stock, and so forth) of the firm, which may fluctuate widely, depending on future earnings prospects of the firm. We cannot generalize on representative costs of capital.

Somewhat related to cost of capital is the computation of the return on total capital (long term debt, capital stock, and retained earnings) actually achieved by firms. *Fortune* magazine, among others, does an annual analysis of the rate of return on total capital. The after-tax rate of return on total capital for individual firms ranges from 0% to about 40% and averages 8%. *Business Week* magazine does a periodic survey of corporate performance. It reports an after-tax rate of return on common stock and retained earnings. We would expect the values to be higher than the rate of return on total capital, and this is the case. The after-tax return on common stock and retained earnings ranges from 0% to about 65% with an average of 14%. None of the rates of return cited above indicates opportunity costs for enterprises or the MARR they require in economic analyses.

When discussing MARR, firms can usually be divided into two general groups. First, there are firms which are struggling along with an inadequate supply of investment capital, or are in an unstable situation or unstable industry. These firms cannot or do not invest money in anything but the most critical projects with very high rates of return and a rapid return of the capital invested. Often these firms use payback period and establish a criterion of one year or less, before income taxes. For an investment project with a five-year life, this corresponds to about a 60% after-tax rate of return. When these firms do rate of return analysis they reduce the MARR to possibly 25% to 30% after income taxes. There is potentially a substantial difference between a one year before-tax payback period and a 30% after-tax MARR, but this apparently does not disturb firms that specify this type of dual criteria.

The second group of firms represents the bulk of all enterprises. They are in a more stable situation and take a longer range view of capital investments. Their greater money supply enables them to invest in capital investment projects that firms in the first group will reject. Like the first group, this group of firms also uses payback and rate of return analysis. When small capital investments (of about $500 or less) are considered, payback period is often the only analysis technique used. The criterion for accepting a proposal may be a before-tax payback period not exceeding one or two years. Larger investment projects are analyzed by rate of return. Where there is a normal level of business risk, an after-tax MARR of 12% to 15% appears to be widely used. The MARR is increased when there is greater risk involved.

In Chapter 9 we saw that payback period is not a proper method for the economic analysis of proposals. Thus industrial use of payback criteria is *not* recommended. Fortunately, the trend in industry is toward greater use of accurate methods and less use of payback period.

The values of MARR given above are approximations. But the values quoted appear to be opportunity costs rather than cost of borrowed money or cost of capital. This indicates that firms cannot or do not obtain money to fund projects whose rates of return are nearer to the cost of borrowed money or cost of capital. While one could make a case that good projects are needlessly being rejected, there may be practical business reasons why firms operate as they do.

One cannot leave this section without noting that the MARR used by enterprises is so much higher than can be obtained by individuals. (Where can you get a 30% after-tax rate of return without excessive risk?) The reason appears to be that businesses are not faced with the intensively competitive situation that confronts an individual. There might be thousands of people in any region seeking a place to invest $2000 with safety, but how many people could (or would) want to invest $500,000 in a business? This diminished competition, combined with a higher risk, appears to explain at least some of the difference.

## SUMMARY

There are four general sources of capital available to an enterprise. The most important one is money generated from the operation of the firm. This has two components. There is the portion of profit that is retained in the business. In addition, a profitable firm generates funds equal to its depreciation charges that are available for reinvestment.

The three other sources of capital are from outside the operation of the enterprise.

1. Borrowed money from banks, insurance companies, and so forth.
2. Longer term borrowing from a lending institution or from the public in the form of mortgage bonds.
3. Sale of equity securities like common or preferred stock.

Retained profits and cash equal to depreciation charges are the primary sources of investment capital for most firms, and the only sources for many enterprises.

In selecting a value of MARR three values are frequently mentioned.

1. Cost of borrowed money.
2. Cost of capital. This is a composite cost of the components of the overall capitalization of the enterprise.
3. Opportunity cost. This refers to the cost of the opportunity foregone. Or stated more simply, opportunity cost is the rate of return on the best investment project that is rejected.

The MARR should be equal to the highest one of these three values.

When there is a risk aspect to the problem (probabilities are known), this can be handled by techniques like expected value and simulation. Where there is uncertainty (probabilities of the various outcomes are not known), there are analytical techniques, but they are less satisfactory. A method commonly used to adjust for risk and uncertainty is to increase the MARR. This method has the effect of distorting the time value of money relationship. The effect is to discount longer term consequences more heavily compared to short term consequences. This may or may not be desirable.

It is difficult to generalize about the values of the MARR used in industry. Some firms operate with a lack of adequate capitalization or in an unstable situation. For these firms major emphasis is given to the payback period rather than the long term rate of return. Typical investment criteria might be a one year before-tax payback period or a 25% to 30% after-tax rate of return.

Most firms operate in a more stable environment and make capital expenditures that would be rejected by less stable firms. For small capital investments (about $500 or less), before-tax payback period is used with a one or two year payback criterion for judging whether or not to make the investment. On larger projects after-tax rate of return is the most common analysis technique with the MARR equal to 12% to 15%. On projects with greater risk or uncertainty, the MARR is adjusted upward.

## PROBLEMS

**15-1** Examine the financial pages of your newspaper (or *The Wall St. Journal*) and determine the current interest rate on the following securities.

(a) U.S. Treasury bond due in five years.

(b) General obligation bond of a municipal district, city or a state due in 20 years.

(c) Corporate debenture bond of a U.S. industrial firm due in 20 years.

Explain why the interest rates are different for these different bonds.

**15-2** Consider four mutually exclusive alternatives.

|                         | A   | B     | C    | D    |
|-------------------------|-----|-------|------|------|
| Initial cost            | 0   | 100   | 50   | 25   |
| Uniform annual benefit  | 0   | 16.27 | 9.96 | 5.96 |
| Computed rate of return | 0%  | 10%   | 15%  | 20%  |

Each alternative has a ten year useful life and no salvage value. Over what range of interest rates is *C* the preferred alternative?   (*Answer:* $4.5\% < i \leqslant 9.6\%$)

**15-3** Frequently we read in the newspaper that one should lease an automobile rather than buying it. For a typical 24-month lease on a car costing $4400, the monthly lease charge is about $125 per month. At the end of the 24 months the car is returned to the lease company (which owns the car). As an alternative the same car could be bought with no down payment and 24 equal monthly payments, with interest at a 12% nominal annual percentage rate. At the end of 24 months the car is fully paid for. The car would then be worth about half of its original cost.

(a) Over what range of nominal before-tax interest rates is leasing the preferred alternative?

(b) What are some of the reasons that would make leasing more desirable than is indicated in (a)?

**15-4** Assume you have $2000 available for investment for a five-year period. You wish to *invest* the money—not just spend it on fun things. There are obviously many alternatives available. You should be willing to assume a modest amount of risk of loss of some or all of the money if this is necessary, but not a great amount of risk (no investments in poker games or at horse races). How would you invest the money? What is your minimum attractive rate of return? Explain.

**15-5** There are many venture capital syndicates that consist of a few (say eight or ten) wealthy people who combine to make investments in small and (hopefully) growing businesses. Typically, the investors hire a young investment manager (often an engineer with a MBA) who seeks and analyzes investment opportunities for the group. Would you estimate that the MARR sought by this group is more or less than 12%? Explain.

# 16
## Economic Analysis
## in
## Government

So far we have considered economic analysis where a firm pays all the investment costs and receives all the benefits. But when we examine the activities of government we find a quite different situation. They receive revenue through various forms of taxation and are expected to do things "in the public interest." Thus the government pays, but it receives few if any of the benefits. The original requirement for economic analysis of federal flood control projects contained what has become a famous line for describing benefits in federal projects* ". . . the benefits to whomsoever they may accrue . . . ." Unlike industrial firms the federal government pays the costs to obtain benefits for whomsoever might be fortunate enough to receive them.

Perhaps it appears desirable to construct a canal across Florida. The costs will be paid by the federal government. But it is clear that the federal government will not receive the benefits of the canal. Instead the benefits will be received by landowners in the area and the people who visit and enjoy the area being enhanced by the canal.

This situation can present all sorts of problems. For one it means that the intended beneficiaries of a federal project will be very anxious to get the project approved and funded. A second problem concerns the measurement of the benefits. It is obviously more difficult to measure actual benefits when they are so widely disseminated. Other difficulties are the selection of an interest rate and from whose viewpoint the analysis should be made. Finally, in benefit-cost ratio analysis there may be a problem deciding what goes in the numerator and what goes in the denominator of the ratio.

*Flood Control Act of 1936.

## FROM WHOSE VIEWPOINT
## SHOULD THE ANALYSIS BE MADE?

When governmental bodies do economic analysis an important question concerns the proper viewpoint of the analysis. A look at industry will help to explain the problem. Industrial economic analysis must also be based on a viewpoint, but in this case there is an obvious answer—a firm pays the costs and counts *its* benefits. Thus both the costs and benefits are measured from the point of view of the firm. Costs and benefits that occur outside of the firm generally are ignored. In years past, the consequences (costs and benefits) of a firm's actions that occurred outside the firm were called external consequences and ignored. (Ask anyone who has lived near a cement plant, a slaughterhouse, a steel mill, and so forth, about external consequences.) More recently government has forced industry to reduce pollution and other undesirable external consequences (see Figure 16-1). The result has been to force firms to take a larger or community viewpoint in evaluating the consequences of their actions.

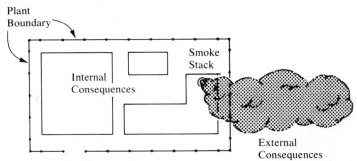

**Figure 16-1.**  Internal and external consequences for an industrial plant.

A small town that levys taxes and spends the money can be expected to take the viewpoint of the town in making decisions. Unless it can be shown that the money can be used effectively, it is unlikely the town council will spend it. But what happens when the money is contributed to the town by the federal government as in "revenue sharing" or some other federal grant? Often the federal government pays a share of project costs varying from 10% to 90%. Example 16-1 illustrates the viewpoint problem that is created.

*EXAMPLE 16-1*

A municipal project will cost $1 million. The federal government will pay 50% of the cost if the project is undertaken. Although the original economic analysis showed the PW of benefits was $1.5 million, a sub-sequent detailed analysis by the town engineer indicates a more realistic estimate of the PW of benefits is $750,000. The town council must decide whether or not to proceed with the project.

From the viewpoint of the town, the project is still a good one. If the town puts up half the cost ($500,000) it will receive all the benefits ($750,000). On the other hand, from an overall viewpoint the revised estimate of $750,000 of benefits does not justify the $1 million expenditure. This illustrates the dilemma caused by the varying viewpoints. For economic efficiency one does not want to encourage the expenditure of money, irrespective of the source, unless the benefits at least equal the costs.

The possible viewpoints are: individual or the firm, town or district in a city, city, state, nation, or international. To avoid suboptimizing, the proper approach is to take a viewpoint at least as broad as those who pay the costs and those who receive the benefits. When the costs and benefits are totally confined to a town, for example, that seems an appropriate viewpoint. But when the money or the benefits go beyond the proposed viewpoint, then the viewpoint should be enlarged to this broader view.

## WHAT INTEREST RATE?

Another thorny problem in economic analysis in government concerns the appropriate interest rate to use in an analysis. For industrial economic analyses some of the alternatives were to use the cost of borrowed money, the cost of capital, or the opportunity cost.

In government most of the money is obtained by taxation and spent about as quickly as it is obtained. There is little time delay between collecting the money from taxpayers and spending it. (Remember, the federal government and many states collect taxes from every paycheck in the form of tax withholding.) The collection of taxes and its disbursement, although based on an annual budget, is actually a continuing process. Using this line of reasoning, some people would argue that there is no interest rate involved because there is little or no time lag between collecting and spending the money. They would advocate a 0% interest rate.

Another approach to the "what interest rate?" question is to recognize that in addition to collecting taxes, most levels of government (federal, state or local) borrow money for capital expenditures. Where money is borrowed for a specific project, one line of reasoning is to use an interest rate equal to the cost of borrowed money. This is less valid an argument for state and local governments than for industrial firms. The reason is that the federal government, through the income tax laws, subsidizes state and local bonded debt. If a state or one of its political subdivisions (like a county, city, or special assessment district) raises money through the sale of bonds, interest paid on the bonds is generally not taxable income to the individual who owns them. This is an important benefit of these bonds (typically called "municipal bonds") with the result that people will naturally require a lower interest rate on these bonds than on similar bonds where the interest is fully taxable. As a rough estimate, when fully taxed bonds yield a 6% interest rate, municipal bonds or other tax-free bonds might have only a $4\frac{1}{2}\%$ interest rate. The difference of $1\frac{1}{2}\%$ represents the effect of the preferred treatment for federal income tax purposes, and hence a form of hidden federal subsidy on tax-free bonds.

Opportunity cost, which is the interest rate on the best opportunity foregone, may take two forms in governmental economic analysis. It may be the opportunity cost within government or the opportunity cost of the taxpayers from whom the money was obtained. We know that for economic efficiency one should select the best projects from among all the prospective projects. We would expect the rate of return on all accepted projects to be higher than the rate of return on any of the rejected projects. This is accomplished by setting the interest rate for use in economic analyses equal to the opportunity cost. If the interest rate is determined solely on alternative governmental use of the money, this might be called the governmental opportunity cost.

Another question is whether or not the money should be taken from the taxpayer for governmental use. Individual taxpayers have varying individual opportunity costs, but the opportunity cost for taxpayers as a group can be approximated. It is economically undesirable to take money from a taxpayer with a 9% opportunity cost, for example, and spend it on a governmental project yielding 6%.

From the foregoing discussion a reasonable conclusion is that in governmental economic analyses the interest rate should be the *largest* one of the following: cost of borrowed money (plus subsidy on tax-free bonds); governmental opportunity cost; or taxpayers opportunity cost.

The tendency in government has been to use relatively low interest rates for economy studies. The structure of typical government projects may partly explain the reason. A highway project, for example, represents a large initial expenditure with only a small continuing maintenance expenditure. The benefits increase over the years as the use of the highway is projected to increase. The present worth of benefits is very sensitive to the interest rate used in the analysis. This is illustrated in Example 16-2.

*EXAMPLE 16-2*

On a public project the annual benefits will be $50,000 beginning two years after start of construction and increasing each year by $50,000 on a uniform gradient. Assume a 30-year analysis period. What capital investment can be justified as of the beginning of construction (a) at a 3% interest rate, (b) at a 9% interest rate?

(a) For a benefit-cost ratio of one,

> Cost (equals PW of cost) = PW of benefits
>
> $\qquad\qquad\qquad\qquad\quad = \$50,000(P/G,3\%,30)$
>
> $\qquad\qquad\qquad\qquad\quad = \$50,000(241.4) = \$12,070,000$

(b) At a 9% interest rate

> Cost $= \$50,000(P/G,9\%,30)$
>
> $\qquad\quad = \$50,000(89.0) = \$4,450,000$

For a benefit-cost ratio of one or more, the construction cost may be as much as \$12,070,000 for a 3% interest rate, but only \$4,450,000 for a 9% interest rate.

## BENEFIT-COST RATIO ANALYSIS

Benefit-cost ratio analysis requires that all the various consequences of a proposed project be classified and placed into either the numerator or denominator of the ratio. At first glance this would appear simply a matter of sorting out the consequences into benefits (for the numerator) or costs (for the denominator). This works satisfactorily when applied to the projects of a firm or an individual. In governmental projects there may be difficulties deciding whether to classify various consequences as items for the numerator or the denominator. Consider Example 16-3.

*EXAMPLE 16-3*

On a proposed governmental project the following consequences have been identified.

1. Initial cost of project to be paid by government is 100.
2. Present worth of future maintenance to be paid by government is 40.
3. Present worth of benefits to the public is 300.
4. Present worth of additional public user costs is 60.

Show the various ways of computing the benefit-cost ratio.

Putting the benefits in the numerator and all the costs in the denominator gives:

$$\text{Benefit-cost ratio} = \frac{\text{all benefits}}{\text{all costs}} = \frac{300}{100 + 40 + 60} = \frac{300}{200} = 1.5$$

An alternate computation is to consider user costs a disbenefit and to subtract them in the numerator rather than adding them in the denominator.

$$\text{Benefit-cost ratio} = \frac{\overset{\text{public}}{\text{benefits}} - \overset{\text{public}}{\text{costs}}}{\text{governmental costs}} = \frac{300 - 60}{100 + 40} = \frac{240}{140} = 1.7$$

Still another variation would be to consider maintenance costs as a disbenefit.

$$\text{Benefit-cost ratio} = \frac{300 - 60 - 40}{100} = \frac{200}{100} = 2.0$$

It should be noted that while three different benefit-cost ratios may be computed, the value of NPW does not change.

$$\text{NPW} = \text{PW of benefits} - \text{PW of costs} = 300 - 60 - 40 - 100 = 100$$

There is no inherently correct way to compute the benefit-cost ratio. For highway projects the authoritative guide is the AASHO "Red Book."* In it the ratio is:

$$\text{Benefit-cost ratio} = \frac{\text{net public benefits}}{\text{government costs}}$$

which corresponds to the second of the three ratios in Example 16-3. Some analysts use the third of the ratios in Example 16-3. This is:

$$\text{Benefit-cost ratio} = \frac{\overset{\text{net public}}{\text{benefits}} - \overset{\text{governmental}}{\text{maintenance costs}}}{\text{governmental investment cost}}$$

The alternate methods of computing the benefit-cost ratio do not change desirable projects (B/C greater than one) into undesirable projects (B/C less than 1) or vice versa. The danger is that by changing the method of its computation one might infer that the resulting higher benefit-cost ratio represents a better project.

## OTHER ASPECTS

Economic analyses made to determine how to spend public monies bring together the full range of analysis difficulties. Since the people who will receive the benefits may pay none of the costs directly (one way or the other, of course, the public *does* pay for public expenditures), they may represent a strong pressure group to obtain the approval and funding of a particular

*American Association of State Highway Officials, *Road User Benefit Analysis For Highway Improvements*, Washington, D.C., 1960.

project. More recently a second and often opposing group of people have become increasingly effective in challenging the wisdom of many development projects. Often the struggle concerns the consequences that are not converted to money and incorporated in the analysis. As a result of the countervailing power of the two groups, it is likely that better decision-making is taking place.

From our examination of benefit-cost ratio analysis in Chapter 9, we know that incremental analysis is necessary in analyzing multiple alternatives with neither input nor output fixed. Unfortunately, some people who use benefit-cost ratio analysis are not aware of the need for incremental analysis.* If one goes to the trouble to make an analysis, it should be a fair and reasonably accurate picture of the situation.

## SUMMARY

Governmental economic analyses use the benefit-cost ratio method largely owing to federal legislation in the 1930's. It has become as familiar to people in government as rate of return is to people in industry.

The viewpoint to take in an analysis is a more difficult problem in government possibly because of the varied and sometimes remote sources of money. When some or all the money is state or federal funds, the appropriate viewpoint should be at this same level. The frequently localized nature of the benefits produces a local viewpoint that considers money from "outside" (Washington, D.C. or the state capital) as "free" because it need not be repaid from the local benefits. From the local viewpoint this seems to be true.

People are not in agreement on "what interest rate" to use. Theory suggests it should be the *largest* one of the following: cost of borrowed money (plus subsidy on tax-free bonds); governmental opportunity cost; or taxpayers opportunity cost. We would expect the opportunity cost to be substantially higher than the cost of borrowed money and therefore recommend that opportunity cost is the proper choice in this situation. The interest rates in actual use (one example is $6\frac{5}{8}\%$) appear to better reflect the cost of borrowed money than the opportunity cost. The typical situation of cost now and benefits later makes projects appear more desirable when lower interest rates are used. This is not unique to public projects, but the effect is probably exaggerated due to the relatively long analysis periods.

Aside from all the other difficulties with benefit-cost ratio analysis, we find that, like payback period, there is more than one way by which it may

---

*The author once almost lost a struggle with a Federal agency over whether or not incremental benefit-cost ratio analysis was required in a particular multiple alternative situation. It demonstrated that being right is no assurance that one will prevail.

be computed. The classification of some costs as disbenefits moves them from the denominator to the numerator. The result will be to increase the value of the ratio. While this has no real effect, it may create problems in deciding which of several projects is best, or otherwise mislead people.

## PROBLEMS

**16-1** Consider the following investment opportunity.

| | |
|---|---|
| Initial cost | $100,000 |
| Additional cost at end of year 1 | 50,000 |
| Benefit at end of year 1 | 0 |
| Annual benefits at end of years 2–10 | 20,000/yr |

With interest at 7%, what is the benefit-cost ratio for this project? (*Answer:* 0.83)

**16-2** A government agency has estimated that a flood control project has costs and benefits that are parabolic, according to the equation:

$$[\text{Present worth of benefits}]^2 - 22[\text{present worth of cost}] + 44 = 0$$

where both benefits and costs are stated in millions of dollars. What is the present worth of cost for the optimal size project?

**16-3** The Highridge Water District needs an additional supply of water from Steep Creek. The engineer has selected two plans for comparison:

*Gravity plan*: Divert water at a point ten miles up Steep Creek and carry it through a pipeline by gravity to the district.

*Pumping plan*: Divert water at a point near the district and pump it through 2 miles of pipeline to the district. The pumping plant can be built in two stages, with half capacity installed initially and the other half ten years later.

Use a 40-year analysis period and 8% interest. Salvage values can be ignored. During the first ten years, the average use of water will be less than during the remaining 30 years. Costs are as follows.

| | *Gravity* | *Pumping* |
|---|---|---|
| Initial investment | $2,800,000 | $1,400,000 |
| Additional investment in tenth year | 0 | 200,000 |
| Operation, maintenance, replacements | 10,000/yr | 25,000/yr |
| Power cost (average first 10 years) | 0 | 50,000/yr |
| Power cost (average next 30 years) | 0 | 100,000/yr |

Select the more economical plan. (*Answer:* Pumping plan)

**16-4** The Federal government proposes to construct a multi-purpose water project. This project will provide water for irrigation and for municipal uses. In addition, there will be flood control benefits and recreation benefits. The estimated project benefits computed for ten-year periods for the next 50 years are given in the table.

| Purpose | 1st decade | 2nd decade | 3rd decade | 4th decade | 5th decade |
|---|---|---|---|---|---|
| Municipal | $ 40,000 | $ 50,000 | $ 60,000 | $ 70,000 | $110,000 |
| Irrigation | 350,000 | 370,000 | 370,000 | 360,000 | 350,000 |
| Flood control | 150,000 | 150,000 | 150,000 | 150,000 | 150,000 |
| Recreation | 60,000 | 70,000 | 80,000 | 80,000 | 90,000 |
| Totals | $600,000 | $640,000 | $660,000 | $660,000 | $700,000 |

The annual benefits may be assumed to be one-tenth of the decade benefits. The operation and maintenance cost of the project is estimated to be $15,000 per year. Assume a 50-year analysis period with no net project salvage value.

(a) If an interest rate of 5% is used, and a benefit-cost ratio of unity, what capital expenditure can be justified to build the water project now?

(b) If the interest rate is changed to 8%, how does this change the justified capital expenditure?

**16-5** The city engineer has prepared two plans for the construction and maintenance of roads in the city park. Both plans are designed to provide the anticipated road and road maintenance requirements for the next 40 years. The minimum attractive rate of return used by the city is 7%. Plan *A* is a three stage development program. $300,000 is to be spent immediately, followed by $250,000 at the end of 15 years and $300,000 at the end of 30 years. Maintenance will be $75,000 per year for the first 15 years, $125,000 per year for the next 15 years, and $250,000 per year for the final 10 years. Plan *B* is a two stage program. $450,000 is required immediately (including money for some special equipment), followed by $50,000 at the end of 15 years. Maintenance will be $100,000 per year for the first 15 years and $125,000 for each of the subsequent years. At the end of 40 years it is believed the equipment may be sold for $150,000.

(a) Determine which plan should be chosen, using benefit-cost ratio analysis.

(b) If you favored plan *B*, what value of MARR would you want to use in the computations? Explain.

**16-6** The state is considering the elimination of a railroad grade crossing by building an overpass. The new structure would cost $900,000. The analysis period is assumed to be 30 years on the theory that either the railroad or the highway above it will be relocated by then. Salvage value of the bridge (actually the net value of the land on either side of the railroad tracks) 30 years hence is estimated to be $100,000. A 6% interest rate is to be used. About 1000 vehicles per day are delayed due to trains at the grade crossing. Trucks represent 40% and 60% are other vehicles. Time for truck drivers is valued at $9 per hour and for other drivers at $1.80 per hour. Average time saving per vehicle will be two minutes if the overpass is built. No time saving occurs for the railroad. The installation will save the railroad an annual expense of $10,000 now spent for crossing guards. During the ten-year preceding period the railroad has paid out $300,000 in settling lawsuits and accident cases related to the grade crossing. The proposed project will entirely eliminate both these expenses. The state estimates that the new overpass will save it about $2000 per year in expenses directly due to the accidents. The overpass, if built, will belong to the state. Should the overpass be built? If the overpass is built, how much should the railroad be asked to contribute to the state as its share of the $900,000 construction cost?

**16-7** An existing two-lane highway between two cities is to be converted to a four-lane divided freeway. The distance between them is ten miles. The average daily traffic (ADT) on the new freeway is forecast to average 20,000 vehicles per day over the next 20 years. Trucks represent 5% of the total traffic. Annual maintenance on the existing highway is $1500 per lane mile. The existing accident rate is 4.58 per million vehicle miles (MVM). Three alternate plans of improvement are now under consideration.

*Plan A*: Improve along the existing development by adding two lanes adjacent to the existing lanes at a cost of $450,000 per mile. It is estimated that this plan will reduce auto travel time by two minutes, and will reduce truck travel time by one minute when compared to the existing highway. The plan *A* estimated accident rate is 2.50 per MVM. Annual maintenance is estimated to be $1250 per lane mile.

*Plan B*: Improve along the existing alignment with grade improvements at a cost of $650,000 per mile. Plan *B* would add two additional lanes, and it is estimated that this plan would reduce auto and truck travel time by three minutes each, when compared to the existing facility. The accident rate on this improved road is estimated to be 2.40 per MVM. Annual maintenance is estimated to be $1000 per lane mile.

*Plan C*: Construct a new freeway on new alignment at a cost of $800,000 per mile. It is estimated that this plan would reduce auto travel time by five minutes, and truck travel time by four minutes when compared to the existing highway. Plan *C* is 0.3 miles longer than *A* or *B*. The estimated accident rate for *C* is 2.30 per MVM. Annual maintenance is estimated to be $1000 per lane mile. Plan *C* includes the abandonment of the existing highway with no salvage value.

Useful data: Operating cost—autos:      6¢ per mile
                     —trucks:     18¢ per mile
          Time saving   —autos:      3¢ per minute
                     —trucks:    15¢ per minute
         Average accident cost: $1200

If a 5% interest rate is used, which of the three proposed plans should be adopted? (*Answer:* Plan *C*)

# 17
# Rationing Capital Among Competing Projects

We have until now dealt with situations where at some interest rate there is an ample amount of money to make all desired capital investments. But the concept of scarcity of resources is fundamental to a free market economy. It is through this mechanism that more economically attractive activities are encouraged at the expense of less desirable activities. We see this situation in industrial firms. There are often more ways of spending money than money available. The result is that from the available alternatives, we must select the more attractive projects and reject, or at least delay, the less attractive projects.

This problem of rationing capital among competing projects is one part of a two-part problem called capital budgeting. An industrial firm in planning its capital expenditures is faced with two questions: where will money for capital expenditures come from and how shall we allocate available money among the various competing projects? In Chapter 14 we discussed the sources of money for capital expenditures as one aspect in deciding on an appropriate interest rate for economic analysis calculations. Thus, the first problem has been examined.

Throughout the book, for any project we have identified the best alternative from two or more feasible alternatives. We have, therefore, identified in each project the attractive alternative. And until now we have looked at these projects in an isolated setting. It has been almost as if the firm had just one project it was considering. In reality, we know that this is not the case. A firm will find that there are a great many projects that are economically attractive. This situation raises two problems not previously considered:

1.  How do you rank projects to show their order of economic attractiveness?

2. What do you do if there is not enough money to pay the costs of all economically attractive projects?

In this chapter we will look at the typical situation faced by a firm— multiple projects and an inadequate money supply to fund all attractive projects. To do this we must review our concept of capital expenditure situations and the available alternatives. Then we can summarize the various techniques that have been presented for determining if an alternative is economically attractive. We will start by screening all alternatives to find those that would merit further consideration. Following this, we will select the best alternative from each project, assuming there is no shortage of money. The next step will be the addition of a budget constraint. When there is not enough money to fund the best alternative from each project, we will have to do what we can with the limited amount of money available. It will become important that we have a technique for accurately ranking the various competing projects in their order of economic attractiveness. All this is designed to answer the question, "How shall we allocate available money among the various competing projects?"

## CAPITAL EXPENDITURE
## PROJECT PROPOSALS

At the beginning of the book we described decision-making as selecting the best alternative to achieve the desired objective in a given situation or problem. By carefully defining our objective, the model, and the choice criteria, the given situation was reduced to one of selecting the best from the feasible alternatives. In this chapter we will call this engineering decision-making process for a given situation or problem a *project proposal*. Associated with the various project proposals are alternatives. For a firm with many project proposals, the following situation may result.

### Capital Expenditure Proposals

*Project 1:* Additional manufacturing facility

Alternative *A*   Lease an existing building.

    *B*   Construct a new building.

    *C*   Contract for the manufacturing to be done overseas.

*Project 2:* Replace old grinding machine

Alternative *A*   Purchase semiautomatic machine.

    *B*   Purchase automatic machine.

*Project 3:* Production of parts for the assembly line

Alternative *A*   Make the parts in the plant.

    *B*   Buy the parts from a subcontractor.

Our task will be to apply economic analysis techniques to this more complex problem.

## Mutually Exclusive Alternatives and Single Project Proposals

Until now we have dealt with mutually exclusive alternatives, that is, where selecting one alternative results in rejecting the other alternatives being considered. Even in the simplest problems encountered, the question was one of selection between alternatives. Should, for example, machine *A* or machine *B* be purchased to perform the necessary task? Clearly, the purchase of one of the machines meant that the other one would not be purchased. Since either machine would perform the task, the selection of one precludes the possibility of selecting the other one as well. Even in the case of multiple alternatives, we have been considering mutually exclusive alternatives. A typical example was: What size pipeline should be installed to supply water to a remote construction site? Only one alternative is to be selected. This is different from the situation for single proposals where only one course of action is outlined. Consider Example 17-1.

### EXAMPLE 17-1

The general manager has received the following project proposals from the various operating departments.

1. The foundry wishes to purchase a new ladle to speed up their casting operation.
2. The machine shop has asked for some new inspection equipment.
3. The painting department reports they must make improvements to the spray booth to conform with new air pollution standards.
4. The office manager wants to buy a larger, more modern safe.

Each project consists of a single course of action. Note that the single project proposals are also independent, for there is no interrelationship or interdependence among them. We could decide to allocate money for none, some, or all of the various project proposals.

*Do-Nothing Alternative.*   We saw that the four project proposals above each had a single course of action. We could, for example, buy the office manager a new safe and buy the inspection equipment for the machine shop. But we could also decide not to buy the office manager a safe or the equipment for the machine shop. There is, then, an alternative to buying the safe for the office manager. We could decide not to buy him the safe—to do nothing. And we could decide to do nothing about the request for the machine shop inspection equipment. In fact, there are do-nothing alternatives for each of the four single project proposals.

This puts a new light on what we have defined as a single project proposal. In fact, a single proposal is really half of a pair of mutually exclusive project alternatives:

1*A*. Purchase the foundry a new ladle.

1*B*. Do nothing. (Do not purchase a new ladle.)

2*A*. Obtain the inspection equipment for the machine shop.

2*B*. Do nothing. (Do not obtain the inspection equipment.)

3*A*. Make improvements to the spray booth in the painting department.

3*B*. Do nothing. (Do not make the improvements.)

4*A*. Buy a new safe for the office manager.

4*B*. Do nothing. (Let him use the old safe!)

The single course of action and its do-nothing alternative are mutually exclusive. One can adopt 1*A* (buy the ladle) or 1*B* (do not buy the ladle), but not both. Alternatives 1*A* and 1*B* are mutually exclusive. We find that what we considered to be a single course of action is really a pair of mutually exclusive alternatives. Even alternative 3 is in this category. The originally stated single proposal was:

"The painting department reports that they must make improvements to the spray booth to conform with new air pollution standards."

Since the painting department reports they *must* make the improvements, is there actually another alternative? Although at first glance we might not think so, the company does have one or more alternatives available. It may be possible to change the paint, or the spray equipment, and thereby solve the air pollution problem without any improvements to the spray booth. In this situation there does not seem to be a practical do-nothing alternative, for failure to comply with the air pollution standards might result in large fines or even shutting down the plant. But if there is not a practical do-nothing alternative, there might be a number of do-something-else alternatives. We conclude that all project proposals may be considered to have mutually exclusive alternatives.

## Identifying and Rejecting Unattractive Alternatives

It is clear that no matter what the circumstances may be, we want to eliminate from further consideration any alternative that fails to meet the minimum level of economic attractiveness, provided one of the other alternatives does meet the criterion. Table 17-1 summarizes five techniques that may be used.

At first glance it appears that many calculations are required, but this is not the situation. *Any* of the five techniques listed in Table 17-1 may be used to determine whether or not to reject an alternative. Each will produce the same decision regarding reject–don't reject.

**Table 17-1.**  Criteria for rejecting unattractive alternatives

| For each alternative compute | Reject alternative when | Do not reject alternative when |
|---|---|---|
| Rate of return, $i$ | $i <$ MARR | $i \geqslant$ MARR |
| Present worth, PW | PW of benefits $<$ PW of costs | PW of benefits $\geqslant$ PW of costs |
| Annual cost, EUAC Annual benefit, EUAB | EUAC $>$ EUAB | EUAC $\leqslant$ EUAB |
| Benefit-cost ratio, B/C | B/C $< 1$ | B/C $\geqslant 1$ |
| Net present worth, NPW | NPW $< 0$ | NPW $\geqslant 0$ |

**Table 17-2.**  Criteria for choosing the best alternative from among mutually exclusive alternatives

| Analysis method | Situation | | |
|---|---|---|---|
| | Fixed input (The cost of each alternative is the same) | Fixed output (The benefits from each alternative are the same) | Neither input nor output fixed (Neither the costs nor the benefits for each alternative are the same) |
| Present worth | Maximize present worth of benefits | Minimize present worth of cost | Maximize net present worth |
| Annual cash flow | Maximize equivalent uniform annual benefits | Minimize equivalent uniform annual cost | Maximize (EUAB − EUAC) |
| Benefit-cost ratio | Maximize benefit-cost ratio | Maximize benefit-cost ratio | Incremental benefit-cost ratio analysis is required |
| Rate of return | Incremental rate of return analysis is required | | |

## Selecting the Best Alternative from Each
## Project Proposal

The task of selecting the best alternative from among two or more mutually exclusive alternatives has been a primary subject of this book. Since a project proposal is this same form of problem, we may use any of the several methods discussed in Chapters 5 through 9. The criteria are summarized in Table 17-2.

# RATIONING CAPITAL BY
# RATE OF RETURN

One way of looking at the capital rationing problem is through the use of rate of return. The technique for selecting from among independent projects may be illustrated by an example.

*EXAMPLE 17-2*

Nine independent projects are being considered. From the following data Figure 17-1 may be prepared.

| Project | Cost | Uniform annual benefit | Useful life | Salvage value | Computed rate of return |
|---|---|---|---|---|---|
| 1 | $100 | $23.85 | 10 yrs | $ 0 | 20% |
| 2 | 200 | 39.85 | 10 | 0 | 15% |
| 3 | 50 | 34.72 | 2 | 0 | 25% |
| 4 | 100 | 20.00 | 6 | 100 | 20% |
| 5 | 100 | 20.00 | 10 | 100 | 20% |
| 6 | 100 | 18.00 | 10 | 100 | 18% |
| 7 | 300 | 94.64 | 4 | 0 | 10% |
| 8 | 300 | 47.40 | 10 | 100 | 12% |
| 9 | 50 | 7.00 | 10 | 50 | 14% |

If a capital budget of $650 is available, which projects should be selected?
Looking at the nine projects, we see that some are expected to produce a larger rate of return than others. It is natural that if we are to select from among them, we will pick those with a higher rate of return. When the projects are arrayed by rate of return, as in Figure 17-1, the choice of project 3, 1, 4, 5, 6, and 2 is readily apparent, and is a correct decision.

In Example 17-2 the rate of return was computed for each project and then the projects were arranged in order of decreasing rate of return. For a fixed amount of money in the capital budget, the projects are selected by going

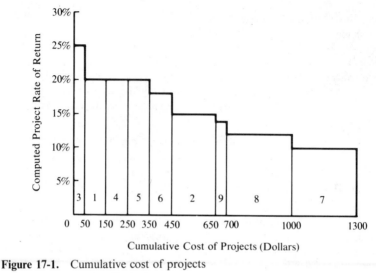

**Figure 17-1.** Cumulative cost of projects
vs rate of return

down the list until the money is exhausted. Using this procedure the point where the money runs out is the point where we cutoff approving projects. The rate of return at this point is called the *cutoff rate of return*. Figure 17-2 illustrates the general situation.

For any set of ranked projects and any capital budget, the rate of return at which the budget is exhausted is called the cutoff rate of return. In Figure 17-2 the cost of the individual projects is small compared to the capital budget. The cumulative cost curve is a relatively smooth curve producing a specific cutoff rate of return. Looking back at Figure 17-1 in Example 17-2, we see the curve is actually a step function. For Example 17-2 the cutoff rate of return is between 14% and 15% for a capital budget of $650.

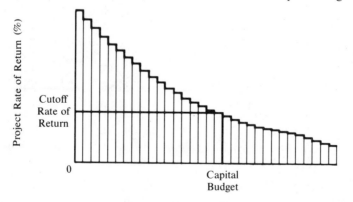

**Figure 17-2.** Location of the cutoff rate of
return

### Significance of the Cutoff Rate of Return

Cutoff rate of return is determined by the comparison of an established capital budget and the available projects. One must examine all the projects and all the money for some period of time (like an annual budget) to compute the cutoff rate of return. It is a computation relating known projects with a known money supply. For this period of time the cutoff rate of return is the opportunity cost (rate of return on the opportunity or project foregone) and also the minimum attractive rate of return. In other words, the minimum attractive rate of return to get a project accomplished is the cutoff rate of return.

$$\text{MARR} = \text{cutoff rate of return} = \text{opportunity cost}$$

We generally use the minimum attractive rate of return to decide whether or not to approve an individual project even though we do not know exactly what other projects will be proposed during the year. In this situation we cannot know if the MARR is equal to the cutoff rate of return. When MARR is different from the cutoff rate of return, incorrect decisions may occur. This will be illustrated in the next section.

## RATIONING CAPITAL BY PRESENT WORTH METHODS

Throughout this book we have choosen from among project alternatives to maximize net present worth. If we can do the same thing for a group of projects, and we do not exceed the available money supply, then the capital budgeting problem is solved.

But more frequently, the capital budgeting problem is one where we will be unable to accept all desirable projects. We, therefore, have a task not previously encountered. We must choose the best from the larger group of acceptable projects.

Lorie and Savage* showed that the proper technique is to use a multiplier $p$ to decrease the attractiveness of an alternative in proportion to its use of the scarce supply of money. The revised criterion is

$$\text{NPW} - p(\text{PW of cost}) \qquad \text{where } p \text{ is a multiplier computed by trial and error.}$$

If a value of $p$ were selected (say 0.1) then some alternatives with a positive NPW will have a negative $\text{NPW} - p(\text{PW of cost})$. This new criterion will reduce the number of favorable alternatives and thereby tend to reduce the cost of the projects meeting this more severe criterion. By trial and error

---

*Lorie, J. and L. Savage, "Three Problems in Rationing Capital," *Journal of Business*, October, 1955, pp. 229–239.

the multiplier *p* is adjusted until the cost of the projects meeting the NPW −
*p*(PW of cost) criterion equals the available money supply (the capital
budget).

## EXAMPLE 17-3

Using the present worth method, determine which of the nine independent projects of Example 17-2 should be included in a capital budget of
$650. The minimum attractive rate of return has been set at 8%.

| Project | Cost | Uniform annual benefit | Useful life | Salvage value | Computed NPW |
|---|---|---|---|---|---|
| 1 | $100 | $23.85 | 10 yrs | $ 0 | $60.04 |
| 2 | 200 | 39.85 | 10 | 0 | 67.40 |
| 3 | 50 | 34.72 | 2 | 0 | 11.91 |
| 4 | 100 | 20.00 | 6 | 100 | 55.48 |
| 5 | 100 | 20.00 | 10 | 100 | 80.52 |
| 6 | 100 | 18.00 | 10 | 100 | 67.10 |
| 7 | 300 | 94.64 | 4 | 0 | 13.46 |
| 8 | 300 | 47.40 | 10 | 100 | 64.38 |
| 9 | 50 | 7.00 | 10 | 50 | 20.13 |

Locating a value of *p* in NPW − *p*(PW of cost) by trial and error:

| Project | Cost | Computed NPW | Trial p = 0.20 NPW − p(PW of cost) | | Trial p = 0.25 NPW − p(PW of cost) | |
|---|---|---|---|---|---|---|
| 1 | $100 | $60.04 | $40.04 | cost $100 | $35.04 | cost $100 |
| 2 | 200 | 67.40 | 27.40 | 200 | 17.40 | 200 |
| 3 | 50 | 11.91 | 1.91 | 50 | −0.59 | |
| 4 | 100 | 55.48 | 35.48 | 100 | 30.48 | 100 |
| 5 | 100 | 80.52 | 60.52 | 100 | 55.52 | 100 |
| 6 | 100 | 67.10 | 47.10 | 100 | 42.10 | 100 |
| 7 | 300 | 13.46 | −46.54 | | −61.54 | |
| 8 | 300 | 64.38 | 4.38 | 300 | −10.62 | |
| 9 | 50 | 20.13 | 10.13 | 50 | 7.63 | 50 |
| | $1300 | | | $1000 | | $650 |

For a value of *p* equal to 0.25 the best set of projects is computed to
be 1, 2, 4, 5, 6, and 9.

This answer does not agree with the solution obtained in Example 17-2.
The difficulty is that the interest rate used in the present worth calculations is not equal to the computed cutoff rate of return. In Example 17-2
the cutoff rate of return was between 14% and 15%, say 14.5%. We will
recompute the present worth solution using MARR = 14.5%.

| Project | Cost | Computed NPW at 14.5% | Cost of projects with positive NPW |
|---|---|---|---|
| 1 | $100 | $22.01 | $100 |
| 2 | 200 | 3.87 | 200 |
| 3 | 50 | 6.81 | 50 |
| 4 | 100 | 21.10 | 100 |
| 5 | 100 | 28.14 | 100 |
| 6 | 100 | 17.91 | 100 |
| 7 | 300 | −27.05 | |
| 8 | 300 | −31.69 | |
| 9 | 50 | −1.28 | |
| | | | $650 |

At a MARR of 14.5% the best set of projects is the same as computed in Example 17-2, namely, projects 1, 2, 3, 4, 5, and 6, and their cost equals the capital budget. One can see that only projects with a rate of return greater than MARR can have a positive NPW at this interest rate. With MARR equal to the cutoff rate of return we *must* obtain the same solution by either the rate of return or present worth methods.

Figure 17-3 outlines the present worth method for the more elaborate case where there are independent projects each with mutually exclusive alternatives.

## EXAMPLE 17-4

A company is preparing its capital budget for next year. The amount has been set at $250 by the Board of Directors. The MARR of 8% is believed to be close to the cutoff rate of return. The following project proposals are being considered.

| Project proposals | Cost | Uniform annual benefit | Salvage value | Useful life | Computed NPW |
|---|---|---|---|---|---|
| Proposal 1 | | | | | |
| Alternative A | $100 | $23.85/yr | $ 0 | 10 yrs | $60.04 |
| B | 150 | 32.20 | 0 | 10 | 66.06 |
| C | 200 | 39.85 | 0 | 10 | 67.40 |
| D | 0 | 0 | | | 0 |
| Proposal 2 | | | | | |
| Alternative A | 50 | 14.92 | 0 | 5 | 9.57 |
| B | 0 | 0 | | | 0 |

| Project proposals | Cost | Uniform annual benefit | Salvage value | Useful life | Computed NPW |
|---|---|---|---|---|---|
| Proposal 3 | | | | | |
| Alternative *A* | 100 | 18.69 | 25 | 10 | 36.99 |
| *B* | 150 | 19.42 | 125 | 10 | 38.21 |
| *C* | 0 | 0 | | | 0 |

Which project alternatives should be selected, based on present worth methods?

### *p = 0.10*

| Project proposals | Cost | NPW | Alternative with largest positive NPW Alt. | Cost | NPW − p(PW of cost) | Alternative with largest positive NPW − p(PW of cost) Alt. | Cost |
|---|---|---|---|---|---|---|---|
| Proposal 1 | | | | | | | |
| Alternative *A* | $100 | $60.04 | | | $50.04 | | |
| *B* | 150 | 66.06 | | | 51.06 | 1*B* | $150 |
| *C* | 200 | 67.40 | 1*C* | $200 | 47.40 | | |
| *D* | 0 | 0 | | | 0 | | |
| Proposal 2 | | | | | | | |
| Alternative *A* | 50 | 9.57 | 2*A* | 50 | 4.57 | 2*A* | 50 |
| *B* | 0 | 0 | | | 0 | | |
| Proposal 3 | | | | | | | |
| Alternative *A* | 100 | 36.99 | | | 26.99 | 3*A* | 100 |
| *B* | 150 | 38.21 | 3*B* | 150 | 23.21 | | |
| *C* | 0 | 0 | | | 0 | | |
| | | | | $400 | | | $300 |

| | |
|---|---|
| Total required money too much, let *p* = 0.10 and recompute | Total still too much, increase *p* to 0.15 and recompute |

### *p = 0.15*

| Project proposals | Cost | NPW | NPW − p(PW of cost) | Alternative with largest positive NPW − p(PW of cost) Alt. | Cost |
|---|---|---|---|---|---|
| Proposal 1 | | | | | |
| Alternative *A* | $100 | $60.04 | $45.04 | 1*A* | $100 |
| *B* | 150 | 66.06 | 43.56 | | |
| *C* | 200 | 67.40 | 37.40 | | |
| *D* | 0 | 0 | 0 | | |
| Proposal 2 | | | | | |
| Alternative *A* | 50 | 9.57 | 2.07 | 2*A* | 50 |
| *B* | 0 | 0 | 0 | | |
| Proposal 3 | | | | | |
| Alternative *A* | 100 | 36.99 | 21.99 | 3*A* | 100 |
| *B* | 150 | 38.21 | 15.71 | | |
| *C* | 0 | 0 | 0 | | |
| | | | | | $250 |

This is the desired set of of projects

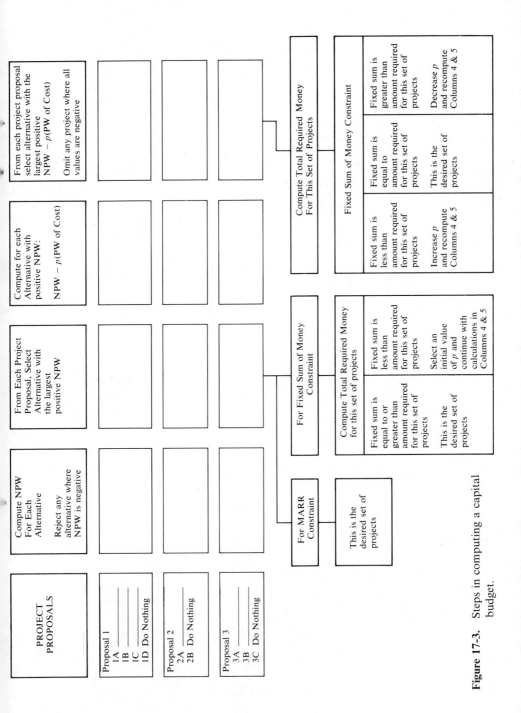

**Figure 17-3.** Steps in computing a capital budget.

## EXAMPLE 17-5

Solve Example 17-4 by rate of return methods. For project proposals with two or more alternatives, incremental rate of return analysis is required. The data from Example 17-4 and the computed rate of return for each alternative and each increment of investment is shown in the tabulation below.

| | | | | | | | *Incremental analysis* | | |
|---|---|---|---|---|---|---|---|---|---|
| | | *Uniform* | | | *Computed* | | *Uniform* | | *Computed* |
| | | *annual* | *Salvage* | *Useful* | *rate of* | | *annual* | *Salvage* | *rate of* |
| | *Cost* | *benefit* | *value* | *life* | *return* | *Cost* | *benefit* | *value* | *return* |
| Proposal 1 | | | | | | | | | |
| A | $100 | $23.85/yr | $0 | 10 yrs | 20.0% | | | | |
| B–A | | | | | | $50 | $8.35/yr | $0 | 10.6% |
| B | 150 | 32.20 | 0 | 10 | 17.0% | | | | |
| C–B | | | | | | 50 | 7.65 | 0 | 8.6% |
| C–A | | | | | | 100 | 16.00 | 0 | 9.6% |
| C | 200 | 39.85 | 0 | 10 | 15.0% | | | | |
| D | 0 | 0 | 0 | | 0% | | | | |
| Proposal 2 | | | | | | | | | |
| A | 50 | 14.92 | 0 | 5 | 15.0% | | | | |
| B | 0 | 0 | 0 | | 0% | | | | |
| Proposal 3 | | | | | | | | | |
| A | 100 | 18.69 | 25 | 10 | 15.0% | | | | |
| B–A | | | | | | 50 | 0.73 | 100 | 8.3% |
| B | 150 | 19.42 | 125 | 10 | 12.0% | | | | |
| C | 0 | 0 | 0 | | 0% | | | | |

The various separable increments of investment may be ranked by rate of return. They are plotted in a cumulative cost vs rate of return graph in Figure 17-4. The ranking of projects by rate of return gives the following.

*Project*

1A

2A

3A

1B in place of 1A

1C in place of 1B

3B in place of 3A

For a budget of $250 the selected projects are 1A, 2A, and 3A. Note that if a budget of $300 were available, 1B would replace 1A, making the proper set of projects 1B, 2A, and 3A. At a budget of $400, 1C would replace 1B and 3B would replace 3A, making the selected projects 1C, 2A, and 3B. These answers agree with the computations in Example 17-4.

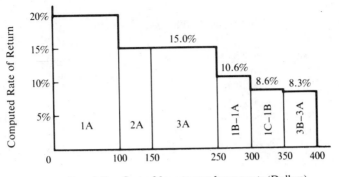

Cumulative Cost of Investment Increments (Dollars)

**Figure 17-4.**  Cumulative cost vs incremental rate of return.

## RANKING PROJECT PROPOSALS

Closely related to the problem of capital budgeting is the matter of ranking project proposals. We will first examine a method of ranking by present worth methods and then show that project rate of return is not a suitable method of ranking projects.

Anyone who has ever bought firecrackers probably used the practical ranking criterion of "biggest bang for the buck" in selecting the fireworks. This same criterion, stated more eloquently, may be used to correctly rank independent projects.

Rank independent projects according to their value of net present worth divided by the present worth of cost. The appropriate interest rate is MARR (as a reasonable estimate of the cutoff rate of return).

Example 17-6 illustrates the method of computation.

*EXAMPLE 17-6*

Rank the following nine independent projects in their order of desirability, based on a 14.5% minimum attractive rate of return. (To facilitate matters the necessary computations are included in the tabulation.)

| Project | Cost | Uniform annual benefit | Useful life | Salvage value | Computed rate of return | Computed NPW at 14.5% | Computed NPW/cost |
|---|---|---|---|---|---|---|---|
| 1 | $100 | $23.85 | 10 yrs | $ 0 | 20% | $22.01 | 0.2201 |
| 2 | 200 | 39.85 | 10 | 0 | 15% | 3.87 | 0.0194 |

| Project | Cost | Uniform annual benefit | Useful life | Salvage value | Computed rate of return | Computed NPW at 14.5% | Computed NPW/cost |
|---------|------|------------------------|-------------|---------------|-------------------------|-----------------------|-------------------|
| 3 | 50 | 34.72 | 2 | 0 | 25% | 6.81 | 0.1362 |
| 4 | 100 | 20.00 | 6 | 100 | 20% | 21.10 | 0.2110 |
| 5 | 100 | 20.00 | 10 | 100 | 20% | 28.14 | 0.2814 |
| 6 | 100 | 18.00 | 10 | 100 | 18% | 17.91 | 0.1791 |
| 7 | 300 | 94.64 | 4 | 0 | 10% | −27.05 | −0.0902 |
| 8 | 300 | 47.40 | 10 | 100 | 12% | −31.69 | −0.1056 |
| 9 | 50 | 7.00 | 10 | 50 | 14% | −1.28 | −0.0256 |

Ranked by NPW/PW of cost, the projects are listed.

| Project | $\dfrac{NPW}{PW \text{ of cost}}$ | Rate of return |
|---------|-----------------------------------|----------------|
| 5 | 0.2814 | 20% |
| 1 | 0.2201 | 20% |
| 4 | 0.2110 | 20% |
| 6 | 0.1791 | 18% |
| 3 | 0.1362 | 25% |
| 2 | 0.0194 | 15% |
| 9 | −0.0256 | 14% |
| 7 | −0.0902 | 10% |
| 8 | −0.1056 | 12% |

The rate of return tabulation illustrates that it is *not* a satisfactory ranking criterion and would have given a different ranking from the present worth criterion.

In Example 17-6 the projects are ranked according to the ratio NPW/PW of cost. In Figure 17-3 the criterion used is NPW − $p$(PW of cost). If one were to compute the value of $p$ at which NPW − $p$(PW of cost) = 0, we would obtain $p$ = NPW/PW of cost. Thus the multiplier $p$ is the ranking criterion at the point where NPW − $p$(PW of cost) = 0.

If independent projects can be ranked in their order of desirability, then selection of projects to be included in a capital budget is a simple task. One may proceed down the list of ranked projects until the capital budget is exhausted. The only difficulty with this scheme occasionally occurs when the capital budget is more than enough for $n$ projects, but too little for $n + 1$ projects. In Example 17-6 a capital budget of $300 is just right to fund the top three projects. But a capital budget of $550 is more than enough for the top five projects (sum = $450) but not enough for the top six projects (sum = $650). When we have this lumpiness problem, one cannot always say with certainty that the best use of a capital budget of $550 is to fund the top five projects. There may be some other set of projects that makes better use of the available $550. While some trial and error computations may indicate the proper set of projects, more elaborate techniques are needed to prove optimality.

As a practical matter, a capital budget probably has some flexibility. If in Example 17-6 the tentative capital budget is $550, then a careful examination of project 2 will dictate whether to expand the capital budget to $650 (to be able to include project 2) or to drop back to $450 (and leave project 2 out of the capital budget).

## SUMMARY

Prior to this chapter we have assumed that all worthwhile projects are approved and implemented. But industrial firms, like individuals and governments, are typically faced with more good projects than there is money available. The task is to select the best projects and reject, or at least delay, the rest.

Mutually exclusive alternatives are those where the acceptance of one alternative effectively prevents the adoption of the other alternatives. This could be because the alternatives perform the same function (like pump *A* vs pump *B*) or occupy the same physical location (like a gas station vs a hamburger stand). If a project has a single alternative of doing something, we know there is likely to be a mutually exclusive alternative of doing nothing—or possibly doing something else. A project proposal may be thought of as having two or more mutually exclusive alternatives. Projects are assumed in this chapter to be independent.

Undesirable alternatives can be located by any of a variety of analysis techniques. Five criteria for rejection are:

Rate of return less than MARR
PW of benefits less than PW of costs
EUAB less than EUAC
Benefit-cost ratio less than 1
NPW less than 0

Selection of the best from among several project alternatives may be accomplished by choosing the alternative by one of the following methods.

For any situation:
  Maximize NPW
  Maximize (EUAB − EUAC)
For either fixed input or fixed output:
  Maximize benefit-cost ratio
 All other situations:
  Incremental rate of return analysis required
  Incremental benefit-cost ratio analysis required

Capital may be rationed among competing investment opportunities by either rate of return or present worth methods. The results may not always be the same for these two methods in many practical situations.

If projects are ranked by rate of return, a proper procedure is to go down the list until the capital budget is exhausted. The rate of return at this point is the cutoff rate of return. This procedure gives the best group of projects, but does not necessarily have them in the proper priority order.

Maximizing NPW is an appropriate present worth selection criterion where the available projects do not exhaust the money supply. But if the amount of money required for the best alternative from each project exceeds the available money, a more severe criterion is imposed: adopt only those alternatives and projects that have a positive $NPW - p(PW$ of cost). The value of the multiplier $p$ is chosen by trial and error until the alternatives and projects meeting the criterion just equal the available capital budget money.

It has been shown in earlier chapters that the usual business objective is to maximize NPW, and this is not necessarily the same as maximizing rate of return. One suitable procedure is to use the ratio NPW/PW of cost to rank the projects. This present worth ranking method will order the projects so that for a limited capital budget, NPW will be maximized. We know that MARR may be adjusted from time to time to reasonably balance the cost of the projects that meet the MARR criterion and the available supply of money. This adjustment of the MARR to equal the cutoff rate of return is essential for the rate of return and present worth methods to yield compatible results.

Another way of ranking is by incremental rate of return analysis. Once a ranking has been made, we can go down the list and accept the projects until the money runs out. There is a theoretical difficulty if the capital budget contains more money than is required for $n$ projects, but not enough for one more, or $n + 1$ projects. As a practical matter capital budgets are seldom inflexible with the result that some additional money may be allocated if the $n + 1$ project looks like it should be included.

## PROBLEMS

**17-1** The following ten independent projects each have a ten-year life and no salvage value.

| Project | Cost | Uniform annual benefit | Computed rate of return |
|---------|------|------------------------|--------------------------|
| 1 | $ 5 thousand | $1.03 thousand | 16% |
| 2 | 15 | 3.22 | 17% |
| 3 | 10 | 1.77 | 12% |
| 4 | 30 | 4.88 | 10% |

| Project | Cost | Uniform annual benefit | Computed rate of return |
|---|---|---|---|
| 5 | 5 | 1.19 | 20% |
| 6 | 20 | 3.83 | 14% |
| 7 | 5 | 1.00 | 15% |
| 8 | 20 | 3.69 | 13% |
| 9 | 5 | 1.15 | 19% |
| 10 | 10 | 2.23 | 18% |

The projects have been proposed by the staff of the Ace Card Company. The minimum attractive rate of return (MARR) of Ace has been 12% for several years.

(a) If there is ample money available, what projects should Ace approve?

(b) Rank order all the acceptable projects in their order of desirability.

(c) If only $55,000 is available, which projects should be approved?

**17-2** At Miami Products four project proposals (three with mutually exclusive alternatives) are being considered. All the alternatives have a ten-year useful life and no salvage value.

| Project proposal | Cost | Uniform annual benefit | Computed rate of return |
|---|---|---|---|
| Project 1 | | | |
| Alternative A | $25 thousand | $4.61 thousand | 13% |
| B | 50 | 9.96 | 15% |
| C | 10 | 2.39 | 20% |
| Project 2 | | | |
| Alternative A | 20 | 4.14 | 16% |
| B | 35 | 6.71 | 14% |
| Project 3 | | | |
| Alternative A | 25 | 5.56 | 18% |
| B | 10 | 2.15 | 17% |
| Project 4 | 10 | 1.70 | 11% |

(a) Using rate of return methods, determine which set of projects should be undertaken if the minimum attractive rate of return is 10%.

(b) Using rate of return methods, which set of projects should be undertaken if the capital budget is limited to $100,000?

(c) For a budget of $100,000, what interest rate should be used in rationing capital by present worth methods? (Limit your answer to a value for which there is a compound interest table available at the back of the book.)

(d) Using the interest rate determined in (c), rank order the eight different investment opportunities by the present worth method.

(e) For a budget of $100,000 and the ranking in (d), which of the investment opportunities should be selected?

**17-3** Al Dale is planning his Christmas shopping as he must buy gifts for seven people. To quantify how much the various people would enjoy receiving a list of prospective gifts, Al has assigned appropriateness units (called "ohs") for each gift if given to each of the seven people. A rating of five ohs represents a gift that the recipient would really like. A rating of four ohs indicates the recipient would like it four-fifths as much; three ohs, three-fifths as much, and so forth. A zero rating indicates an inappropriate gift that cannot be given to that person. These data are tabulated below.

*"Oh" rating of gift if given to various family members*

|  | *Prospective gift* | *Father* | *Mother* | *Sister* | *Brother* | *Aunt* | *Uncle* | *Cousin* |
|---|---|---|---|---|---|---|---|---|
| 1. | $10 box of candy | 4 | 4 | 2 | 1 | 5 | 2 | 3 |
| 2. | $6 box of cigars | 3 | 0 | 0 | 1 | 0 | 1 | 2 |
| 3. | $8 necktie | 2 | 0 | 0 | 3 | 0 | 3 | 2 |
| 4. | $10 shirt or blouse | 5 | 3 | 4 | 4 | 4 | 1 | 4 |
| 5. | $12 sweater | 3 | 4 | 5 | 4 | 3 | 4 | 2 |
| 6. | $15 camera | 1 | 5 | 2 | 5 | 1 | 2 | 0 |
| 7. | $3 calendar | 0 | 0 | 1 | 0 | 1 | 0 | 1 |
| 8. | $8 magazine subscription | 4 | 3 | 4 | 4 | 3 | 1 | 3 |
| 9. | $9 book | 3 | 4 | 2 | 3 | 4 | 0 | 3 |
| 10. | $8 game | 2 | 2 | 3 | 2 | 2 | 1 | 2 |

The objective is to select the most appropriate set of gifts for the seven people (that is, maximize total ohs) that can be obtained with the selected budget.

(a) How much will it cost to buy the seven gifts the people would like best, if there is ample money for Christmas shopping?

(b) If the Christmas shopping budget is set at $56, which gifts should be purchased, and what is their total appropriateness rating in ohs?

(c) If the Christmas shopping budget must be cut to $45, which gifts should be purchased, and what is their total appropriateness rating in ohs?
(*Answer:* (a) $84)

**The following facts are to be used in solving Problems 17-4 through 17-7.**

In assembling data for the Peabody Company annual capital budget, five independent projects are being considered. Detailed examination by the staff has resulted in the identification of from three to six mutually exclusive do-something alternatives for each project. In addition, each project has a do-nothing alternative. The projects and their alternatives are listed below.

|  | *Cost* | *Uniform annual benefit* | *Useful life* | *End of useful life salvage value* | *Computed rate of return* |
|---|---|---|---|---|---|
| **Project 1** | | | | | |
| Alternative *A* | $40 thousand | $13.52 thousand | 2 yrs | $20 thousand | 10% |
| *B* | 10 | 1.87 | 16 | 5 | 18% |
| *C* | 55 | 18.11 | 4 | 0 | 12% |
| *D* | 30 | 6.69 | 8 | 0 | 15% |
| *E* | 15 | 3.75 | 2 | 15 | 25% |
| **Project 2** | | | | | |
| Alternative *A* | 10 | 1.91 | 16 | 2 | 18% |
| *B* | 5 | 1.30 | 8 | 0 | 20% |
| *C* | 5 | 0.97 | 8 | 2 | 15% |
| *D* | 15 | 5.58 | 4 | 0 | 18% |

| | Cost | Uniform annual benefit | Useful life | End of useful life salvage value | Computed rate of return |
|---|---|---|---|---|---|
| *Project 3* | | | | | |
| Alternative *A* | 20 | 2.63 | 16 | 10 | 12% |
| *B* | 5 | 0.84 | 16 | 0 | 15% |
| *C* | 10 | 1.28 | 16 | 0 | 10% |
| *D* | 15 | 2.52 | 16 | 0 | 15% |
| *E* | 10 | 3.50 | 4 | 0 | 15% |
| *F* | 15 | 2.25 | 16 | 15 | 15% |
| *Project 4* | | | | | |
| Alternative *A* | 10 | 2.61 | 8 | 0 | 20% |
| *B* | 5 | 0.97 | 16 | 0 | 18% |
| *C* | 5 | 0.90 | 16 | 5 | 18% |
| *D* | 15 | 3.34 | 8 | 0 | 15% |
| *Project 5* | | | | | |
| Alternative *A* | 5 | 0.75 | 8 | 5 | 15% |
| *B* | 10 | 3.50 | 4 | 0 | 15% |
| *C* | 15 | 2.61 | 8 | 5 | 12% |

Each project concerns operations at the St. Louis brewery. The plant was leased from another firm many years ago and the lease expires 16 years from now. For this reason, the analysis period for all projects is 16 years. Peabody considers 12% to be the minimum attractive rate of return.

In solving the Peabody Co. problem an important assumption concerns the situation at the end of the useful life of an alternative when the alternative has a useful life less than the 16-year analysis period. Two replacement possibilities are listed.

*Assumption one*: When an alternative has a useful life less than 16 years, it will be replaced by a new alternative with the same useful life as the original. This may need to occur more than once. The new alternative will have a 12% computed rate of return, and hence a NPW = 0 at 12%.

*Assumption two*: When an alternative has a useful life less than 16 years, it will be replaced at the end of its useful life by an identical alternative (one with the same cost, uniform annual benefit, useful life and salvage value as the original alternative).

**17-4** For an unlimited supply of money, and replacement assumption one, which project alternatives should Peabody select? Solve the problem by present worth methods. (*Answer:* Project alternatives 1*B*, 2*A*, 3*F*, 4*A*, and 5*A*)

**17-5** For an unlimited supply of money, and replacement assumption two, which project alternatives should Peabody select? Solve the problem by present worth methods.

**17-6** For an unlimited supply of money, and replacement assumption two, which project alternatives should Peabody select? Solve the problem by rate of return methods. (*Hint*: By careful inspection of the alternatives you should be able to reject about half of them. Even then the problem requires lengthy calculations.)

**17-7** For a capital budget of $55,000, and replacement assumption two, which project alternatives should Peabody select? (*Answer:* Project alternatives 1*E*, 2*A*, 3*F*, 4*A*, and 5*A*.)

# 18

# Digital
# Computer
# Applications

As long as one is solving simple problems to illustrate particular principles, the calculations are usually quite simple. But as problems become more realistic, each additional complexity adds substantially to the length of the calculations. Depreciation, income taxes, estimates of fluctuating future events, sensitivity analysis—all these combine to increase the complexity of the calculations.

In many situations the digital computer may be used to perform the numerical calculations. Three separate topics will be used to illustrate practical applications of the digital computer.

1.  *Computation of rate of return for a cash flow.* Because rate of return computations can be both lengthy and complex, only a superficial solution was provided in earlier chapters. In this chapter we will make a more careful and precise analysis of the computation of rate of return.

2.  *Capital expenditure analysis by simulation.* In Chapter 14, simulation was shown to be a practical way of solving problems where the probabilities were known for some of the future costs and/or benefits. A computer program will be presented to solve these problems by simulation.

3.  *Computing specialized compound interest tables.* Economic analysis calculations depend on being able to find equivalencies for money at different points in time. The tables provided at the back of this book provide compound interest factors for a range of values of *n* and *i*. Additional tables of compound interest factors may be readily generated on a computer.

## ANOTHER LOOK AT THE COMPUTATION OF RATE OF RETURN

In Chapter 7 the rate of return was defined for an investment as follows.

Rate of return is the interest rate earned on the unrecovered investment such that the payment schedule makes the unrecovered investment equal to zero at the end of the life of the investment.

In a borrowing situation the definition was given.

Rate of return is the interest rate paid on the unpaid balance of a loan such that the payment schedule makes the unpaid loan balance equal to zero when the final payment is made.

In the actual calculation of rate of return we wrote one equation relating costs and benefits (for example, present worth of cost = present worth of benefits) and solved the equation for the unknown rate of return. This works fine if the situation represents either a pure investment or a pure borrowing situation and there is a single positive rate of return.

Unfortunately, there are times when these two conditions are not met. In Chapter 7 a remedy was suggested. By application of an external interest rate, the cash flow was adjusted until the number of sign changes was reduced to one. The cash flow rule of signs then tells us there is either none or one positive rate of return. And with only one sign change in the cash flow, the situation must be one of either pure investment or of pure borrowing. This remedy works. But as we will see in this chapter, we may be adjusting the cash flow when no adjustment is necessary, and even when an adjustment is required, we may be making too large an adjustment. The resulting computed rate of return is thus affected by adjustments which are either unnecessary or too much.

A cash flow may represent any kind of situation. It may be a pure borrowing, pure investment, or a mixture of the two. This is illustrated by Table 18-1. In Example *A* the typical investment situation is represented by the investment of 50 at year 0, followed by the return of the resulting benefits in years 1 through 4. Example *B* represents a borrowing situation. Fifty is received in year 0, followed by four payments of 15 each in years 1 through 4 to repay the loan. Note that *A* and *B* are mirror images of one another. That is, changing the sign of all the cash flows in *A* gives us *B*, and vice versa.

Example *C* represents a mixed situation. Initially there is a receipt of benefits (like a borrowing situation) followed by investments in years 1 and 2. Finally, there are benefits in years 3 and 4. The result is a mixed situation

**Table 18-1.**   Examples of different cash flow situations

|  | Example A | Example B | Example C |
|---|---|---|---|
|  | *pure investment* | *pure borrowing* | *mixed borrowing and investment* |
| *Year* | *Cash flow* | *Cash flow* | *Cash flow* |
| 0 | − $50 | + $50 | + $50 |
| 1 | + 15 | − 15 | − 30 |
| 2 | + 15 | − 15 | − 30 |
| 3 | + 15 | − 15 | + 15 |
| 4 | + 15 | − 15 | + 15 |

with years 0, 1, and 2 looking like a borrowing situation and at the same time years 1, 2, 3, and 4 look like an investment situation.

In computing the rate of return for a cash flow one must carefully decide what he wishes the number to represent. Is it to be the internal rate of return earned on invested money, the external rate of return paid on borrowed money, or what? As this is a book on capital expenditure analysis, the view here is that we want to determine the internal rate of return, that is, the rate of return earned on the money invested in the project while it is actually in the project. Money that is invested outside of the project will generally be assumed to earn some established external rate of return.

**Analysis of a Cash Flow as
an Investment Situation**

A good deal can be learned about the desirability of a cash flow as an investment situation. The goal is to produce a single value that accurately portrays the profitability of the investment opportunity reflected by the cash flow. We do not want multiple rates of return. (After all, how can an investment have both a 20% and a 60% rate of return at the same time?) And we do not want a rate of return that assumes that one must temporarily invest money outside of the investment project at an unrealistic interest rate. If a suitable external investment rate is 6%, for example, then the computations must not be based on some other rate. In short, we want a a single, realistic value representing the rate of return (note that this could also be called the "profitability rate") on the investment.

There are four tests that will help us to understand the investment situation represented by any cash flow. They are:

1. Cash flow rule of signs.
2. Accumulated cash flow sign test.
3. Algebraic sum of the cash flow.
4. Net investment conditions.

These will be examined one by one.

## Cash Flow Rule of Signs

There may be as many positive rates of return as there are sign changes in the cash flow.

If we let $a_i$ represent the cash flow in year $i$, then the entire cash flow could be represented as follows:

| Year | Cash flow |
|------|-----------|
| 0 | $a_0$ |
| 1 | $a_1$ |
| 2 | $a_2$ |
| 3 | $a_3$ |
| . | . |
| . | . |
| . | . |
| $n$ | $a_n$ |

As described in detail in Chapter 7, a sign change is where successive terms in the cash flow (ignoring zeros) have different signs.

| Number of sign changes in cash flow | Number of positive rates of return |
|-------------------------------------|------------------------------------|
| 0 | 0 (or rate of return = infinity) |
| 1 | 1, or 0 |
| 2 | 2, 1, or 0 |
| 3 | 3, 2, 1, or 0 |
| . | . |
| . | . |

## Accumulated Cash Flow Sign Test

The accumulated cash flow is the algebraic sum of the cash flow to that point in time. If we let $a_i$ represent the cash flow in a year and $A_i$ to represent the accumulated cash flow, the situation is:

| Year | Cash flow | Accumulated cash flow |
|------|-----------|------------------------|
| 0 | $a_0$ | $A_0 = a_0$ |
| 1 | $a_1$ | $A_1 = a_0 + a_1$ |
| 2 | $a_2$ | $A_2 = a_0 + a_1 + a_2$ |
| 3 | $a_3$ | $A_3 = a_0 + a_1 + a_2 + a_3$ |
| . | . | . |
| . | . | . |
| . | . | . |
| $n$ | $a_n$ | $A_n = a_0 + a_1 + a_2 + a_3 + \cdots + a_n$ |

The sequence of accumulated cash flows $(A_0, A_1, A_2, A_3, \ldots, A_n)$ is examined to determine the number of sign changes. As before, a sign change is where successive terms in the accumulated cash flow (ignoring zeros) have different signs.

Norstrøm[†] proved that sufficient (but not necessary) conditions for a single positive rate of return are:

1.  The accumulated cash flow in the $n$th year is not equal to zero ($A_n \neq 0$).
2.  There is exactly one sign change in the sequence of accumulated cash flows.

## Algebraic Sum of the Cash Flow

$$\text{Algebraic sum} = \sum_{i=0}^{n} a_i$$

or looking at the accumulated cash flow,

$$\text{Algebraic sum} = A_n$$

A positive algebraic sum ($A_n > 0$) suggests a positive rate of return and an algebraic sum equal to zero ($A_n = 0$) suggests a 0% rate of return. In either case, however, there may be multiple positive rates of return.

A negative algebraic sum tells that the costs exceed the benefits of the project. This would seem to immediately mean that a positive rate of return

†Norstrøm, Carl J., "A Sufficient Condition for a Unique Nonnegative Internal Rate of Return," *Journal of Financial and Quantitative Analysis,* VII (June, 1972), pp. 1835–9.

would be impossible in this situation. This is usually true. It has been shown by Merrett and Sykes,† however, that where there are two or more sign changes in the cash flow, and a substantial outlay near the end of the life of a project, the result may be one or more positive rates of return.

## Net Investment Conditions

Two conditions have been shown‡ to be sufficient (but not necessary) to establish that a computed positive rate of return $i^*$ is the only positive rate of return. The conditions are as follows.

1. The cash flow in the $n$th year must be positive ($a_n > 0$).
2. Given that a positive rate of return $i^*$ has been computed, there is a net investment throughout the life of the project until the end of the $n$th year when the net investment becomes zero. Mathematically, this is:

$$\text{Net investment in any year } k = \sum_{j=0}^{k} a_j(1 + i^*)^{k-j} \quad \text{for } k = 0,1,2,\ldots,n$$

When either of these two conditions are not met, we know a net investment does not exist throughout the life of the project. This means there are one or more periods when the project has a net outflow of money *which will later be required to be returned to the project*. This money can be put into an external investment until such time as it is needed in the project. The interest rate of the external investment ($e^*$) will be the interest rate at which the money can in fact be invested outside the project. The external interest rate ($e^*$) is unrelated to the internal rate of return ($i^*$) on the project. If there is no external investment, then no value of $e^*$ is required in the computations.

## Application of the Four Tests of a Cash Flow

The four tests can be used to produce a chart for computing the rate of return for any investment project. Table 18-2 is such a chart. The required computations may be computed by hand or on a digital computer. A com-

---

†Merrett, A. J. and Allen Sykes. *The Finance and Analysis of Capital Projects*, 2nd ed. London: Longman, 1973, p. 135.

‡Soper, C. S., "The Marginal Efficiency of Capital: A Further Note," *Economic Journal* LXIX, pp. 174–7.

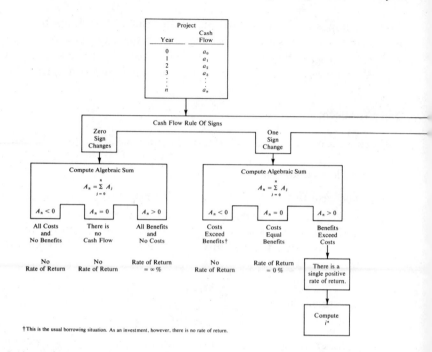

† This is the usual borrowing situation. As an investment, however, there is no rate of return.

puter program called ROR has been written to perform all the computations. A listing of the program is given in the Appendix.

The computation of the rate of return for any investment project may be
illustrated by the following example problems.

## EXAMPLE 18-1

Given the following cash flow, compute the internal rate of return $i^*$.
If needed, use an external interest rate $e^*$ of 6%.

| Year | Cash flow |
|------|-----------|
| 0 | $-\$2000$ |
| 1 | $+1200$ |
| 2 | $+400$ |
| 3 | $+400$ |
| 4 | $-200$ |
| 5 | $+400$ |

First, we will write out the cash flow and accumulated cash flow and then
determine the sign changes in both along with the algebraic sum of the
cash flow.

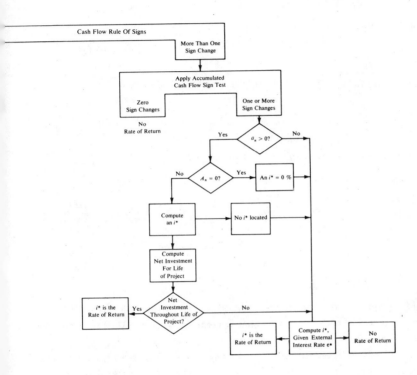

|  | *Cash flow* | *Accumulated cash flow* |
|---|---|---|
| *Year* | $a_i$ | $A_i$ |
| 0 | − $2000 | − $2000 |
| 1 | + 1200 | − 800 |
| 2 | + 400 | − 400 |
| 3 | + 400 | 0 |
| 4 | − 200 | − 200 |
| 5 | + 400 | + 200 = $A_n$ |
|  | + $200 |  |
| Sign changes = | 3 | = 1 |

The cash flow rule of signs indicates there may be as many as three positive rates of return. Since there is one sign change in the accumulated cash flow and $A_n \neq 0$, these are sufficient conditions for a single positive rate of return. It may be computed in the usual manner.

$$\text{PW of cost} = \text{PW of benefits}$$

$$2000 + 200(P/F,i\%,4) = 1200(P/F,i\%,1) + 400[(P/F,i\%,2)$$
$$+ (P/F,i\%,3) + (P/F,i\%,5)]$$

Try $i = 5\%$.

$$2000 + 200(0.8227) = 1200(0.9524) + 400[0.9070 + 0.8638 + 0.7835]$$

$$2165 = 2165$$

Thus $i^* = 5\%$.

Compute the net investment for the project at each time period.

| Year | Cash flow | | Net investment |
|------|-----------|---|---------------|
| 0 | − $2000 | | − $2000.0 |
| 1 | + 1200 | − 2000(1 + 0.05) + 1200 = | − 900.0 |
| 2 | + 400 | − 900(1 + 0.05) + 400 = | − 545.0 |
| 3 | + 400 | − 545(1 + 0.05) + 400 = | − 172.3 |
| 4 | − 200 | − 172.3(1 + 0.05) − 200 = | − 380.9 |
| 5 | + 400 | − 380.9(1 + 0.05) + 400 = | 0 |

The fact that the net investment becomes zero at the end of the fifth year proves that 5% is a rate of return for the cash flow. We also see that there is a continuing net investment in the project for all time periods until the end of the project. This means there is no external investment of money and hence no need for an external interest rate. We conclude that $i^* = 5\%$ is a correct measure of the profitability of the project represented by the cash flow.

## EXAMPLE 18-2

Given the following cash flow, compute the internal rate of return $i^*$. If needed, use an external interest rate $e^*$ of 6%.

| Year | Cash flow |
|------|-----------|
| 0 | − $100 |
| 1 | + 50 |
| 2 | + 50 |
| 3 | + 50 |
| 4 | + 50 |
| 5 | − 73.25 |

As in the previous example, determine the sign changes in the cash flow and the accumulated cash flow and compute the sum of the cash flow.

| Year | Cash flow $a_i$ | Accumulated cash flow $A_i$ |
|---|---|---|
| 0 | $-\$100$ | $-\$100$ |
| 1 | $+50$ | $-50$ |
| 2 | $+50$ | $0$ |
| 3 | $+50$ | $+50$ |
| 4 | $+50$ | $+100$ |
| 5 | $-73.25$ | $+26.75 = A_n$ |
| | $+\$26.75$ | |

Sign changes $= 2$          $= 1$

With one sign change in the accumulated cash flow and $A_n \neq 0$, we know there is a single positive rate of return. The problem was devised with a 20% interest rate as is shown:

$$\text{PW of cost} = \text{PW of benefits}$$
$$100 + 73.25(P/F,20\%,5) = 50(P/A,20\%,4)$$
$$100 + 73.25(0.4019) = 50(2.589)$$
$$129.45 = 129.45$$

Thus $i^* = 20\%$. Check net investment.

| Year | Cash flow | | Net investment |
|---|---|---|---|
| 0 | $-\$100$ | | $-\$100$ |
| 1 | $+50$ | $-100(1+0.20) + 50$ $=$ | $-70$ |
| 2 | $+50$ | $-70(1+0.20) + 50$ $=$ | $-34$ |
| 3 | $+50$ | $-34(1+0.20) + 50$ $=$ | $+9.20$ |
| 4 | $+50$ | $+9.20(1+0.20) + 50$ $=$ | $+61.04$ |
| 5 | $-73.25$ | $+61.04(1+0.20) - 73.25$ $=$ | $0$ |

Note that the fact that net investment becomes zero at the end of the last year confirms that 20% is a rate of return for the cash flow. It is also clear that there is not net investment throughout the life of the project. At the end of year 3 there is in fact $+9.20$ that must be invested outside of the project represented by the cash flow. And in year 4 the amount for external investment has increased to $+61.04$. The money invested externally must be returned to the project in year 5 to provide the needed final disbursement of 73.25. In the table above the money has been assumed to be invested at a 20% interest rate. More importantly, this same assumption of a 20% external investment is implicit in the PW of cost = PW of benefits calculation that also indicated that $i^* = 20\%$.

The problem statement specifies that external investment may be expected to earn a 6% interest rate. Since this is less than the 20% in the calculations above, this means that a benefit (the external interest earned) of the cash flow is reduced. For $e^* = 6\%$, we can see that $i^* < 20\%$. The calculation must be done by trial and error. Try $i = 15\%$.

| Year | Cash flow | | | Net investment |
|------|-----------|--|--|----------------|
| 0 | $-\$100$ | | | $-\$100.00$ |
| 1 | $+50$ | $-100(1+0.15) + 50$ | $=$ | $-65.00$ |
| 2 | $+50$ | $-65(1+0.15) + 50$ | $=$ | $-24.75$ |
| 3 | $+50$ | $-24.75(1+0.15) + 50$ | $=$ | $+21.54$ |
| 4 | $+50$ | $+21.54(1+0.06) + 50$ | $=$ | $+72.83$ |
| 5 | $-73.25$ | $+72.83(1+0.06) - 73.25$ | $=$ | $+3.95$ |

The net investment does not equal zero at the end of the project. This indicates our trial $i = 15\%$ is in error. The remaining positive net investment signals that $i$ should be increased. Further trials would guide us to $i = 16.5\%$. For $e^* = 6\%$ and $i = 16.5\%$ we have:

| Year | Cash flow | | | Net investment |
|------|-----------|--|--|----------------|
| 0 | $-\$100$ | | | $-\$100.00$ |
| 1 | $+50$ | $-100.00(1+0.165) + 50$ | $=$ | $-66.50$ |
| 2 | $+50$ | $-66.50(1+0.165) + 50$ | $=$ | $-27.47$ |
| 3 | $+50$ | $-27.47(1+0.165) + 50$ | $=$ | $+17.99$ |
| 4 | $+50$ | $+17.99(1+0.06) + 50$ | $=$ | $+69.07$ |
| 5 | $-73.25$ | $+69.07(1+0.06) - 73.25$ | $=$ | $-0.03$ |

We see that with $e^* = 6\%$, $i^*$ is very close to 16.5%.

The results of Example 18-2 are a little startling. The two sign changes in the cash flow warned us that there might be as many as two positive rates of return. From the accumulated cash flow sign test we proved that in reality there was one positive rate of return. We then showed that 20% was the single positive rate of return. One would be tempted to stop at this point, satisfied that $i^* = 20\%$ is a proper measure of the profitability of the project represented by the cash flow.   Examination of the net investment throughout the life of the project reveals that external investments occur at the end of year 3 and year 4. Using an external interest rate $e^*$, we found the true profitability of the project was $i^* = 16.5\%$. The surprising discovery is that where there are multiple sign changes in an investment project cash flow, there may be a unique positive rate of return, but the existance of external investment may mean that the unique positive rate of return is not a suitable measure of project profitability.

## EXAMPLE 18-3

Given the following cash flow, compute the internal rate of return $i^*$. If needed, use an external interest rate $e^*$ of 6%.

| Year | Cash flow |
|------|-----------|
| 0 | −$100 |
| 1 | +330 |
| 2 | −362 |
| 3 | +132 |

We will begin the solution by writing the accumulated cash flow and checking the sign changes.

| Year | Cash flow | Accumulated cash flow |
|------|-----------|-----------------------|
| 0 | −$100 | −$100 |
| 1 | +330 | +230 |
| 2 | −362 | −132 |
| 3 | +132 | $0 = A_n$ |
| | 0 | |
| Sign changes = 3 | | = 2 |

Careful examination of these data reveal the following.

1. There may be as many as three positive rates of return.
2. One rate of return is 0%.
3. The accumulated cash flow does not meet the sufficient conditions for a single positive rate of return.
4. At $i^* = 0\%$ there is external investment at the end of year 1. (The accumulated cash flow sequence becomes the net investment sequence at 0%.)

At this point we have two choices. We can search to find out how many positive rates of return there are and their numerical values. As an alternative we can seek $i^*$ for the given $e^* = 6\%$.

This particular cash flow was devised by picking the desired roots of 0%, 10%, and 20% for a third order polynomial as follows: let $x = 1 + i$. For $i = 0.00, 0.10,$ and $0.20,$

$$(x - 1.0)(x - 1.1)(x - 1.2) = 0$$

Multiplying we get:

$$x^3 - 3.3x^2 + 3.62x - 1.32 = 0$$

Multiplying by $-100$ gives:

$$-100x^3 + 330x^2 - 362x + 132 = 0$$
$$-100(1 + i)^3 + 330(1 + i)^2 - 362(1 + i) + 132 = 0$$

This represents the future worth of the following cash flow.

| Year | Cash flow |
|------|-----------|
| 0 | $-\$100$ |
| 1 | $+330$ |
| 2 | $-362$ |
| 3 | $+132$ |

Thus the three positive rates of return for the cash flow are 0%, 10%, and 20%. Knowing that there are three positive rates of return, and their values, has not helped us in our search for a true measure of project profitability.

A trial and error search is made for $i^*$, given that $e^* = 6\%$. For a trial $i = 5.98\%$

| Year | Cash flow | | | Net investment |
|------|-----------|---|---|----------------|
| 0 | $-100$ | | | $-100$ |
| 1 | $+330$ | $-100(1 + 0.0598) + 330$ | $=$ | $+224.02$ |
| 2 | $-362$ | $+224.02(1 + 0.06) - 362$ | $=$ | $-124.54$ |
| 3 | $+132$ | $-124.54(1 + 0.0598) + 132$ | $=$ | $+0.01$ |

For $e^* = 6\%$, $i^*$ is about 5.98%. We see that the exterior investment is more desirable than the internal project.

## EXAMPLE 18-4

Given the following cash flow, compute the internal rate of return $i^*$. If needed, use an external interest rate $e^*$ of 6%.

| Year | Cash flow | Accumulated cash flow |
|------|-----------|-----------------------|
| 0 | $-\$200$ | $-\$200$ |
| 1 | $+150$ | $-50$ |
| 2 | $+50$ | $0$ |
| 3 | $+50$ | $+50$ |
| 4 | $+50$ | $+100$ |
| 5 | $+50$ | $+150$ |
| 6 | $+50$ | $+200$ |
| 7 | $+50$ | $+250$ |
| 8 | $-285$ | $-35$ |
|  | $-\$35$ |  |
| Sign changes $= 2$ | | $= 2$ |

Although the algebraic sum of the cash flow is negative, the surprising fact is that there are two positive rates of return for the cash flow (at about 8.7% and 17.4%). The accumulated cash flow shows that an important benefit is the substantial external investment of money for relatively long periods of time. If we compute the external investment at $e* = 6\%$ (rather than 8.7% or 17.4%), the benefits will be substantially reduced and the computed internal rate of return will decline. The internal rate of return is computed by trial and error. Try $i = 4\%$.

| Year | Cash flow | | | Net investment |
|---|---|---|---|---|
| 0 | − $200 | | | − $200.0 |
| 1 | + 150 | − 200(1.04) + 150 | = | − 58.0 |
| 2 | + 50 | − 58.0(1.04) + 50 | = | − 10.3 |
| 3 | + 50 | − 10.3(1.04) + 50 | = | + 39.3 |
| 4 | + 50 | + 39.3(1.06) + 50 | = | + 91.7 |
| 5 | + 50 | + 91.7(1.06) + 50 | = | + 147.2 |
| 6 | + 50 | + 147.2(1.06) + 50 | = | + 206.0 |
| 7 | + 50 | + 206.0(1.06) + 50 | = | + 268.4 |
| 8 | − 285 | + 268.4(1.06) − 285 | = | − 0.5 |

With $e* = 6\%$, $i*$ is about 4%.

## EXAMPLE 18-5

Consider the following project cash flow. The external interest rate $e* = 6\%$.

| Year | Cash flow |
|---|---|
| 0 | − $500 |
| 1 | + 300 |
| 2 | + 300 |
| 3 | + 300 |
| 4 | − 600 |
| 5 | + 200 |
| 6 | + 135.66 |

(a) Solve for the project internal rate of return ($i*$) using the methods described in Chapter 7.

There are three sign changes in the cash flow. We wish to reduce this to one sign change. We will eliminate the − 600 at the end of year 4 by accumulating money at the external interest rate ($e* = 6\%$) to equal the required $600. The computations are as follows.

| Year | Cash flow | External investment | | Transformed cash flow |
|------|-----------|---------------------|---|-----------------------|
| 0 | − $500 | | | − $500 |
| 1 | + 300 | | | + 300 |
| 2 | + 300 | $x(1.06)^2$   $x = 250.98$ | | + 49.02 |
| 3 | + 300 | + 300(1.06) ⤵   ↓ | | 0 |
| 4 | − 600 | + 318   + 282 | | 0 |
| 5 | + 200 | | | + 200 |
| 6 | + 135.66 | | | + 135.66 |

As described in the computations above, all the money at year 3 is assumed to be invested externally at 6%. The + $300 increases to + $318 at the end of one year and can be brought back to the cash flow at the end of year 4. This makes the internal investment cash flow have a zero for year three and − 600 + 318 = − $282 for year 4. A further alteration of the cash flow is needed to reduce the number of sign changes to one. Part of the year 2 cash flow must be set aside in an external investment so that its accumulated sum at the end of two years will be + 282. This can be brought back to the cash flow at the end of year 4 with the result that the internal investment cash flow will have a zero for year 4. The amount to invest externally in year 2 is $x = +282/(1.06)^2 = +250.98$. The result is the transformed cash flow. The transformed cash flow has one sign change, and therefore cannot have more than one positive rate of return. Using the technique described in Chapter 7, we will see that $i^*$ is very close to 11.1%.

| Year | Transformed cash flow | Present worth at 11% | Present worth at 12% | Present worth at 11.1% |
|------|----------------------|---------------------|---------------------|------------------------|
| 0 | − $500 | − $500 | − $500 | − $500 |
| 1 | + 300 | + 270.27 | + 267.86 | + 270.03 |
| 2 | + 49.02 | + 39.79 | + 39.08 | + 39.71 |
| 3 | 0 | 0 | 0 | 0 |
| 4 | 0 | 0 | 0 | 0 |
| 5 | + 200 | + 118.69 | + 113.49 | + 118.16 |
| 6 | + 135.66 | + 72.53 | + 68.73 | + 72.14 |
| | + $184.68 | + $1.28 | − $10.86 | + $0.04 |

(b) Solve for the project internal rate of return ($i^*$) using the methods described in this chapter.

The procedure will be to solve the cash flow for a rate of return, check net investment, and proceed using $e^*$, if necessary.

Some preliminary computations, not shown here, lead us to a 15%

rate of return. This may be verified, and net investment computed for
$i* = 15\%$.

| Year | Cash flow | Present worth at 15% | | | Net investment at 15% |
|------|-----------|---------------------|---|---|----------------------|
| 0 | − $500 | − $500 | | | − $500.00 |
| 1 | + 300 | + 260.87 | − 500(1.15) + 300 | = | − 275.00 |
| 2 | + 300 | + 226.84 | − 275(1.15) + 300 | = | − 16.25 |
| 3 | + 300 | + 197.25 | − 16.25(1.15) + 300 | = | + 281.31 |
| 4 | − 600 | − 343.05 | + 281.31(1.15) − 600 | = | − 276.49 |
| 5 | + 200 | + 99.44 | − 276.49(1.15) + 200 | = | − 117.96 |
| 6 | + 135.66 | + 58.65 | − 117.96(1.15) + 135.66 | = | 0 |
| | + $135.66 | $0 | | | |

The computation of net investment at 15% indicates there is not net
investment throughout the life of the project. At the end of year 3 there
is an external investment of $281.31. The computation is based on
$e* = i* = 15\%.$

For $e* = 6\%$, we must compute $i*$ by trial and error.

For $e* = 6\%$, and trial $i = 13\%$.

| Year | Cash flow | | | Net investment |
|------|-----------|---|---|----------------|
| 0 | − $500 | | | − $500.00 |
| 1 | + 300 | − 500.00(1.13) + 300 | = | − 265.00 |
| 2 | + 300 | − 265.00(1.13) + 300 | = | + 0.55 |
| 3 | + 300 | + 0.55(1.06) + 300 | = | + 300.58 |
| 4 | − 600 | + 300.58(1.06) − 600 | = | − 281.38 |
| 5 | + 200 | − 281.38(1.13) + 200 | = | − 117.96 |
| 6 | + 135.66 | − 117.96(1.13) + 135.66 | = | + 2.36 |

The end of year 6 net investment is not zero, indicating that $i*$ is not
equal to our 13% trial $i$. Select 13.2% as another trial $i$ and repeat the
computation.

| Year | Cash flow | | | Net investment |
|------|-----------|---|---|----------------|
| 0 | − $500 | | | − $500.00 |
| 1 | + 300 | − 500.00(1.132) + 300 | = | − 266.00 |
| 2 | + 300 | − 266.00(1.132) + 300 | = | − 1.11 |
| 3 | + 300 | − 1.11(1.132) + 300 | = | + 298.74 |
| 4 | − 600 | + 298.74(1.06) − 600 | = | − 283.33 |
| 5 | + 200 | − 283.33(1.132) + 200 | = | − 120.73 |
| 6 | + 135.66 | − 120.73(1.132) + 135.66 | = | − 1.01 |

Results of the two trials:

|            | *End of year 6* |
|------------|-----------------|
| *Trial i*  | *Net investment* |
| 13.0%      | +2.36           |
| 13.2%      | −1.01           |

We conclude that for $e^* = 6\%$, $i^*$ is approximately 13.13%. At 13.13% the net investment is:

| Year | Cash flow |                                  |     | Net investment |
|------|-----------|----------------------------------|-----|----------------|
| 0    | − $500    |                                  |     | − $500.00      |
| 1    | + 300     | − 500.00(1.1313) + 300           | =   | − 265.65       |
| 2    | + 300     | − 265.65(1.1313) + 300           | =   | − 0.53         |
| 3    | + 300     | − 0.53(1.1313) + 300             | =   | + 299.40       |
| 4    | − 600     | + 299.40(1.06) − 600             | =   | − 282.64       |
| 5    | + 200     | − 282.64(1.1313) + 200           | =   | − 119.75       |
| 6    | + 135.66  | − 119.75(1.1313) + 135.66        | =   | − 0.19         |

(c) Explain the difference in the results obtained in parts (a) and (b).

The two computation methods transform the project cash flow into different amounts of internal and external investment. This is illustrated by the following tabulation.

|      |           | Part (a) | | Part (b) | |
|------|-----------|---------------------|----------------------|---------------------|----------------------|
| Year | Cash flow | Internal investment | External investment | Internal investment | External investment |
| 0    | − $500    | − $500.00           |                      | − $500.00           |                      |
| 1    | + 300     | + 300               |                      | + 300.00            |                      |
| 2    | + 300     | + 49.02             | + $250.98            | + 300.00            |                      |
| 3    | + 300     |                     | + 300.00             | + 0.60              | + $299.40            |
| 4    | − 600     |                     | − 600.00             | − 282.64            | − 317.36             |
| 5    | + 200     | + 200.00            |                      | + 200.00            |                      |
| 6    | + 135.66  | + 135.66            |                      | + 135.66            |                      |

The tabulation shows that the methods outlined in this chapter do not necessarily assume as much external investment as is required by the methods of Chapter 7. We now see clearly that the criterion on how much to alter a cash flow is properly based on maintaining a net investment—not on reducing the number of sign changes in the cash flow.

Based on net investment a smaller portion of the benefits are assumed in an external investment at $e^* = 6\%$. For values of $i^*$ greater than $e^*$, we will find the part (b) $i^*$ to be larger than the part (a) $i^*$.

From what has been said, plus the five example problems, there are some important conclusions to be noted. When there are multiple positive rates of return for a cash flow, in general, none of them are a suitable measure of project profitability. Even in the situation where there is only one positive rate of return, that value still may not be a good indicator of project profitability. The critical question is whether the rate of return is exclusively the return on funds invested in the project or is the return on the combination of funds in the project and funds temporarily invested outside the project. If the project cash flow reflects both internal and external investments, the one or more positive rates of return assume that the internal rate of return $i^*$ equals the external rate of return $e^*$. This is seldom a valid assumption. Where it is not, an external rate of return $e^*$ should be selected and $i^*$ computed, given $e^*$.

In a situation where an initial investment is made, followed by subsequent benefits (only one sign change in the cash flow), there is no temporary external investment and no difficulty in computing a suitable $i^*$. There will be other situations where in the later years of a cash flow an additional net investment is required. This may or may not result in multiple positive rates of return. It will, however, mean that an external interest rate $e^*$ is needed to compute a suitable $i^*$.

## RATE OF RETURN COMPUTER PROGRAM

A FORTRAN program called ROR has been written to analyze a cash flow and compute a rate of return on the internal investment ($i^*$) using, if necessary, a preselected external interest rate, $e^*$. A complete listing of the program is given in the Appendix. Its use may be illustrated by an example.

### EXAMPLE 18-6

For the cash flow in Example 18-5, solve for $i^*$, given $e^* = 6\%$, by use of the ROR computer program.

```
                    RATE OF RETURN COMPUTATION
      FROM NEWNAN   -   ENGINEERING ECONOMIC ANALYSIS


                                ACCUMULATED
      YEAR        CASH FLOW       CASH FLOW
       0           -500.00        -500.00
       1            300.00        -200.00
       2            300.00         100.00
       3            300.00         400.00
       4           -600.00        -200.00
       5            200.00            0
       6            135.66         135.66


          SUM =   135.66
          NUMBER OF SIGN CHANGES IN CASH FLOW = 3
                    IN ACCUMULATED CASH FLOW = 3
```

```
CASH FLOW RULE OF SIGNS
     THERE MAY BE AS MANY POSITIVE RATES OF RETURN AS THERE
     ARE SIGN CHANGES IN THE CASH FLOW.  FOR THIS CASH FLOW
     POSSIBLE NUMBER OF POSITIVE RATES OF RETURN
     BETWEEN 0 AND 3.
SIGN CHANGES IN ACCUMULATED CASH FLOW
     TWO CONDITIONS THAT ARE SUFFICIENT (BUT NOT NECESSARY)
     FOR A SINGLE POSITIVE RATE OF RETURN ARE
          THE ACCUMULATED CASH FLOW IN YEAR N IS NOT EQUAL TO 0.
     AND
          THERE IS ONE SIGN CHANGE IN THE ACCUMULATED CASH FLOW.

FOR THIS CASH FLOW
     ACCUMULATED CASH FLOW IN YEAR N = 135.66
     AND NUMBER OF SIGN CHANGES = 3
NO CONCLUSIONS CAN BE REACHED FROM THIS CALCULATION.

ALGEBRAIC SUM OF CASH FLOW
     A NEGATIVE SUM INDICATES THAT COSTS EXCEED BENEFITS.
     GENERALLY AN UNSATISFACTORY INVESTMENT SITUATION.

     A POSITIVE SUM INDICATES ONE OR MORE POSITIVE RATES OF
     RETURN.

     A ZERO SUM FREQUENTLY INDICATES A ZERO PERCENT RATE OF
     RETURN.

     FOR THIS CASH FLOW, SUM = 135.66

USING ABOVE DATA, PROCEED WITH ANALYSIS.


  TRIAL INTERNAL RATE OF RETURN = 0.1316
        EXTERNAL INTEREST RATE = 0.0600
```

| YEAR | CASH FLOW | COMPUTATION | | | INTERNAL INVESTMENT | EXTERNAL INVESTMENT |
|------|-----------|-------------|---|---|---------------------|---------------------|
| 0 | -500.00 | | | | -500.00 | 0 |
| 1 | 300.00 | -500.00(1+0.132) + | 300.00 = | -265.80 | 300.00 | 0 |
| 2 | 300.00 | -265.80(1+0.132) + | 300.00 = | -0.78 | 300.00 | 0 |
| 3 | 300.00 | -0.78(1+0.132) + | 300.00 = | 299.11 | 0.89 | 299.11 |
| 4 | -600.00 | 299.11(1+0.06) + | -600.00 = | -282.94 | -282.94 | -317.06 |
| 5 | 200.00 | -282.94(1+0.132) + | 200.00 = | -120.18 | 200.00 | 0 |
| 6 | 135.66 | -120.18(1+0.132) + | 135.66 = | -0.33 | 135.66 | 0 |

```
NEXT TRIAL
TRIAL INTERNAL RATE OF RETURN = 0.13104
        EXTERNAL INTEREST RATE = 0.0600
```

| YEAR | CASH FLOW | COMPUTATION | | | INTERNAL INVESTMENT | EXTERNAL INVESTMENT |
|------|-----------|-------------|---|---|---------------------|---------------------|
| 0 | -500.00 | | | | -500.00 | 0 |
| 1 | 300.00 | -500.00(1+0.131) + | 300.00 = | -265.52 | 300.00 | 0 |
| 2 | 300.00 | -265.52(1+0.131) + | 300.00 = | -0.31 | 300.00 | 0 |
| 3 | 300.00 | -0.31(1+0.131) + | 300.00 = | 299.65 | 0.35 | 299.65 |
| 4 | -600.00 | 299.65(1+0.06) + | -600.00 = | -282.37 | -282.37 | -317.63 |
| 5 | 200.00 | -282.37(1+0.131) + | 200.00 = | -119.38 | 200.00 | 0 |
| 6 | 135.66 | -119.38(1+0.131) + | 135.66 = | 0.64 | 135.02 | 0.64 |

```
CALCULATION INDICATES   FUE INTERNAL RATE OF RETURN VERY
CLOSE TO 0.1313.   THIS  LUE CONSIDERED THE ANSWER.

RESULTS OF RATE OF RETURN COMPUTATION WITH EXTERNAL INTEREST
   RATE = 0.0600
   EXTERNAL INTEREST RATED NEEDED IN COMPUTATION.
   COMPUTED INTERNAL RATE OF RETURN EQUALS 0.1313
ANALYSIS IS COMPLETED.
```

## CAPITAL EXPENDITURE ANALYSIS BY SIMULATION

In Chapter 14 we saw that a useful technique for analyzing investment proposals is simulation. Simulation is a very good analysis tool, but the extreme number of calculations makes computation by hand methods impractical. A digital computer program may be written to do the myriad of calculations for us. Capital Expenditure Analysis Program, CEAP, is such a simulation program. Basically, 11 components of an investment project must be specified.

1. Investment.
2. Investment timing.
3. Analysis period.
4. Depreciation method.
5. End-of-life salvage value.
6. Borrowed money (if any) used to finance investment.
7. Expensed costs (Annual operation and maintenance costs, and so forth).
8. Income tax rate.
9. Schedule of project benefits.
10. Trend (increase or decrease) of benefits, if any.
11. Interest rate on any external investment.

A detailed documentation of CEAP is in the Appendix. Two example problems illustrate how CEAP functions.

### EXAMPLE 18-7

Apply CEAP to Example 14-12. The problem statement for Example 14-12 is as follows. If a more accurate scales is installed on a production line it will reduce the error in computing postage charges and save $250 a year. The useful life of the scales is believed to be uniformly distributed and range from 12 to 16 years. The initial cost of the scales is estimated to be normally distributed with a mean of $1500 and a standard deviation of $150.

The CEAP program requires four basic data cards. In this problem one additional data card is required to load the cumulative life distribution (LCD). The format of the input data cards is as follows:

```
            Column
                    11111111112222222222233333333334444444
Data card   12345678901234567890123456789012345678901234 56
   1        14  16 5                     0 2      1500       150
   2        1    1                       7 0 1
   3        1                            1         .060
   4              250                     1          200   1
```

The LCD distribution card contains 25 values as follows:

12 12 12 12 12 13 13 13 13 13 14 14 14 14 14 15 15 15 15 15 16 16 16 16 16

The output from the computer program contains three elements.

1.  The input data are printed out for verification of the data cards and to document the results.
2.  Some summary results of the simulation are printed out in one line for each iteration.
3.  A plot of rate of return vs relative frequency is printed, together with values for the computed mean, standard deviation, variance (standard deviation squared) and the number of iterations of the simulation program ($N$).

The output for this problem follows.

```
            CAPITAL EXPENDITURE ANALYSIS PROGRAM
                        PROJECT 14
        FROM NEWNAN - ENGINEERING ECONOMIC ANALYSIS

INVESTMENT
        NORMAL DISTRIBUTION WITH MEAN = 1500 AND STD DEV = 150
    INVESTMENT TIMING
        TOTAL INVESTMENT MADE AT BEGINNING OF YEAR 1 (YEAR 0)
    ANALYSIS PERIOD
        MAX USEFUL LIFE = 16
        CUMULATIVE DISTRIBUTION
                USEFUL LIFE   CUMULATIVE PROBABILITY
                    12                0.04
                    12                0.08
                    12                0.12
                    12                0.16
                    12                0.20
                    13                0.24
                    13                0.28
                    13                0.32
```

|    |      |
|----|------|
| 13 | 0.36 |
| 13 | 0.40 |
| 14 | 0.44 |
| 14 | 0.48 |
| 14 | 0.52 |
| 14 | 0.56 |
| 14 | 0.60 |
| 15 | 0.64 |
| 15 | 0.68 |
| 15 | 0.72 |
| 15 | 0.76 |
| 15 | 0.80 |
| 16 | 0.84 |
| 16 | 0.88 |
| 16 | 0.92 |
| 16 | 0.96 |
| 16 | 1.00 |

```
DEPRECIATION METHOD
      DEPRECIABLE LIFE = 0 YEARS
      BEFORE-TAX COMPUTATION.  DEPRECIATION NOT COMPUTED.
END-OF-LIFE SALVAGE VALUE
      NO SALVAGE VALUE
BORROWED MONEY
      NO BORROWED MONEY
EXPENSED COSTS
      NO EXPENSED ANNUAL COST
INCOME TAX
      INCOME TAXES IGNORED
BENEFIT SCHEDULE
      UNIFORM BENEFIT = 250.
BENEFIT TREND
      NO TREND
EXTERNAL INVESTMENT
      INTEREST RATE EARNED ON ANY EXTERNAL INVESTMENT = 0.060
MODEL ITERATIONS = 200
```

|    | ITER | CAPITALIZED | TOTAL | TOTAL | TOTAL | SUM NET | RATE OF |
|----|------|-------------|-------|-------|-------|---------|---------|
| YR | NO | INVESTMENT | BENEFITS | EXP COST | TAXES | CASH FLOW | RETURN |
| 15 | 1 | 1682. | 3750. | 0 | 0 | 2068. | 0.124 |
| 14 | 2 | 1276. | 3500. | 0 | 0 | 2224. | 0.177 |
| 16 | 3 | 1458. | 4000. | 0 | 0 | 2542. | 0.155 |
| 15 | 4 | 1212. | 3750. | 0 | 0 | 2538. | 0.193 |
| 12 | 5 | 1713. | 3000. | 0 | 0 | 1287. | 0.098 |
| 15 | 6 | 1557. | 3750. | 0 | 0 | 2193. | 0.138 |
| 12 | 7 | 1273. | 3000. | 0 | 0 | 1727. | 0.165 |
| 13 | 8 | 1720. | 3250. | 0 | 0 | 1530. | 0.107 |
| 12 | 9 | 1506. | 3000. | 0 | 0 | 1494. | 0.127 |
| 15 | 10 | 1561. | 3750. | 0 | 0 | 2189. | 0.138 |
| . | . | . | . | . | . | . | . |
| . | . | . | . | . | . | . | . |
| . | . | . | . | . | . | . | . |
| 12 | 191 | 1216. | 3000. | 0 | 0 | 1784. | 0.175 |
| 15 | 192 | 1422. | 3750. | 0 | 0 | 2328. | 0.155 |
| 14 | 193 | 1685. | 3500. | 0 | 0 | 1815. | 0.118 |
| 13 | 194 | 1470. | 3250. | 0 | 0 | 1780. | 0.137 |
| 14 | 195 | 1211. | 3500. | 0 | 0 | 2289. | 0.189 |
| 14 | 196 | 1622. | 3500. | 0 | 0 | 1878. | 0.124 |
| 16 | 197 | 1600. | 4000. | 0 | 0 | 2400. | 0.135 |
| 14 | 198 | 1239. | 3500. | 0 | 0 | 2261. | 0.182 |
| 15 | 199 | 1424. | 3750. | 0 | 0 | 2326. | 0.155 |
| 16 | 200 | 1590. | 4000. | 0 | 0 | 2410. | 0.138 |

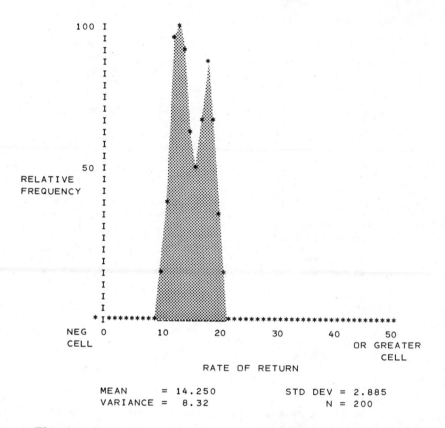

```
100 I              *
    I             *
    I            *
    I          
    I                    *
    I
    I
    I
    I
    I                    *
    I                  *
    I               *
    I              *
 50 I
    I
RELATIVE I        *
FREQUENCY I
    I                    *
    I
    I
    I
    I       *            *
    I
    I
    I
  * I***********         **************************
    NEG  0      10      20      30      40      50
    CELL                                  OR GREATER
                                             CELL
                    RATE OF RETURN
```

```
MEAN      =  14.250         STD DEV = 2.885
VARIANCE  =   8.32               N = 200
```

The plot of rate of return vs relative frequency may be compared with Figure 14-15 and 14-16. The results are substantially the same, but the computer generated results took about $3 of computer time, compared to several hours of hand calculations for Figure 14-16.

### EXAMPLE 18-8

Compute a graph of rate of return vs relative frequency for a proposed investment in machinery, based on the CEAP computer program with 200 iterations.

The machinery at year 0 has a normally distributed cost with a $10,000 mean and a $200 standard deviation. The machinery has a normally distributed useful life that will average 20 years with a two-year standard deviation, but will not exceed 30 years. The salvage value at the end of the machine's useful life is $200. The analysis period is the useful life of the equipment.

Sum-of-years digits depreciation is to be used for tax purposes, based on a 20-year depreciable life. The combined tax rate (federal and state) for this corporation is 52%. Half of the initial cost of the machinery will be borrowed at $6\frac{1}{2}\%$ interest and will be repaid in ten uniform annual end-of-year payments. (Interest paid on the loan is part of the deductible expensed cost of the corporation.) Annual operation and maintenance expense is $200.

The benefits from buying the machinery are normally distributed with mean of $2000 per year and a $450 standard deviation. The distribution of annual benefits does not shift during the analysis period (no benefit trend). In computing rate of return, any external investment is assumed to be at 6%.

If ISWCH is set equal to 1, the detailed output will give a cash flow table and complete rate of return computation (ROR) for each iteration. For 200 iterations this might produce 1000 pages of computer output. The following portions of the computer output are reproduced here:

1. Documentation of the input data.
2. Detailed cash flow table for the first iteration.
3. Summary results for a portion of the 200 iterations.
4. The graph of rate of return vs relative frequency.

```
        CAPITAL EXPENDITURE ANALYSIS PROGRAM
                    PROJECT 1
    FROM NEWNAN - ENGINEERING ECONOMIC ANALYSIS

INVESTMENT
        NORMAL DISTRIBUTION WITH MEAN = 10,000
        AND STD DEV = 200.
    INVESTMENT TIMING
        TOTAL INVESTMENT MADE AT BEGINNING OF YEAR 1 (YEAR 0)
    ANALYSIS PERIOD
        MAX USEFUL LIFE = 30
        NORMAL DISTRIBUTION WITH MEAN = 20 AND STD DEV = 2.
    DEPRECIATION METHOD
        DEPRECIABLE LIFE = 20 YEARS
        SUM-OF-YEARS DIGITS
    END-OF-LIFE SALVAGE VALUE
        SALVAGE VALUE FIXED = 200.
    BORROWED MONEY
        UNIFORM REPAYMENT SCHEDULE
        TERM OF LOAN = 10 YEARS
        PROPORTION OF YEAR 0 INVESTMENT BORROWED = 0.50
        NOMINAL INTEREST RATE PER ANNUM = 0.0650
    EXPENSED COSTS
        EXPENSED COSTS FIXED = 200.
    INCOME TAX
        UNIFORM TAX RATE
        TAX RATE = 0.52
    BENEFIT SCHEDULE
        NORMAL DISTRIBUTION WITH MEAN = 2000
        AND STD DEV = 450.
    BENEFIT TREND
        NO TREND
    EXTERNAL INVESTMENT
        INTEREST RATE EARNED ON ANY EXTERNAL INVESTMENT = 0.060
MODEL ITERATIONS = 200
```

| YR | CUMULATIVE INVESTMENT | ANNUAL INVESTMENT | DEPREC SCHEDULE | BORROWED MONEY | LOAN PAYMT PRIN + INT | LOAN INTEREST |
|---|---|---|---|---|---|---|
| 0 | 10328. | 10328. | | 5164. | | |
| 1 | 10328. | 0 | 965. | | 718. | 336. |
| 2 | 10328. | 0 | 916. | | 718. | 311. |
| 3 | 10328. | 0 | 868. | | 718. | 284. |
| 4 | 10328. | 0 | 820. | | 718. | 256. |
| 5 | 10328. | 0 | 772. | | 718. | 226. |
| 6 | 10328. | 0 | 723. | | 718. | 194. |
| 7 | 10328. | 0 | 675. | | 718. | 160. |
| 8 | 10328. | 0 | 627. | | 718. | 124. |
| 9 | 10328. | 0 | 579. | | 718. | 85. |
| 10 | 10328. | 0 | 531. | | 718.' | 44. |
| 11 | 10328. | 0 | 482. | | 0 | 0 |
| 12 | 10328. | 0 | 434. | | 0 | 0 |
| 13 | 10328. | 0 | 386. | | 0 | 0 |
| 14 | 10328. | 0 | 338. | | 0 | 0 |
| 15 | 10328. | 0 | 289. | | 0 | 0 |
| 16 | 10328. | 0 | 241. | | 0 | 0 |
| 17 | 10328. | 0 | 193. | | 0 | 0 |

| YR | -INVESTMENT | + PROJECT BENEFITS | - EXPENSED COST | - INCOME TAX | EQUALS | NET CASH FLOW |
|---|---|---|---|---|---|---|
| 0 | 10328. | | | 0 | | -10328. |
| 1 | 0 | 1327. | 200. | +90. | | 1217. |
| 2 | 0 | 1289. | 200. | +72. | | 1161. |
| 3 | 0 | 2277. | 200. | -481. | | 1596. |
| 4 | 0 | 1137. | 200. | +72. | | 1009. |
| 5 | 0 | 1483. | 200. | -148. | | 1135. |
| 6 | 0 | 2128. | 200. | -525. | | 1402. |
| 7 | 0 | 2171. | 200. | -591. | | 1380. |
| 8 | 0 | 2091. | 200. | -593. | | 1298. |
| 9 | 0 | 1806. | 200. | -490. | | 1116. |
| 10 | 0 | 2660. | 200. | -980. | | 1479. |
| 11 | 0 | 1513. | 200. | -432. | | 881. |
| 12 | 0 | 2538. | 200. | -990. | | 1348. |
| 13 | 0 | 2184. | 200. | -831. | | 1153. |
| 14 | 0 | 2891. | 200. | -1224. | | 1467. |
| 15 | 0 | 2857. | 200. | -1231. | | 1426. |
| 16 | 0 | 1220. | 200. | -405. | | 615. |
| 17 | 0 | 2805. | 200. | -1104. | | 1701. |
| 17 | SALVAGE VALUE = 200. | | | | | |

| YR | ITER NO | CAPITALIZED INVESTMENT | TOTAL BENEFITS | TOTAL EXP COST | TOTAL TAXES | SUM NET CASH FLOW | RATE OF RETURN |
|---|---|---|---|---|---|---|---|
| 17 | 1 | 10328. | 34376. | 3400. | 9791. | 11057. | 0.095 |
| 20 | 2 | 9724. | 42659. | 4000. | 14161. | 14973. | 0.131 |
| 16 | 3 | 9595. | 30790. | 3200. | 8486. | 9709. | 0.106 |
| 22 | 4 | 9580. | 45857. | 4400. | 15706. | 16371. | 0.140 |
| 20 | 5 | 10354. | 38746. | 4000. | 11735. | 12857. | 0.105 |
| 23 | 6 | 10315. | 46384. | 4600. | 15419. | 16250. | 0.117 |
| 17 | 7 | 9684. | 35178. | 3400. | 10608. | 11686. | 0.111 |
| 21 | 8 | 10456. | 41764. | 4200. | 13138. | 14171. | 0.112 |
| 23 | 9 | 10189. | 45533. | 4600. | 15055. | 15889. | 0.104 |
| 23 | 10 | 10295. | 44475. | 4600. | 14439. | 15341. | 0.107 |
| . | . | . | . | . | . | . | . |
| . | . | . | . | . | . | . | . |
| . | . | . | . | . | . | . | . |

| 22 | 191 | 9580. | 45857. | 4400. | 15706. | 16371. | 0.140 |
| 20 | 192 | 10354. | 38746. | 4000. | 11735. | 12857. | 0.105 |
| 23 | 193 | 10315. | 46384. | 4600. | 15419. | 16250. | 0.117 |
| 17 | 194 | 9684. | 35178 | 3400. | 10608. | 11686. | 0.111 |
| 21 | 195 | 10456. | 41764. | 4200. | 13138. | 14171. | 0.112 |
| 23 | 196 | 10189. | 45533. | 4600. | 15055. | 15889. | 0.104 |
| 23 | 197 | 10295. | 44475. | 4600. | 14439. | 15341. | 0.107 |
| 18 | 198 | 10080. | 36017. | 3600. | 10694. | 11842. | 0.107 |
| 22 | 199 | 10456. | 41211. | 4400. | 12745. | 13809. | 0.111 |
| 19 | 200 | 10293. | 41239. | 3800. | 13173. | 14173. | 0.120 |

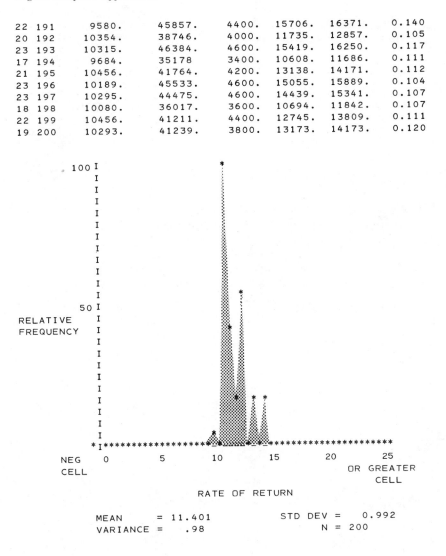

## COMPUTING SPECIALIZED
## COMPOUND INTEREST TABLES

One problem frequently encountered is that the available compound interest tables do not contain a particular desired compound interest factor. What are the values, for example, of $(A/P,55\%,3\text{yrs})$ and $(F/G,6.4\%,12\text{yrs})$? Neither factor is in the tables at the back of this book. When the need arises, it is a relatively simple matter to derive an expression for the compound interest factor and then compute a table of values on a computer.

In Chapter 4, equations for the two single payment factors with continuous compounding were derived.

$$P = Fe^{-rn}$$
$$F = Pe^{rn}$$

The four uniform series factors for continuous compounding are:

$$A = F\left[\frac{e^r - 1}{e^{rn} - 1}\right] \qquad F = A\left[\frac{e^{rn} - 1}{e^r - 1}\right]$$

$$A = P\left[\frac{e^r - 1}{1 - e^{-rn}}\right] \qquad P = A\left[\frac{1 - e^{-rn}}{e^r - 1}\right]$$

If desired, a set of compound interest tables could be computed for continuous compounding. Table 18-3 is an example for a nominal 8% interest rate.

**Table 18-3.** Compound interest tables with continuous compounding nominal interest rate = 8% (effective interest rate = 8.3287%)

| | SINGLE PAYMENT | | UNIFORM PAYMENT SERIES | | | |
|---|---|---|---|---|---|---|
| | Compound Amount Factor | Present Worth Factor | Sinking Fund Factor | Capital Recovery Factor | Compound Amount Factor | Present Worth Factor |
| | Find $F$ Given $P$ | Find $P$ Given $F$ | Find $A$ Given $F$ | Find $A$ Given $P$ | Find $F$ Given $A$ | Find $P$ Given $A$ |
| $n$ | $F/P$ | $P/F$ | $A/F$ | $A/P$ | $F/A$ | $P/A$ |
| 1 | 1.083 | .9231 | 1.0000 | 1.0833 | 1.000 | .923 |
| 2 | 1.174 | .8521 | .4800 | .5633 | 2.083 | 1.775 |
| 3 | 1.271 | .7866 | .3071 | .3903 | 3.257 | 2.562 |
| 4 | 1.377 | .7261 | .2208 | .3041 | 4.528 | 3.288 |
| 5 | 1.492 | .6703 | .1693 | .2526 | 5.905 | 3.958 |
| 6 | 1.616 | .6188 | .1352 | .2185 | 7.397 | 4.577 |
| 7 | 1.751 | .5712 | .1109 | .1942 | 9.013 | 5.148 |
| 8 | 1.896 | .5273 | .0929 | .1762 | 10.764 | 5.676 |
| 9 | 2.054 | .4868 | .0790 | .1623 | 12.660 | 6.162 |
| 10 | 2.226 | .4493 | .0680 | .1512 | 14.715 | 6.612 |
| 11 | 2.411 | .4148 | .0590 | .1423 | 16.940 | 7.027 |
| 12 | 2.612 | .3829 | .0517 | .1350 | 19.351 | 7.409 |
| 14 | 3.065 | .3263 | .0403 | .1236 | 24.792 | 8.089 |
| 15 | 3.320 | .3012 | .0359 | .1192 | 27.857 | 8.390 |
| 16 | 3.597 | .2780 | .0321 | .1154 | 31.177 | 8.668 |
| 17 | 3.896 | .2567 | .0288 | .1120 | 34.774 | 8.925 |
| 18 | 4.221 | .2369 | .0259 | .1091 | 38.670 | 9.162 |
| 20 | 4.953 | .2019 | .0211 | .1044 | 47.463 | 9.583 |
| 22 | 5.812 | .1720 | .0173 | .1006 | 57.781 | 9.941 |
| 25 | 7.389 | .1353 | .0130 | .0963 | 76.711 | 10.382 |
| 26 | 8.004 | .1249 | .0119 | .0952 | 84.100 | 10.507 |
| 30 | 11.023 | .0907 | .0083 | .0916 | 120.345 | 10.917 |
| 40 | 24.533 | .0408 | .0035 | .0868 | 282.547 | 11.517 |
| 50 | 54.598 | .0183 | .0016 | .0848 | 643.535 | 11.787 |
| 60 | 121.510 | .0082 | .0007 | .0840 | 1446.928 | 11.908 |
| 70 | 270.426 | .0037 | .0003 | .0836 | 3234.913 | 11.962 |
| 80 | 601.845 | .0017 | .0001 | .0834 | 7214.145 | 11.987 |

*EXAMPLE 18-9*

An end-of-year deposit of $100 is made each year for 25 years in a bank savings account. How much will be in the account at the end of the 25 years if annual interest of (a) 8% is compounded annually? (b) 8% is compounded continuously? (c) 8.3287% is compounded annually?

(a) Using the regular compound interest tables,
$$F = A(F/A,i\%,n) = 100(F/A,8\%,25\text{yrs}) = 100(73.106) = \$7310.60$$

(b) For continuous compounding Table 18-3 is applicable.
$$F = A(F/A,r\%,n) = 100(76.711) = \$7671.10$$

(c) For $i = 8.3287\%$ we can compute the future sum from the equation

$$F = A\left[\frac{(1 + i)^n - 1}{i}\right] = \$100\left[\frac{1.0832872^5 - 1}{0.083287}\right] = 100(76.711)$$

$$= \$7671.10$$

The effective interest rate for 8% compounded continuously is shown in Table 4-3 to be 8.3287%. Thus the computation in (b) is also applicable for (c).

## SUMMARY

Although this chapter is entitled Digital Computer Applications, an important topic discussed is the computation of rate of return for an investment project. The situation is far more complex than was stated in Chapter 7.

The cash flow rule of signs is the simple test of the number of changes of sign in the project cash flow. Zero sign changes is the unusual case of all benefits or all disbursements. This is immediately seen as either good or a disaster. One sign change is the conventional situation. This is the case that has been discussed throughout most of this book. It generally leads to a single positive rate of return that is a valid measure of project profitability. Multiple sign changes are a warning sign that a valid measure of project profitability may be difficult to locate.

The accumulated cash flow sign test is a test of the accumulated cash flows $(A_0, A_1, A_2, \ldots, A_n)$. The algebraic sum of the cash flow is simply the sum of the project cash flow and equals the accumulated cash flow, $A_n$. If there is one sign change in the sequence and $A_n \neq 0$, these are sufficient (but not necessary) conditions for a single positive rate of return. When there is more than one sign change in the cash flow rule of signs, the existance of a single positive rate of return does not necessarily give us a suitable measure of project profitability. The critical test is net investment.

Net investment is the computation to determine whether or not there is a continuing net investment throughout the life of the project for a value of $i^*$. Thus the first step in computing net investment is to locate a value of $i^*$. Then net investment is computed year-by-year with the net investment earning $i^*$ each year. A cash flow will either have net investment throughout the life of the project, or it will not have it. A cash flow that does not have net investment is one where there is a net outflow of money from the project which will later be required to be returned to the project. At this point we need to invest the outflow of money someplace else (at external interest rate $e^*$) until it is required back in the project.

If there is net investment throughout the life of the project, we know this is sufficient (but not necessary) for a single positive rate of return, and this $i^*$ is a suitable measure of project profitability. When there is not net investment throughout the life of the project, another rate of return $i^*$ must be computed, with an external interest rate $e^*$ assumed for project money temporarily invested externally. This revised $i^*$ is an appropriate measure of project profitability. As we have seen, the calculation of $i^*$, using an external interest rate $e^*$, can be lengthy. A FORTRAN computer program ROR can be used to perform all the required computations.

Simulation is another analysis tool that has lengthy calculations. The logic and fundamental techniques were described in Chapter 14. Here a computer program CEAP is provided to perform simulation.

Finally, additional compound interest tables may be readily constructed with a computer.

## A CLOSING COMMENT

Engineering economic analysis, as you have seen, is a practical subject. And it is a lot more than just understanding compound interest computations. The techniques described are suitable for use in one's own personal life as well as one's professional career.

The test is not whether the calculations are accurate and technically correct, but whether a good decision is made. If you will work at it, and help others around you to do the same, then better decisions will result and the ultimate purpose of this book will have been accomplished.

## PROBLEMS

**18-1**  Consider the following situation.

| Year | Cash flow |
|---|---|
| 0 | -$500 |
| 1 | +2000 |
| 2 | -1200 |
| 3 | -300 |

(a) For the cash flow, what information can be learned from
  1. Cash flow rule of signs.
  2. Accumulated cash flow sign test.
  3. Algebraic sum of the cash flow.
  4. Net investment conditions.

(b) Compute the internal rate of return for the cash flow. If there is external investment, assume it is made at the same interest rate as the internal rate of return.

(c) Compute the internal rate of return for the cash flow. This time assume a 6% external interest rate.

**18-2** Repeat Problem 18-1 for the following cash flow.

| Year | Cash flow |
|------|-----------|
| 0 | − $500 |
| 1 | + 200 |
| 2 | − 500 |
| 3 | + 1200 |

(*Answer:* (b) 21.1%, (c) 21.1%)

**18-3** Repeat Problem 18-1 for the following cash flow.

| Year | Cash flow |
|------|-----------|
| 0 | − $500 |
| 1 | + 200 |
| 2 | − 500 |
| 3 | + 200 |

**18-4** Repeat Problem 18-1 for the following cash flow.

| Year | Cash flow |
|------|-----------|
| 0 | − $100 |
| 1 | + 360 |
| 2 | − 570 |
| 3 | + 360 |

**18-5** Given the following cash flow.

| Year | Cash flow |
|------|-----------|
| 0 | − $200 |
| 1 | + 100 |
| 2 | + 100 |
| 3 | + 100 |
| 4 | − 300 |
| 5 | + 100 |
| 6 | + 200 |
| 7 | + 200 |
| 8 | − 124.5 |

Compute the internal rate of return assuming

(a) External interest rate equals the internal rate of return.

(b) External interest rate equals 6%.

(*Answer:* (a) 20% (b) 18.9%)

**18-6**   In Examples 7A-1 and 7A-2 (in the appendix of Chapter 7) an analysis was made of the proposed sale of four planes from Going Aircraft Company to Interair. In Example 7A-2 the rate of return was computed to be 8.4%, given a 6% external interest rate. Using the methods described in this chapter, compute $i^*$, given $e^* = 6\%$. Explain why your answer is or is not different from the 8.4% computed in Example 7A-2.

**18-7**   Calculate an equation for the uniform series present worth factor, assuming uniform *beginning of period* payments or disbursements. Write a computer program to generate a table of values for a 6% interest rate and values of $n$ from 1 to 50.

# Appendix A
## Computer Programs

ROR.     Rate of return. Given a cash flow, compute a rate of return on the internal investment $i*$ using, if necessary, a preselected external interest rate $e*$.

CEAP.    Capital expenditure analysis program. A capital investment simulation program.

          CEAP documentation.

          Listing of CEAP program.

## RATE OF RETURN PROGRAM

```
      PROGRAM ROR
C     RATE OF RETURN COMPUTATION
C     FROM NEWNAN - ENGINEERING ECONOMIC ANALYSIS
      DIMENSION CASH (50), TEMP(51)
      IRD = 60
      IPR = 61
    1 READ(IRD,2) NR,N,EXT,CASHO,(CASH(I),I=1,N)
    2 FORMAT(2I2,F4.3,F12.2/(6F12.2))
C     UNLESS RESET, PRINT TABULATION AND RESULTS IN SUBROUTINE
      ISW = 2
C     UNLESS RESET, EXTERNAL I = EXT
      MTH = 2
C     NR = 1 ONLY 1 PROBLEM TO SOLVE, OR THIS IS LAST ONE.
C        = 2 ANOTHER SEPARATE PROBLEM FOLLOWS THIS ONE.
C     PROGRAM COMPUTES NCHGS = NO. OF SIGN CHGS IN CASH FLOW.
C                       SUM  = ALGEBRAIC SUM OF CASH FLOW.
C                       NCASH = NO. SIGN CHGS IN ACCUM CASH FL.
      WRITE(IPR,3)
    3 FORMAT(///25X,26HRATE OF RETURN COMPUTATION/11X,
    1 43HFROM NEWNAN - ENGINEERING ECONOMIC ANALYSIS//)
      EXT2 = EXT
      TEMP(1) = CASHO
      SUM = CASHO
      DO 10 I=1,N
      TEMP(I+1) = CASH(I)
   10 SUM = SUM + CASH(I)
      CALL SIGNS(TEMP,N+1,NCHGS)
      DO 20 I=1,N
```

```
   20 TEMP(I+1) = TEMP(I) + CASH(I)
      CALL SIGNS(TEMP,N+1,NCASH)
C     OUTPUT THE RESULTS OF THESE COMPUTATIONS
      WRITE(IPR,21) CASHO,CASHO,(I,CASH(I),TEMP(I+1), I=1,N)
   21 FORMAT(35X,11HACCUMULATED/10X,4HYEAR,6X,9HCASH FLOW,7X,
     1 9HCASH FLOW//12X,1H0,4X,F12.2,5X,F12.2/(11X,I2,4X,
     2 F12.2,5X,F12.2))
      WRITE(IPR,22) SUM,NCHGS,NCASH
   22 FORMAT(/12X,5HSUM =,F12.2//10X,22HNUMBER OF SIGN CHANGE
     1S/22X,14HIN CASH FLOW =,I3/10X,26HIN ACCUMULATED CASH F
     2LOW =,I3///)
      WRITE(IPR,30) NCHGS
   30 FORMAT(5X,23HCASH FLOW RULE OF SIGNS/10X,35HTHERE MAY B
     1E AS MANY POSITIVE RATES/10X,35HOF RETURN AS THERE ARE
     2SIGN CHANGES/10X,17HIN THE CASH FLOW.//10X,18HFOR THIS
     3CASH FLOW/10X,43HPOSSIBLE NUMBER OF POSITIVE RATES OF R
     4ETURN/10X,13HBETWEEN 0 AND,I3///)
      WRITE (IPR,35)
   35 FORMAT(5X,37HSIGN CHANGES IN ACCUMULATED CASH FLOW/10X,
     1 39HTWO CONDITIONS THAT ARE SUFFICIENT (BUT/10X,36HNOT
     2NECESSARY) FOR A SINGLE POSITIVE/10X,18HRATE OF RETURN
     3ARE//15X,35HTHE ACCUMULATED CASH FLOW IN YEAR N/20X,
     4 21HIS NOT EQUAL TO ZERO.//15X,3HAND//15X,31HTHERE IS O
     5NE SIGN CHANGE IN THE/20X,22HACCUMULATED CASH FLOW.//)
      WRITE(IPR,36) TEMP(N+1),NCASH
   36 FORMAT(10X,18HFOR THIS CASH FLOW/10X,33HACCUMULATED CAS
     1H FLOW IN YEAR N =,F12.2/10X,28HAND NUMBER OF SIGN CHAN
     2GES =,I3//)
      IF (TEMP(N+1).NE.0..AND.NCASH.EQ.1) GO TO 41
      WRITE(IPR,40)
   40 FORMAT(10X,52HNO CONCLUSIONS CAN BE REACHED FROM THIS C
     1ALCULATION./)
      GO TO 43
   41 WRITE(IPR,42)
   42 FORMAT(10X,41HTHERE IS A SINGLE POSITIVE RATE OF RETURN
     1//)
C     ALGEBRAIC SUM OF CASH FLOW
   43 WRITE(IPR,50) SUM
   50 FORMAT(5X,26HALGEBRAIC SUM OF CASH FLOW/10X,42HA NEGATI
     1VE SUM INDICATES THAT COSTS EXCEED/10X,48HBENEFITS. GEN
     2ERALLY AN UNSATISFACTORY INVESTMENT/10X,10HSITUATION.//
     3 10X,45HA POSITIVE SUM INDICATES ONE OR MORE POSITIVE
     4 1X,16HRATES OF RETURN.//10X,33HA ZERO SUM FREQUENTLY I
     5NDICATES A/10X,27HZERO PERCENT RATE OF RETURN/10X,
     6 24HFOR THIS CASH FLOW SUM =,F12.2//)
C     USING ABOVE DATA PROCEED WITH ANALYSIS
      WRITE(IPR,61)
   61 FORMAT(5X,39HUSING ABOVE DATA, PROCEED WITH ANALYSIS//)
      INDEX = NCHGS -1
      IF(INDEX) 65,75,150
C     ZERO SIGN CHANGES
   65 IF(SUM) 66,70,72
   66 WRITE(IPR,67)
   67 FORMAT(5X,25HALL COSTS AND NO BENEFITS/5X,26HTHERE IS N
     1O RATE OF RETURN//5X,22HANALYSIS IS COMPLETED.)
      GO TO 125
   70 WRITE(IPR,71)
   71 FORMAT(5X,21HTHERE IS NO CASH FLOW/5X,21HAND NO RATE OF
     1 RETURN//5X,22HANALYSIS IS COMPLETED.)
      GO TO 125
```

```
   72 WRITE(IPR,73)
   73 FORMAT(5X,25HALL BENEFITS AND NO COSTS/5X,25HRATE OF RE
     1TURN = INFINITY//5X,22HANALYSIS IS COMPLETED.)
      GO TO 125
C     ONE SIGN CHANGE
   75 IF(SUM) 76,80,85
   76 WRITE(IPR,77)
   77 FORMAT(5X,26HTHERE IS NO RATE OF RETURN//
     1 5X,22HANALYSIS IS COMPLETED.)
      GO TO 125
   80 WRITE(IPR,81)
   81 FORMAT(5X,20HCOSTS EQUAL BENEFITS/5X,26HRATE OF RETURN
     1= 0 PERCENT//5X,22HANALYSIS IS COMPLETED.)
      GO TO 125
   85 WRITE(IPR,86)
   86 FORMAT(5X,26HTHERE IS A SINGLE POSITIVE/5X,14HRATE OF R
     1ETURN//)
      MTH = 1
      CALL RORIE(IPR,CASHO,CASH,N,EXT,ISW,RATE,MTH)
C     REPORT RESULTS OF RATE OF RETURN COMPUTATION
C     WITH INTERNAL = EXTERNAL INTEREST RATE.
      WRITE(IPR,88)
   88 FORMAT(5X,37HRESULTS OF RATE OF RETURN COMPUTATION,5X,
     1 29HWITH EXTERNAL = INTERNAL RATE//)
      GO TO 112
C     GREATER THAN ONE SIGN CHANGE
  150 IF (NCASH-1) 89,91,93
C     NO SIGN CHANGES IN ACCUMULATED CASH FLOW
   89 WRITE(IPR,90)
   90 FORMAT(5X,40HNO SIGN CHANGES IN ACCUMULATED CASH FLOW)
      GO TO 76
C     ONE SIGN CHANGE IN ACCUMULATED CASH FLOW
   91 WRITE(IPR,92)
   92 FORMAT(5X,40HONE SIGN CHANGE IN ACCUMULATED CASH FLOW)
      GO TO 93
C     MORE THAN ONE SIGN CHANGE IN ACCUMULATED CASH FLOW
   93 IF (CASH(N)) 94,94,140
C     LAST TERM IN CASH FLOW NEGATIVE (OR ZERO).
   94 WRITE(IPR,95)
   95 FORMAT(5X,41HLAST TERM IN CASH FLOW NEGATIVE (OR ZERO))
  101 CALL RORIE(IPR,CASHO,CASH,N,EXT,ISW,RATE,MTH)
C     REPORT RESULTS OF RATE OF RETURN COMPUTATION
C     USING EXTERNAL INTEREST RATE.
      WRITE(IPR,111) EXT
  111 FORMAT(5X,37HRESULTS OF RATE OF RETURN COMPUTATION/5X,
     1 29HWITH EXTERNAL INTEREST RATE =,F5.3//)
  112 IF (MTN-4) 113,115,121
  113 WRITE(IPR,114)
  114 FORMAT(5X,45HEXTERNAL INTEREST RATE NEEDED IN COMPUTATI
     1ON.//)
      GO TO 123
  115 WRITE(IPR,116)
  116 FORMAT(5X,35HEXTERNAL INTEREST RATE NOT NEEDED.//)
      GO TO 123
  121 WRITE(IPR,122)
  122 FORMAT(//5X,25HNO RATE OF RETURN LOCATED/5X,
     1 22HANALYSIS IS COMPLETED.)
      GO TO 125
  123 WRITE(IPR,124) RATE
  124 FORMAT(5X,32HCOMPUTED INTERNAL RATE OF RETURN/5X,
```

```
    1 6HEQUALS,F7.4//5X,22HANALYSIS IS COMPLETED.)
C     CHECK TO SEE IF THERE IS ANOTHER PROBLEM TO BE SOLVED.
  125 IF (NR.GT.1) 130,200
  130 WRITE(IPR,131)
  131 FORMAT(/1X,37HTHERE IS ANOTHER PROBLEM TO BE SOLVED///)
      GO TO 1
C     LAST TERM IN CASH FLOW GREATER THAN ZERO
C     CHECK TO SEE IF ALGEBRAIC SUM = 0
  140 IF (SUM.EQ.0.) 141,145
C     ALGEBRAIC SUM = 0
  141 WRITE(IPR,142)
  142 FORMAT(5X,38HALGEBRAIC SUM OF CASH FLOW EQUALS ZERO/5X,
    1 42HAN INTERNAL RATE OF RETURN IS ZERO PERCENT)
      GO TO 101
C     COMPUTE INTERNAL RATE OF RETURN.  EXTERNAL = INTERNAL
  145 MTH = 1
      CALL RORIE(IPR,CASHO,CASH,N,EXT,ISW,RATE,MTH)
      WRITE(IPR,96) RATE
   96 FORMAT(5X,33HFOR EXTERNAL = INTERNAL, COMPUTED/5X,
    1 16HRATE OF RETURN =,F7.4//)
      UNINV = CASHO
      JJ = N-1
      DO 100 I=1,JJ
      UNINV = UNINV + CASH(I)*(1.+RATE)**(-I)
      IF (UNINV) 100,100,105
  100 CONTINUE
      WRITE(IPR,155) RATE
  155 FORMAT(5X,31HTHERE IS UNRECOVERED INVESTMENT/5X,39HFOR
    1ALL PERIODS PRIOR TO THE NTH PERIOD//5X,36HTHE COMPUTED
    2 INTERNAL RATE OF RETURN/5X,35HIS THE ONLY POSITIVE ONE
    3 AND EQUALS,F7.4//5X,22HANALYSIS IS COMPLETED.)
      GO TO 125
  105 WRITE(IPR,106)
  106 FORMAT(5X,35HTHERE IS NOT UNRECOVERED INVESTMENT/5X,
    1 39HFOR ALL PERIODS PRIOR TO THE NTH PERIOD//)
      MTH = 2
      EXT = EXT2
      GO TO 101
  200 CONTINUE
      END
C****************************************************************
C****************************************************************
      SUBROUTINE SIGNS(TEMP,M,NO)
      DIMENSION TEMP(51)
      NO = 0
      A = TEMP(1)
      DO 50 I=2,M
      IF (TEMP(I-1).NE.0) A = TEMP(I-1)
      IF (TEMP(I).EQ.0.OR.A.EQ.0.) GO TO 50
      Y = A*TEMP(I)
      IF (Y.LT.0.) NO = NO + 1
   50 CONTINUE
      RETURN
      END
C****************************************************************
C****************************************************************
      SUBROUTINE RORIE(OUT,CASHO,CASH,N,EXT,ISW,RATE,MTH)
      INTEGER OUT
FOR LISTING SEE CAPITAL EXPENDITURE ANALYSIS PROGRAM (CEAP).
```

## CAPITAL EXPENDITURE ANALYSIS
## PROGRAM DOCUMENTATION

*Section 1.* The logic of the capital expenditure analysis model is essentially linear and proceeds from Section 1 to GRAPH. In the initial part of the program, the required dimensioning of the various vectors is accomplished. The essential common cards are added, and reader and printer computer device numbers are assigned. The program then proceeds into Section 1, which reads in the fixed data of the problem. There are four data cards, and these four cards are required in each and every program. Their description is provided on the input data card format sheets. The only other element treated in Section 1 is to establish an integer representation of the term of debt.

*Section 2.* In Section 2 the seed for the pseudorandom number generator RANG is defined. This section also is used to initially load all vectors with zeros. This may or may not be necessary, depending upon the computer on which the program is to be run.

*Section 3.* The four data cards in Section 1 provided an input of some control digits, so that the user might indicate whether the parameters are to be determined internally by built-in distributions, or whether exogeneous data are to be loaded at this time.

If established by the control digits, the following are read:

TP       Investment timing distribution

BRS     Loan repayment schedule

DEP     Depreciation schedule

CCD     Cumulative expensed cost distribution

CBS     Cumulative annual benefit distribution

At this point in the program, the economic analysis problem has been completely defined.

*Section 4.* This section serves the single purpose of echoing the input data. The problem has been described in terms of eleven elements and in Section 4 these elements are output on the printer.

*Section 5.* In this section, the main processing DO loop begins. The first task in Section 5 will be to re-initialize all of those variables which will be used in each individual iteration.

*Section 6.* The analysis period is determined in this section. A useful life control digit is analyzed by means of a computed GO TO. Alternatives available are: (a) useful life is fixed; (b) useful life is normally distributed; (c) useful life has a log normal distribution; (d) the useful life uses Erlang; (if the exponential distribution is desired, then it is obtained from Erlang simply by setting the value of $K$, here called KERL1, $= 1$); and (e) cumulative life distribution LCD is used and was input in Section 3. The value of useful life $N$ has now been determined. The value is compared to a pre-set maximum useful life NMAX, and in the event that $N$ exceeds NMAX, it is reset equal to NMAX. Thus, the analysis period $=$ asset useful life $= N$.

*Section 7.*    In this section the capitalized investment is determined. It follows the same general pattern followed in Section 6, that is, the capital control investment digit IFC is analyzed by means of a computed GO TO with the following options provided: (a) capitalized investment is fixed; (b) capitalized investment normally distributed; (c) log normal distribution; and (d) Erlang.

*Section 8.*    The salvage or terminal value of the asset (SV) is determined in Section 8. Here the options are: (a) no salvage value; (b) salvage value fixed; (c) salvage value is a linear decline from salvage value SVMAX to salvage value SVLOW; and (d) salvage value is determined from cumulative salvage value distribution CSV.

*Section 9.*    This section addresses itself to the relatively complex problem of investment timing. Here the options provided are: (a) total investment is assumed to occur at the beginning of year one, which we describe in this program as year zero; (b) investment is equally distributed over TN years. Here the first investment is assumed to take place at the beginning of year one (year zero) and the last investment occurs at the beginning of year TN. Since the program uses the end-of-year convention, the last investment is at the end of year TN-1. In the event the useful life of the asset is less than TN years, the investment is re-distributed to occur over $N$ years rather than TN years. The last option is (c) investment timing is provided by means of the investment timing schedule TP, which was input in Section 3. The investment timing schedule permits the investment to be distributed over ten years. It is again possible that the investment useful life is less than ten, so the investment timing schedule TN may need to be altered, and if necessary this is accomplished in the vector REBTP (rebalanced TN).

*Section 10.*    This section determines the deductible interest charge which results from borrowed money (if any) used in the project. The borrowed money control digit allows four options: (a) no borrowed money; (b) uniform repayment schedule; (c) equal principal payment schedule, plus annual interest; and (d) repayment schedule is provided from Section 3.

The only function of this section is to develop the interest charge in the event the economic analysis is to be done on an after-tax basis. If this is true, then the interest paid in any year is a tax deductible item. As in prior sections, there must be an examination here to ensure that the borrowed money repayment schedule does not exceed the useful life in any iteration. If it does, then the repayment schedule is accelerated so that the debt is extinguished by the end of the useful life $N$.

*Section 11.* This section computes the depreciation schedule for the asset based upon an established, depreciable life which is provided from the input data cards. Actual computation is done in SUBROUTINE DEPRC. By means of a control digit, the subroutine permits seven options: (a) straight-line depreciation; (b) sum-of-years digits method; (c) double declining balance; (d) double declining balance with conversion to straight line; (e) declining balance at 150% of straight line; (f) depreciation schedule provided by Section 3; and (g) depreciation is ignored. This latter event would occur, presumably, where a before-tax computation is to be made.

In the event that the asset is not depreciated down to its salvage value, by the time the useful life $N$ is reached in any iteration, the actual result for tax purposes would be a loss on disposal. This is assumed to occur at the end of the useful life if the book value of the asset is more than the salvage value. The program then makes the customary accounting computations. The tax adjustment itself is made in Section 14.

*Section 12.* In this section, the before-tax benefit schedule BENF is compiled. The benefit schedule control digit IBS provides for five options: (a) benefits are fixed; (b) benefits are normally distributed; (c) benefit distribution is log normal; (d) benefits follow Erlang distribution; and (e) benefits are determined from the cumulative annual benefit distribution input in Section 3. The distribution is sampled by means of the random number generator. Following the compilation of the benefit schedule, the program provides an option for introducing a trend into the benefit schedule. The benefit control digit ITB allows two options: (a) no trend; and (b) trend equals plus or minus the decimal percentage per year. The benefit schedule is adjusted for the benefit trend.

*Section 13.* The expensed cost schedule EXPC is computed in this section. There are six options: (a) no expensed annual cost; (b) expensed annual cost is fixed; (c) expensed cost is normally distributed; (d) expensed cost follows log normal distribution; (e) expensed cost is obtained from Erlang distribution; and (f) expensed cost distribution was provided in Section 3 and is sampled using the random number generator.

*Section 14.* This section computes the income tax consequences schedule TXS. In this program we define taxable income as equal to annual benefit, minus the depreciation allowed or allowable, minus the interest cost for the year, minus the expensed cost for the year. This is in agreement with customary accounting practice. The income tax control digit provides for three options: (a) computation is to be a before-tax computation (hence income taxes are ignored); (b) uniform tax rate is applied; and (c) uniform tax rate is used together with an investment tax credit allowance. It is assumed that the investment tax benefit occurs at the time of the investment; thus, an investment at year 0 would similarly have its investment tax credit during that year, and would, therefore, be recorded as a saving in year 0.

In the event the asset is retired prior to the end of its depreciable life, there may be a loss on disposal to take into account. The loss would be the depreciated book value, minus the salvage value. If necessary, this adjustment is made.

*Section 15.*   The after-tax cash flow schedule is computed in this section. Here, after-tax cash flow is defined as benefits, minus expensed costs, minus taxes, minus the year's investment. If the parameter ISWCH is set to something other than zero, the entire cash flow table for each iteration is printed at this point. If ISWCH is equal to zero, printing is suppressed and the section is concluded.

*Section 16.*   This section does the preliminary calculations necessary to compute a rate of return for the after-tax cash flow, ATCH. The number of sign changes in the cash flow is computed, along with the sum of the cash flow.

For no sign changes in the cash flow
    and sum is negative or zero:        Rate of return set equal to 0%
    and sum is positive:              Rate of return set equal to 100% (The true rate of return is infinity.)

For one sign change in the cash flow
    and sum is negative or zero:        Rate of return set equal to 0%
    and sum is positive:              Rate of return computed by subroutine RORIE.

For more than one sign change in the cash flow, the rate of return is computed by subroutine RORIE.

*Section 17.*   This section prints an output heading and then prints one line of summary statistics at the end of each iteration. As a result there will be ITER lines of output. This output is not controlled by ISWCH and always appears.

*Section 18.*   This section takes the trial rate of return for the iteration and assigns it to one cell of the vector NR, which will be used to print the rate of return frequency distribution. At this point the range of the main iteration DO loop, labeled 610, is reached. Control is returned to the beginning of Section 5. When the main iteration loop is satisfied, Section 18 calls the rate of return plot routine GRAPH. Following this, the program terminates.

*Subroutine RORIE.* Printing in this subroutine is controlled by ISWCH.

If ISWCH = 0,   ISW = 1      printing is suppressed
   ISWCH = 1,   ISW = 2      the tabulation and results are printed

The algebraic sum of the cash flow, ASUM, is computed. The "original book" method of computing an approximate rate of return is:

$$\text{Rate of return} = \frac{\text{average annual profit after depreciation}}{\text{original investment}}$$

Thus an estimate of the rate of return is roughly

$$\text{TRIAL} = \frac{\text{SUM/N}}{\text{CASHO}} = \frac{\text{SUM}}{\text{CASHO} * \text{N}}$$

In the program, TRIAL is purposely set 10% higher in an attempt to make TRIAL slightly higher than the answer being sought.

The search area is limited to the range 0% to 100%. If trial is computed to be outside the search area, it is set to either 5% or 80%. The search begins with the lower boundary TLOW equal to 0% and the upper boundary THI equal to 100%.

The trial internal rate of return equals TRIAL. The external interest rate depends on the method of computation MTH specified.

| *MTH* | *External interest rate* |
|:---:|:---:|
| 1 | TRIAL |
| 2 | EXT |

The subroutine computes the period-by-period results for the cash flow. If there is net internal investment then in the next period compound interest will increase it by $(1+\text{TRIAL})$. If there is an external investment, it will be increased by $(1+\text{EXT})$ or $(1+\text{TRIAL})$ depending on MTH.

If the TRIAL internal rate of return is correct, the net investment will equal zero at the end of the $N$th year. If instead there is an unrecovered investment (a negative sum) it means that TRIAL is too high. The lower boundary TLOW is increased to TRIAL and a new trial halfway between the last TRIAL and THI is used.

TLOW  = TRIAL
TRIAL = (TLOW + THI)/2.

A new period-by-period computation is done.

If the TRIAL internal rate of return had, instead, produced a positive sum at the end of the $N$th year, we know that TRIAL is too low. This time THI would be reduced to TRIAL and the binary search continued in the diminished interval.

THI    = TRIAL
TRIAL = (TLOW + THI)/2.

When the interval between TLOW and TRIAL is less than or equal to 0.002, TRIAL is considered sufficiently close to the exact solution and the computations are halted. The search should be completed within ten trials. If it is not, the internal rate of return is probably not within the search area. After ten trials the search is stopped.

*Subroutine RANG.*   To minimize the difficulty of implementing the program on various computers, a simple pseudorandom number generator is provided by subroutine RANG. The random numbers produced are greater than 0.0000 and less than or equal to 0.9999. On systems where a random number generator is available as a library function, subroutine RANG can be changed to call the library function.

*Subroutine NORML.*   This subroutine generates random normal numbers, with mean equal to zero and standard deviation equal to one.

*Subroutine DEPRC.*   This subroutine is called in Section 11. It provides for computation of the depreciation schedule by any of the usual methods.

*Subroutine GRAPH.*   GRAPH is arranged to compute a rate of return plot for one of two ranges on the $x$-axis. The range may be 0 to +25%, if the variable SCX is set equal to two. It will be 0 to +50% if SCX is equal to one. The rate of return obtained in each iteration was loaded into the vector NR in section 18. The frequency in the various categories is examined and the graph is then scaled so that the maximum frequency is given relative frequency 100 and all other values are assigned equivalent frequency values. The graph is then output on the printer. The appropriate labeling on the axes of the table and the summary statistics (mean, variance, standard deviation, and $N$) are printed. The value $N$, in this case, indicates the number of iterations used in compiling the data on the chart.

## Data Card 1.

| | Format | Columns | |
|---|---|---|---|
| JOB | I3 | 1–3 | *Job number* or other identifying digits. |
| NMAX | I3 | 5–7 | *Maximum useful life* (analysis period) NMAX ≤ 100 |
| IL | I1 | 9 | *Useful life control digit* |

    1   Useful life is fixed = NMAX

    2   Normally distributed useful life

| | | | |
|---|---|---|---|
| AVEN | F5.2 | 11–15 | Mean life |
| SDN | F5.2 | 17–21 | Standard deviation of life |

    3   Log normal distribution

$\alpha$ = AVEN

$\beta$ = SDN

    4   Erlang distribution

Mean of underlying exponential distribution = AVEN

For Erlang: Resulting mean = KERL1 * AVEN

| | | | |
|---|---|---|---|
| KERL1 | I2 | 23–24 | $K$ for Erlang. When $K = 1$ the exponential distribution is obtained. |

    5   Cumulative useful life distribution is to be provided (LCD)

| | | | |
|---|---|---|---|
| ISWCH | I1 | 26 | *Cash flow table switch* |

    1   With ISWCH = 1 a complete cash flow table is printed each iteration with the result that ITER tables will be printed

    0   Printing of cash flow tables is suppressed

| | | | |
|---|---|---|---|
| IFC | I1 | 28 | *Capitalized investment control digit* |

    1   Capitalized investment fixed

Fixed Investment = AVEI

    2   Normally distributed capitalized investment

| | | | |
|---|---|---|---|
| AVEI | F8.0 | 30–37 | Mean capitalized investment |
| SDI | F8.0 | 39–46 | Standard deviation of capitalized investment |

    3   Log normal distribution

$\alpha$ = AVEI

$\beta$ = SDI

    4   Erlang distribution

Mean of underlying exponential distribution = AVEI

Resulting mean = KERL2 * AVEI

| | | | |
|---|---|---|---|
| KERL2 | I2 | 48–49 | $K$ for Erlang. When $K = 1$, the exponential distribution is obtained |

*Data Card 2.*

|     | Format | Columns |
|-----|--------|---------|
| IT  | I1     | 1       |

    *Investment timing control digit*

      1  Total investment made at beginning of year 1 (year 0)

      2  Investment equally distributed over lesser of TN or $N$ years

| TN  | F3.0   | 3–5     |

         Years over which investment distributed

      3  Percentage distribution of investment timing to be provided (TP)

| IB  | I1     | 7       |

    *Borrowed money control digit*

      1  No borrowed money

      2  Uniform repayment schedule

      3  Equal principal payments plus annual interest

      4  Repayment schedule to be provided (BRS)

| BI  | F5.4   | 9–13    |

    *Effective interest rate per annum charged on borrowed money, stated as a decimal* (for example, $5\% = 0.0500$)

| BN  | F3.0   | 15–17   |

    *Year in which last portion of borrowed money is repaid;* that is, the term of the debt which began at beginning of year 1 (year 0), BN $\leqslant$ 99

| BPC | F5.3   | 19–23   |

    *Portion of year 0 investment that is borrowed, stated as a decimal* (for example, $50\% = 0.500$); BPC may be greater than 1.000

| ID  | I1     | 25      |

    *Depreciation schedule control digit*

      1  Straight line depreciation

      2  Sum-of-years digits depreciation

      3  Double declining balance depreciation

      4  Double declining balance depreciation with conversion to straight line

      5  Declining balance depreciation at 150% of straight line rate

      6  Depreciation schedule to be provided, (DEP)

      7  Depreciation schedule not computed (this option would be used when a before income tax analysis is to be made)

| ND  | I2     | 27–28   |

    *Life of capital asset* (years) to be used in calculation of depreciation schedule. If the value of ND exceeds the analysis period ($N$) in any iteration, the depreciation schedule is computed using $N$ years rather than ND years as the depreciable life.

| ISV | I1 | 30 | *Salvage (Terminal) value control digit* |
|------|------|-------|---|
| | | | 1   Salvage value = 0 |
| AVESV | F8.0 | 32–39 | 2   Salvage value is fixed = AVESV |
| | | | 3   Salvage value declines uniformly from |
| SVMAX | F8.0 | 41–48 |      SVMAX to SVLOW. |
| SVLOW | F8.0 | 50–57 |      At year 0       SV = SVMAX |

                                           At year NMAX     SV = SVMIN

                                  Linear interpolation between 0 and NMAX

                          4   Cumulative salvage value distribution is to be provided (CSV)

## Data Card 3.

| | Format | Columns | |
|------|--------|---------|---|
| IC | I1 | 1 | *Expensed cost control digit* |
| | | | 1   No expensed annual cost |
| AVEC | F8.0 | 3–10 | 2   Expensed cost is fixed = AVEC |
| | | | 3   Normally distributed expensed cost |
| | | |      Mean = AVEC |
| SDC | F8.0 | 12–19 |      Std deviation = SDC |
| | | | 4   Log normal distribution |
| | | |      $\alpha$ = AVEC |
| | | |      $\beta$ = SDC |
| | | | 5   Erlang distribution |
| | | |      Mean of underlying exponential |
| | | |         distribution = AVEC |
| | | |      Resulting mean = KERL3 * AVEC |
| KERL3 | I2 | 21–22 |      $K$ for Erlang; when $K = 1$, the exponential distribution is obtained |
| | | | 6   Cumulative expensed annual cost distribution to be provided (CCD) |
| ITX | I1 | 24 | *Income tax control digit* |
| | | | 1   Income taxes not computed |
| | | | 2   Uniform income tax rate (TAX) stated as a |
| TAX | F3.2 | 26–28 |      decimal (for example, 52% = 0.52). |
| | | | 3   Uniform income tax rate (TAX) and invest- |
| | | |      ment tax credit (TXCD); tax credit stated |
| TXCD | F3.2 | 30–32 |      as a decimal (for example, 7% = 0.07) |
| EXT | F4.3 | 34–37 | *External investment interest rate (EXT)* |
| | | | Stated as a decimal (for example, 6% = 0.060) |

## Data Card 4.

| | Format | Columns | |
|------|--------|---------|---|
| IBS | I1 | 1 | *Benefit schedule (before taxes) control digit* |
| AVEB | F8.0 | 3–10 | 1   Before-tax benefits fixed = AVEB |

| | | | |
|---|---|---|---|
| | | | 2 Normally distributed before-tax benefits |
| | | | Mean = AVEB |
| SDB | F8.0 | 12–19 | Standard deviation = SDB |
| | | | 3 Log normal distribution |
| | | | $\alpha$ = AVEB |
| | | | $\beta$ = SDB |
| | | | 4 Erlang distribution |
| | | | Mean of underlying exponential |
| | | | distribution = AVEB |
| KERL4 | I2 | 21–22 | Resulting mean = KERL4 * AVEB |
| | | | $K$ for Erlang; when $K = 1$, the expo- |
| | | | nential distribution is obtained |
| | | | 5 Cumulative annual benefits schedule (before |
| | | | taxes) distribution to be provided (CBS) |
| ITB | I1 | 24 | *Trend of benefit schedule control digit* |
| | | | 1 No trend adjustment of before-tax benefit |
| | | | schedule |
| | | | 2 Before-tax benefit schedule to be adjusted by |
| | | | a uniform gradient |
| TPC | F5.3 | 26–30 | Annual percentage change per year (TPC) is |
| | | | stated as a signed decimal (a 4% declining |
| | | | gradient = $-0.04$) |
| ITER | I6 | 32–37 | *Number of iterations of the simulation model desired* |
| SCX | F3.0 | 39–41 | *Rate of return graph scale digit* |
| | | | 1 0 to 50 percent plotted range |
| | | | 2 0 to 25 percent plotted range |

## Schedules Specified in Fixed Data Cards.

*Listed in order of loading*

1. LCD   *Cumulative life distribution*
   25 integer values of useful life with a cumulative discrete probability of
   0.04(I); I = 1,25
   FORMAT (25I3)
   Example:                                                    NMAX
   001   005   008   010   012   016   020   024   $\cdots$   xxx

2. TP   *Investment timing distribution*
   Ten percentage values (stated in decimal form) as distribution of invest-
   ment over initial ten-year period; sum = 1.00
   FORMAT (10F3.2)
   Example:
   0.02   0.25   0.30   0.25   0.10   0.08   0.00   0.00   0.00   0.00

3. **BRS**    *Borrowed money repayment schedule*
   Dollar amount of each year's repayment provided for the term of the
   debt; number of elements = BN (=IBN)    $BN_{max} = 99$
   FORMAT (10F8.0) with as many cards as required to read IBN values
   Example:
   0005000.   0010000.   0020000.   . . .   and so forth

4. **DEP**    *Depreciation schedule*
   Dollar amount of each year's depreciation provided for the depreciable
   life of the asset: ND years
   FORMAT (10F8.0) with as many cards as required to read ND values
   Example:
   0002000.   0052000.   0046000.   . . .   and so forth

5. **CSV**    *Cumulative salvage value distribution*
   25 values of salvage value with a cumulative discrete probability of
   0.04(I) I = 1,25; dollar amounts are input
   FORMAT (5F8.0)
   Example:
   0002000.   0002300.   0002800.   0003000.   0003800.
   +4 additional cards

6. **CCD**    *Cumulative expensed cost distribution*
   25 values of expensed cost with a cumulative discrete probability of
   0.04(I) I = 1,25. Dollar amounts are input.
   FORMAT (5F8.0)
   Example:
   0002000.   0002100.   0002300.   0002450.   0002600.
   +4 additional cards

7. **CBS**    *Cumulative annual benefit distribution*
   25 values of annual benefit with a cumulative discrete probability of
   0.04*I, I = 1,25; dollar amounts are input.
   FORMAT (5F8.0)
   Example:
   0002500.   0003000.   0003500.   0004000.   0004500.
   +4 additional cards

# CAPITAL EXPENDITURE ANALYSIS PROGRAM

```
      PROGRAM CEAP
C     CAPITAL EXPENDITURE ANALYSIS PROGRAM
C     FROM NEWNAN - ENGINEERING ECONOMIC ANALYSIS
      DIMENSION TP(50),DEP(50),LCD(25),BRS(50),CSV(25),
     1CCD(25),REBTP(50),CBS(25),BENF(50),CFC(50),BMP(50),
     2BINT(50),FC(50),TXS(50),EXPC(50)
C     REBTP COULD BE DIMENSIONED AS 10 BUT IS 50 FOR EASY
C     RESETTING IN SECTION 5.
      COMMON NR(53),X,SUM1,SUM2,IRP,IRD,TRR,CFCZ,ATCH(50),SCX
      IRD = 60
      IPR = 61
C*****************************************************************
C     SECTION 1 - READ IN FIXED DATA CARDS
C*****************************************************************
      READ(IRD,1)JOB,NMAX,IL,AVEN,SDN,KERL1,ISWCH,IFC,AVEI,
     1 SDI,KERL2
    1 FORMAT(I3,1X,I3,1X,I1,1X,F5.2,1X,F5.2,1X,I2,1X,I1,1X,
     1 I1,1X,F8.0,1X,F8.0,1X,I2)
      READ(IRD,2)IT,TN,IB,BI,BN,BPC,ID,ND,ISV,AVESV,SVMAX,
     1 SVMIN
    2 FORMAT(I1,1X,F3.0,1X,I1,1X,F5.4,1X,F3.0,1X,F5.3,1X,
     1 I1,1X,I2,1X,I1,1X,F8.0,1X,F8.0,1X,F8.0)
      READ(IRD,3)IC,AVEC,SDC,KERL3,ITX,TAX,TXCD,EXT
    3 FORMAT(I1,1X,F8.0,1X,F8.0,1X,I2,1X,I1,1X,F3.2,1X,F3.2,
     1 1X,F4.3)
      READ(IRD,4) IBS,AVEB,SDB,KERL4,ITB,TPC,ITER,SCX
    4 FORMAT(I1,1X,F8.0,1X,F8.0,1X,I2,1X,I1,1X,F5.3,1X,I6,
     1 1X,F3.0)
      IBN=BN
C*****************************************************************
C     SECTION 2 - INITIALIZE
C*****************************************************************
      SEED = 1
 9001 NR(I) = 0
      X = 0.
      SUM1=0.
      SUM2=0.
      DO 320 I=1,10
  320 TP(I)=0.
      DO 321 I=1,25
```

```
      LCD(I)  =0
      CSV(I)  =0.
      CCD(I)  =0.
  321 CBS(I)  =0.
      DO 322 I=1,50
      BENF(I) =0.
  322 DEP(I)  =0.
C*****************************************************************
C     SECTION 3 - READ IN SCHEDULES SPECIFIED IN FIXED DATA
C     CARDS.  CUM LIFE DISTRIBUTION (LCD) CONTAINS 25 INTEGER
C     VALUES OF USEFUL LIFE WITH CUM DISCRETE PROBABILITY OF
C     0.04(I), I=1,25.  FORMAT = 25I3.
C*****************************************************************
      IF(IL-5)8,6,8
    6 READ(IRD,7) (LCD(I), I=1,25)
    7 FORMAT(25I3)
C     INVESTMENT TIMING DIST (TP)
    8 IF(IT-3)201,202,201
  202 READ(IRD,203) (TP(I), I=1,10)
  203 FORMAT(10F3.2)
C     REPAYMENT SCHEDULE (BRS)
  201 IF(IB-4)204,205,204
  205 READ(IRD,206) (BRS(I), I=1,IBN)
  206 FORMAT(10F6.0)
C     DEPRECIATION SCHEDULE (DEP)
  204 IF(ID-6)10,9,10
   10 READ(IRD,11) (DEP(I), I=1,ND)
   11 FORMAT(10F8.0)
C     CUM SALVAGE VALUE (CSV)
    9 IF(ISV-6)207,208,207
  208 READ(IRD,209) (CSV(I), I=1,25)
  209 FORMAT(5F8.0)
C     CUM EXPENSED COST DISTRIBUTION (CCD)
  207 IF(IC-6)210,211,210
  211 READ(IRD,209) (CCD(I), I=1,25)
C     CUM ANNUAL BENEFIT DISTRIBUTION (CBS)
  210 IF(IBS-5)212,213,212
  213 READ(IRD,209) (CBS(I), I=1,25)
  212 CONTINUE
C*****************************************************************
C     SECTION 4 - DOCUMENT THE PROBLEM
C*****************************************************************
      WRITE(IPR,20)
   20 FORMAT(25X,36HCAPITAL EXPENDITURE ANALYSIS PROGRAM)
      WRITE(IPR,21) JOB
   21 FORMAT(37X,7HPROJECT,I5,//)
      WRITE(IPR,23)
```

*418*

```
23 FORMAT(11X,43HFROM NEWNAN - ENGINEERING ECONOMIC ANALYS
1IS/)
   WRITE(IPR,22)
22 FORMAT(11H INVESTMENT)
   GO TO (24,25,26,27),IFC
24 WRITE(IPR,28) AVEI
28 FORMAT(10X,19HINVESTMENT FIXED AT,F10.0)
   GO TO 36
25 WRITE(IPR,29) AVEI, SDI
29 FORMAT(10X,31HNORMAL DISTRIBUTION WITH MEAN =,F10.0,2X,
1 13HAND STD DEV =,F10.0)
   GO TO 36
26 WRITE(IPR,30) AVEI
30 FORMAT(10X,34HLOGNORMAL DISTRIBUTION WITH MEAN = ,F10.0)
   GO TO 36
27 WRITE(IPR,31) KERL2,AVEI
31 FORMAT(10X,22HERLANG DISTRIBUTION K=,I2,5X,6HMEAN =,
1 F10.0)
   GO TO 36
36 WRITE(IPR,32)
32 FORMAT(/5X,17HINVESTMENT TIMING)
   GO TO (33,34,35),IT
33 WRITE(IPR,37)
37 FORMAT(10X,22HTOTAL INVESTMENT MADE ,31HAT BEGINNING OF
1 YEAR 1(YEAR 0) )
   GO TO 43
34 WRITE(IPR,38) TN
38 FORMAT(10X,35HINVESTMENT EQUALLY DISTRIBUTED OVER,F6.0,
1 7H YEARS)
   GO TO 43
35 WRITE(IPR,39)
39 FORMAT(10X,38HPERCENT DISTRIBUTION OF INVESTMENT FOR,
1 16H TEN YEAR PERIOD)
   WRITE(IPR,40)
40 FORMAT(20X,4HYEAR,5X,7HPERCENT)
   DO 41 I=1,10
41 WRITE(IPR,42) I,TP(I)
42 FORMAT(21X,I2,6X,F4.2)
43 WRITE(IPR,44)
44 FORMAT(/5X,15HANALYSIS PERIOD)
   WRITE(IPR,45) NMAX
45 FORMAT(10X,16HMAX USEFUL LIFE=,I4)
   GO TO (46,47,48,49,101),IL
46 WRITE(IPR,102) NMAX
102 FORMAT(10X,13HLIFE FIXED AT,I4,6H YEARS)
   GO TO 103
47 WRITE(IPR,450) AVEN, SDN
450 FORMAT(10X,31HNORMAL DISTRIBUTION WITH MEAN =,F6.2,2X,

1 13HAND STD DEV =,F6.2)
   GO TO 103
48 WRITE(IPR,30) AVEN
   GO TO 103
49 WRITE(IPR,31) KERL1, AVEN
   GO TO 103
101 WRITE(IPR,104)
104 FORMAT(10X,23HCUMULATIVE DISTRIBUTION)
   WRITE(IPR,105)
105 FORMAT(20X,11HUSEFUL LIFE,24H    CUMULATIVE PROBABILITY)
   DO 106 I=1,25
   P=I
   P=P*0.04
106 WRITE(IPR,107)   LCD(I),P
107 FORMAT(25X,I4,12X,F5.2)
103 WRITE(IPR,108)
108 FORMAT(/5X,19HDEPRECIATION METHOD)
   WRITE(IPR,700) ND
700 FORMAT(10X,18HDEPRECIABLE LIFE =,I4,6H YEARS)
   GO TO (109,110,111,112,113,114,129),ID
109 WRITE(IPR,115)
115 FORMAT(10X,13HSTRAIGHT LINE)
   GO TO 124
110 WRITE(IPR,116)
116 FORMAT(10X,19HSUM-OF-YEARS DIGITS)
   GO TO 124
111 WRITE(IPR,117)
117 FORMAT(10X,24HDOUBLE DECLINING BALANCE)
   GO TO 124
112 WRITE(IPR,118)
118 FORMAT(10X,36HDBL DECLINING BAL W/CONV TO STR LINE)
   GO TO 124
113 WRITE(IPR,119)
119 FORMAT(10X,44HDECLINING BALANCE AT 150 PERCENT OF STR
1LINE)
   GO TO 124
129 WRITE(IPR,130)
130 FORMAT(10X,50HBEFORE TAX COMPUTATION   DEPRECIATION NOT
1COMPUTED)
   GO TO 124
114 WRITE(IPR,120)
120 FORMAT(10X,27HDEPRECIATION SCHEDULE INPUT)
   WRITE(IPR,121)
121 FORMAT(20X,4HYEAR,17H   DEPREC FOR YEAR)
   DO 122 I=1,ND
122 WRITE(IPR,123) I, DEP(I)
123 FORMAT(21X,I2,4X,F9.0)
124 WRITE(IPR,125)
```

```
125  FORMAT(/5X,25HEND-OF-LIFE SALVAGE VALUE)
     GO TO (126,127,128,131),ISV
126  WRITE(IPR,132)
132  FORMAT(10X,16HNO SALVAGE VALUE)
     GO TO 138
127  WRITE(IPR,133) AVESV
133  FORMAT(10X,21HSALVAGE VALUE FIXED =,F8.0)
     GO TO 138
128  SVGRAD = (SVMAX-SVMIN)/AVEN
     WRITE(IPR,81) SVMAX,SVMIN, SVGRAD
81   FORMAT(3X,10HMAX S.V. =,F8.0,3X,10HMIN S.V. =,F8.0,3X,
    1 15HS.V. GRADIENT =,F8.0)
     GO TO 138
131  WRITE(IPR,104)
     WRITE(IPR,134)
134  FORMAT(20X,13HSALVAGE VALUE,23H CUMULATIVE PROBABILITY)
     DO 135 I=1,25
     P=I
     P=P*4./100.
135  WRITE(IPR,136) CSV(I),P
136  FORMAT(21X,F9.0,5X,F5.2)
138  WRITE(IPR,139)
139  FORMAT(/5X,14HBORROWED MONEY)
     GO TO (140,141,142,143),IB
140  WRITE(IPR,144)
144  FORMAT(10X,17HNO BORROWED MONEY)
     GO TO 154
141  WRITE(IPR,145)
145  FORMAT(10X,26HUNIFORM REPAYMENT SCHEDULE)
     GO TO 154
142  WRITE(IPR,146)
146  FORMAT(10X,42HEQUAL PRIN PAYMENTS   PLUS ANNUAL INTEREST)
     GO TO 154
143  WRITE(IPR,147)
147  FORMAT(10X,24HREPAYMENT SCHEDULE INPUT)
     WRITE(IPR,148)
148  FORMAT(20X,4HYEAR,5X,17HPRINCIPAL PAYMENT)
     DO 149 I=1,IBN
149  WRITE(IPR,150) I, BRS(I)
150  FORMAT(21X,I3,8X,F9.0)
154  GO TO (151,152,152),IB
152  WRITE(IPR,155) BN
155  FORMAT(10X,14HTERM OF LOAN =,F5.0,6H YEARS)
     WRITE(IPR,156) BPC
156  FORMAT(10X,42HPROPORTION OF YEAR 0 INVESTMENT BORROWED
    1 =, F8.3)
     WRITE(IPR,157) BI
157  FORMAT(10X,29HNOMINAL INTEREST RATE/ANNUM =, F8.4)

151  WRITE(IPR,158)
158  FORMAT(/5X,14HEXPENSED COSTS)
     GO TO (199,159,160,161,162,163),IC
199  WRITE(IPR,200)
200  FORMAT(10X,23HNO EXPENSED ANNUAL COST)
     GO TO 169
159  WRITE(IPR,164) AVEC
164  FORMAT(10X,22HEXPENSED COSTS FIXED =,F8.0)
     GO TO 169
160  WRITE(IPR,29) AVEC, SDC
     GO TO 169
161  WRITE(IPR,30) AVEC
     GO TO 169
162  WRITE(IPR,31) KERL3, AVEC
     GO TO 169
163  WRITE(IPR,104)
     WRITE(IPR,165)
165  FORMAT(20X,14HEXPENSED COSTS,23H CUMULATIVE PROBABILITY)
     DO 167 I=1,25
     P=I
     P=P*4./100.
167  WRITE(IPR,168) CCD(I),P
168  FORMAT(25X,F9.0,5X,F5.2)
169  WRITE(IPR,170)
170  FORMAT(/5X,10HINCOME TAX)
     GO TO (171,172,173),ITX
171  WRITE(IPR,174)
174  FORMAT(10X,20HINCOME TAXES IGNORED)
     GO TO 180
172  WRITE(IPR,175)
175  FORMAT(10X,16HUNIFORM TAX RATE)
     GO TO 179
173  WRITE(IPR,176)
176  FORMAT(10X,38HUNIFORM TAX RATE PLUS INVEST TX CREDIT)
     WRITE(IPR,177) TXCD
177  FORMAT(10X,12HTAX CREDIT =, F5.2)
179  WRITE(IPR,178) TAX
178  FORMAT(10X,10HTAX RATE =,F5.2)
180  WRITE(IPR,181)
181  FORMAT(/5X,16HBENEFIT SCHEDULE)
     GO TO (182,183,184,185,186),IBS
182  WRITE(IPR,187) AVEB
187  FORMAT(10X,17HUNIFORM BENEFIT =,F9.0)
     GO TO 191
183  WRITE(IPR,29) AVEB,SDB
     GO TO 191
184  WRITE(IPR,30) AVEB
     GO TO 191
```

```
C **********************************************************
C    SECTION 6 - DETERMINE ANALYSIS PERIOD (USEFUL LIFE)
C **********************************************************
      GO TO (301,302,303,304,305),IL
      USEFUL LIFE IS FIXED = NMAX
  301 N=NMAX
      GO TO 306
C     USEFUL LIFE IS NORMALLY DISTRIBUTED
  302 CALL NORML(SEED,RND)
      N=AVEN+SDN*RND+0.5
      GO TO 306
C     USEFUL LIFE HAS LOG NORMAL DISTRIBUTION
  303 CALL RLOGN(SEED,RLGN)
      N=AVEN+SDN*RLGN+0.5
      GO TO 306
C     USEFUL LIFE USES ERLAND FOR DISTRIBUTION
  304 CALL ERLAN(AVEN,KERL1,SEED,ERLAN1)
      N = ERLAN1 +0.5
      GO TO 306
C     USEFUL LIFE BASED ON CUM LIFE DIST PROVIDED (LCD(25))
  305 CALL RANG(SEED,R)
      IXX=R*25.+1.
      N=LCD(IXX)
C     IF COMPUTED N GREATER THAN NMAX, SET N=NMAX
  306 IF(N-NMAX) 307,307,308
  308 N=NMAX
  307 CONTINUE
C **********************************************************
C     SECTION 7 - COMPUTE CAPITALIZED INVESTMENT
C **********************************************************
      GO TO (331,332,333,334),IFC
C     CAPITALIZED INVESTMENT IS FIXED = AVEI
  331 CI=AVEI
      GO TO 335
C     CAPITALIZED INVESTMENT NORMALLY DISTRIBUTED
  332 CALL NORML(SEED,RND)
      CI=AVEI+SDI*RND
      GO TO 335
C     CAPITALIZED INVESTMENT LOG NORMAL DISTRIBUTION
  333 CALL RLOGN(SEED,RLGN)
      CI=AVEI+SDI*RLGN
      GO TO 335
C     CAPITALIZED INVESTMENT USING ERLANG
  334 CALL ERLAN(AVEI,KERL2,SEED,ERLAN2)
      CI = ERLAN2
  335 CONTINUE
C **********************************************************
C     SECTION 8 - COMPUTE SALVAGE VALUE (SV)
C **********************************************************

  185 WRITE(IPR,31) KERL4, AVEB
      GO TO 191
  186 WRITE(IPR,104)
      WRITE(IPR,188)
  188 FORMAT(20X,32HBENEFIT    CUMULATIVE PROBABILITY)
      DO 189 I=1,25
      P=I
      P=P*4./100.
  189 WRITE(IPR,190) CBS(I),P
  190 FORMAT(17X,F10.0,10X,F5.2)
  191 WRITE(IPR,192)
  192 FORMAT(/5X,13HBENEFIT TREND)
      GO TO (193,194),ITB
  193 WRITE(IPR,195)
  195 FORMAT(10X,8HNO TREND)
      GO TO 197
  194 WRITE(IPR,196) TPC
  196 FORMAT(10X,7HTREND =,F6.3,9H PER YEAR)
  197 WRITE(IPR,198) EXT, ITER
  198 FORMAT(/5X,19HEXTERNAL INVESTMENT/10X,49HINTEREST RATE
     1EARNED ON ANY EXTERNAL INVESTMENT =,F5.3,//
     2 19H MODEL ITERATIONS =,I6)
C **********************************************************
C     SECTION 5 - THE MAIN PROCESSING LOOP STARTS HERE
C     RESET VARIABLES USED IN EACH ITERATION
C **********************************************************
      DO 610 IK=1,ITER
      SUM = 0.
      TEMP1 = 0.
      TEMP3 = 0.
      TEMP4 = 0.
      TBENF = 0.
      TEXPC = 0.
      TTXS = 0.
      FCZ = 0.
      BMZ = 0.
      CFCZ = 0.
      TAXZ = 0.
      ATCHZ = 0.
      DO 600 I=1,50
      ATCHI(I) = 0.
      REBTP(I) = 0.
      BINT(I) = 0.
      EXPC(I) = 0.
      BMP(I) = 0.
      CFC(I) = 0.
      FC(I) = 0.
  600 TXS(I) = 0.
```

421

```
C***********************************************
      GO TO (401,402,403,406),ISV
C     NO SALVAGE VALUE
401   SV = 0.
      GO TO 407
C     SALVAGE VALUE FIXED = SVF
402   SV = AVESV
      GO TO 407
C     SALVAGE VALUE IS A LINEAR DECLINE FROM SVMAX TO SVLOW
C     USING THE GRADIENT.
403   IF(NMAX-N) 1100,1100,1101
1100  SV = SVMIN
      GO TO 407
1101  SV=((SVMAX-SVMIN)*(NMAX-N))/NMAX+SVMIN
      GO TO 407
C     SALVAGE VALUE BASED ON CUM SV DIST PROVIDED. CSV(I)
C     VECTOR INPUT = DOLLAR SUM(F8.0) FOR EACH 0.04
406   CALL RANG(SEED,R)
      IXY = R*25.+1.
      SV = CSV(IXY)
407   CONTINUE
C***********************************************
C     SECTION 9 - INVESTMENT TIMING + CALCULATION OF FC(I)
C***********************************************
      NO = N
      IF(ND.GT.N) NO=ND
      GO TO (351,352,353),IT
C     TOTAL INVESTMENT AT BEGINNING OF YEAR 1 (YEAR 0)
351   FCZ = CI
      CFCZ = CI
      DO 61 I=1,NO
61    CFC(I) = CI
      GO TO 363
C     INVESTMENT EQUALLY DISTRIBUTED OVER (TN) YEARS
C     FIRST INVESTMENT AT BEGINNING OF YEAR 1 (YEAR 0)
C     LAST INVESTMENT AT BEGINNING OF YEAR TN (YEAR TN-1)
352   ITN = TN
      IF(N-ITN) 355,354,354
      IF TN GREATER THAN N, SET TN = N
355   ITN = N
354   FCZ = CI/TN
      CFCZ = FCZ
      FC(1) = CI/TN
      CFC(1) = CFCZ + FC(1)
      ITN = ITN -1
      DO 356 I=2,ITN
356   FC(I) = CI/TN
      DO 65 I=2,NO

65    CFC(I) = CFC(I-1) + FC(I)
      GO TO 363
C     INVESTMENT TIMING SCHEDULE INPUT
C     VECTOR TP(I) I=1,10 CONTAINS PERCENTAGE STATED AS DECIML
C     TYP ELEMENT (0.20) INPUT AS F3.2 SUM OF ELEMENTS = 1.00
C     IF N LESS THAN 10, A REBALANCE OF TP(I) MAY BE NECESSARY
353   IF(N-10) 357,358,358
357   DO 359 I=N,9
359   SUM = SUM + TP(I+1)
      IF (SUM) 358,358,360
360   FN = N
      DO 361 I=1,N
361   REBTP(I) = TP(I) + SUM/FN
C     REBALANCING OF TP(I) IF NEEDED, COMPLETED.
C     LOADING OF FC(I) VECTOR.
      FC(1) = REBTP(1)*CI
      CFC(1) = FC(1)
      DO 362 I=2,NO
362   FC(I) = REBTP(I)*CI
      DO 601 I=2,NO
601   CFC(I) = CFC(I-1) + FC(I)
      GO TO 363
C     N EQUAL OR GREATER THAN 10
358   FC(1) = TP(1)*CI
      CFC(1) = FC(1)
      DO 461 I=2,10
461   FC(I) = TP(I)*CI
      DO 602 I=2,NO
602   CFC(I) = CFC(I-1) + FC(I)
363   CONTINUE
C***********************************************
C     SECTION 10 - BORROWED MONEY AND INTEREST CHARGE
C     CONTROL DIGIT (IB) = 1   NO BORROWED MONEY
C                          2   UNIFORM REPAYMENT SCHEDULE
C                          3   EQUAL PRIN PAYMENTS + ANNUAL INT
C                          4   REPAYMENT SCHEDULE PROVIDED
C     TEMP1 = AMOUNT OF BORROWED MONEY REPAID
C     BMZ = TOTAL BORROWED MONEY
C***********************************************
C     CALCULATION OF INTEREST CHARGE VECTOR BINT(I)
      BMZ = BPC*FCZ
      GO TO (381,382,382,384),IB
C     NO BORROWED MONEY
381   GO TO 396
C     CHECK THAT TERM OF DEBT DOES NOT EXCEED USEFUL LIFE
382   IF(N-IBN) 387,388,388
C     TERM OF DEBT GREATER THAN N, SO ADJUST IBN=BN=N
387   IBN = N
```

```
      BN = N
  388 IF (IB-2) 389,389,383
C     UNIFORM REPAYMENT SCHEDULE
  389 CR=BI*(1.+BI)**BN/(((1.+BI)**BN)-1.)
      DO 385 I=1,IBN
      BMPI(I) = BMZ*CR
      BINT(I) = (BMZ - TEMP1)*BI
  385 TEMP1 = BMP(I) - BINT(I) + TEMP1
      GO TO 396
C     EQUAL PRINCIPAL PAYMENTS PLUS ANNUAL INTEREST
  383 P = BMZ/BN
      DO 386 I=1,IBN
      BINT(I) = (BMZ-TEMP1)*BI
      BMP(I) = BINT(I) + P
  386 TEMP1 = TEMP1 + P
      GO TO 396
C     REPAYMENT SCHEDULE PROVIDED BRS(I) I=1,IBN
C     FORM OF INPUT = DOLLAR SUM (F8.0)
C     CHECK THAT REPAYMENT SCHEDULE DOES NOT EXCEED N
  384 IF(N-IBN) 390,391,391
C     REPAYMENT SCHEDULE GREATER THAN N
C     DISTRIBUTE AMT BEYOND N EQUALLY TO TERMS 1,IBN
  390 ITEM = N+1
      DO 392 I=ITEM,100
  392 TEMP3 = TEMP3 + BRS(I)
      FN = N
      DO 393 I=1,N
  393 BRS(I) = BRS(I) + TEMP3/FN
      IBN = N
C     REPAYMENT SCHEDULE DOES NOT EXCEED N OR HAS BEEN
C     REALLOCATED TO EQUAL N
  391 DO 394 I=1,IBN
      TEMP4 = TEMP4 + BRS(I)
  394 DO 395 I=1,IBN
      BINT(I) = TEMP4*BI
      BMP(I) = BINT(I) + BRS(I)
  395 TEMP4 = TEMP4 - BRS(I)
  396 CONTINUE
C     *************************************************************
C     SECTION 11 - COMPUTE DEPRECIATION SCHEDULE. ND = LIFE OF
C     ASSET FOR USE IN ALLOWABLE DEPRECIATION CALCULATION.
C     IF ND EXCEEDS THE USEFUL LIFE (ANALYSIS PERIOD) N, A
C     LOSS ON DISPOSAL TAX ADJUSTMENT MAY BE REQUIRED.
C     SUBROUTINE DEPRC CALCULATES DEPRECIATION SCHEDULE FOR
C     ND LIFE.  TAX ADJUSTMENT MADE IN SECTION 14.
C     *************************************************************
      CALL DEPRC (ND,ID,SV,CFC,DEP)
C     *************************************************************

C     SECTION 12 - COMPUTE BENEFITS SCHEDULE BEFORE TAXES BENF(I)
C     *************************************************************
      DO 419 I=1,N
      GO TO (411,412,413,414,415),IBS
C     BENEFITS FIXED = AVEB
  411 BENF(I) = AVEB
      GO TO 419
C     BENEFITS NORMALLY DISTRIBUTED
  412 CALL NORML(SEED,RND)
      BENF(I) = AVEB + SDB*RND
      GO TO 419
C     BENEFITS LOG NORMAL DISTRIBUTION
  413 CALL RLOGN(SEED,RLGN)
      BENF(I) = AVEB+SDB*RLGN
      GO TO 419
C     BENEFITS USING ERLANG
  414 CALL ERLAN(AVEB,KERL4,SEED,ERLAN4)
      BENF(I) = ERLAN4
      GO TO 419
C     CUM ANNUAL BENEFIT DISTRIBUTION PROVIDED CBS(I)
C     PROBABILITY OF ELEMENT = 0.04I.   DOLLAR SUM (F8.0)
  415 CALL RANG(SEED,R)
      IXY = R*25.+1.
      BENF(I) = CBS(IXY)
      TEMP1 = 0.
  419 TBENF = TBENF + BENF(I)
C     TREND OF BENEFIT ADJUSTMENT
      GO TO (421,422),ITB
C     NO TREND -  CONTINUE IF ITB = 1, OTHERWISE COMPUTE.
  422 TBENF = 0.
      DO 421 I=1,N
      FI = I
      BENF(I) = BENF(I)*(1.+TPC*FI)
      TBENF = TBENF + BENF(I)
  421 CONTINUE
C     *************************************************************
C     SECTION 13 - COMPUTE EXPENSED COST SCHEDULE EXP(I)
C     *************************************************************
      IF(IC-1) 470,470,471
  471 DO 470 I=1,N
      GO TO (470,472,473,474,475,476),IC
C     EXPENSED COST FIXED
  472 EXPC(I) = AVEC
      GO TO 470
C     EXPENSED COST NORMALLY DISTRIBUTED
  473 CALL NORML(SEED,RND)
      EXPC(I) = AVEC+SDC*RND
      GO TO 570
```

423

```
C     EXPENSED COST USING LOG NORMAL DISTRIBUTION
474   CALL RLOGN(SEED,RLGN)
      EXPC(I) = AVEC + SDC*RLGN
      GO TO 570
C     EXPENSED COST USING ERLANG
475   CALL ERLAN(AVEC,KERL3,SEED,ERLAN3)
      EXPC(I) = ERLAN3
      GO TO 570
C     CUMULATIVE ANNUAL EXPENSED COST DISTRIBUTION PROVIDED.
C     CCD(I). INPUT = DOLLAR SUM(F8.0) FOR EACH 0.04
476   CALL RANG (SEED,R)
      IXX = R*25.+1.
      EXPC(I) = CCD(IXX)
570   IF (EXPC(I)) 571,470,470
571   EXPC(I) = 0.
C     IF EXPC(I) IS NEGATIVE, RESET EXPC(I) = 0.
470   TEXPC = TEXPC + EXPC(I)
C***********************************************************
C     SECTION 14 - COMPUTE INCOME TAX SCHEDULE TXS(I)
C     TAXABLE INCOME = BENEFITS - DEPREC - INTEREST COSTS
C                                        -EXPENSED COST.
C     IF ITX = 1, INCOME TAXES IGNORED.
C***********************************************************
      GO TO (1500,481,481),ITX
481   DO 482 I=1,N
C     UNIFORM TAX RATE
482   TXS(I) = TAX*(BENF(I)-DEP(I)-BINT(I)-EXPC(I))
      IF ITX = 3, ADD INVESTMENT TAX CREDIT EFFECT.
      IF(ITX-3) 480,483,483
C     INVESTMENT SUBJECT TO TAX CREDIT
483   K=1
      TAXZ= -FCZ*TXCD
      DO 484 I=1,N
      IF(CFC(I)-SV) 484,484,485
485   IF(K-1) 486,486,487
486   TXS(1) = TXS(1) - (CFC(1) - FCZ)*TXCD
      K = K+1
      GO TO 484
487   TXS(I) = TXS(I)-(CFC(I)-CFC(I-1))*TXCD
484   CONTINUE
C     IF ASSET RETIRED PRIOR TO END OF ITS DEPRECIABLE LIFE
C     THERE WILL BE A LOSS ON DISPOSAL. LOSS IS DEPREC BOOK
C     VALUE - SALVAGE VALUE (SV).
      IF(N-ND) IS NEGATIVE, THERE IS A LOSS
      IF(N-ND) 490,492,492
480   IXX = N+1
      SUM = 0.
      DO 491 I=IXX,ND
491   SUM = SUM + DEP(I)
      TXS(N) = TXS(N) - TAX*SUM
492   DO 618 I=1,N
618   TTXS = TTXS + TXS(I)
      TTXS = TTXS + TAXZ
C***********************************************************
C     SECTION 15 - COMPUTE AFTER-TAX CASH FLOW SCHEDULE ATCH(I)
C     ATCH = BENEFITS - EXP COST - TAXES -INVESTMENT FOR YR.
C***********************************************************
1500  DO 495 I=1,N
495   ATCH(I) = BENF(I) - EXPC(I) - TXS(I) - FC(I)
      ATCH(N) = ATCH(N) + SV
      ATCHZ = -FCZ - TAXZ
C     OUTPUT OF COMPUTED SCHEDULES
      IF (ISWCH.EQ.0) GO TO 612
      WRITE(IPR,496)
496   FORMAT(1H1,3X,10HCUMULATIVE,5X,6HANNUAL,5X,6HDEPREC,3X,
     1 8HBORROWED,1H LOAN PAYMT,7X,4HLOAN,8X,7H-INV +,
     2 10HPROJECT -,8HEXPENSED,3X,10H- INCOME =,6X,3HNET,/
     3  1H ,35HYR INVESTMENT INVESTMENT  SCHEDULE,6X,5HMONEY
     4  11H PRIN + INT,3X,8HINTEREST,14X,8HBENEFITS,6X,5HCOSTS
     5  8X,14HTAX CASH FLOW)
C     PRINT YEAR 0 LINE
      TAXZ = -TAXZ
      WRITE(IPR,1001) CFCZ,FCZ,BMZ,TAXZ,ATCHZ
1001  FORMAT(2X,1H0,2F11.0,11X,F11.0,55X,2F11.0)
      TAXZ = -TAXZ
      DO 498 I=1,N
      TXS(I) = -TXS(I)
      WRITE(IPR,605) I,CFC(I),FC(I),DEP(I),BMP(I),BINT(I),
     1 BENF(I),EXPC(I),TXS(I),ATCH(I)
605   FORMAT(IX,I2,3F11.0,11X,2F11.0,11X,4F11.0)
      TXS(I) = -TXS(I)
498   CONTINUE
      WRITE(IPR,1000) N,SV
1000  FORMAT(1X,I2,64X,13HSALVAGE VALUE,F11.0)
C***********************************************************
C     SECTION 16 - RATE OF RETURN COMPUTATION
C***********************************************************
612   L = 0
C     SET METHOD OF COMPUTATION IN RORIE.  EXTERNAL I = EXT
      MTH = 2
      K = N-1
C     IF ATCHZ IS MINUS, COMPARE WITH ATCH(1) FOR POSSIBLE
C     SIGN CHANGE.
      IF (ATCHZ) 1009,1010,1010
1009  IF (ATCH(1).GT.0.) L=1
1010  FLOW = ATCHZ + ATCH(1)
```

424

```
C        PRINT RATE OF RETURN GRAPH
         CALL GRAPH
         WRITE(IPR,951)
951      FORMAT(4H END)
         STOP
         END
C**********************************************************
C**********************************************************
         SUBROUTINE DEPRC(ND,ID,SV,CFC,DEP)
C        CONTROL DIGIT (ID) =1 STRAIGHT LINE DEPRECIATION
C                           2 SUM-OF-YEARS DIGITS DEPRECIATION
C                           3 DOUBLE DECLINING BALANCE
C                           4 DDB WITH CONVERSION TO STR LINE
C                           5 150% DECLINING BALANCE
C                           6 DEPREC SCHEDULE PROVIDED DEP(I)
C                           7 DEPRECIATION IGNORED
C
C        ND=DEPREC LIFE,  SV=SALVAGE VALUE, CFC(I)=CUM INVESTMENT
         COMMON NR(53),X,SUM1,SUM2,IPR,IRD,TRR,CFCZ,ATCH(50),SCX
         DIMENSION CFC(50),DEP(50)
         CDEP = 0.
C        7  DEPRECIATION NOT  CALCULATED
         IF (ID-7) 56,50,50
56       DO 50 I=1,ND
         A=I
         B=ND
         GO TO (51,52,53,54,55,50),ID
C        1 STRAIGHT LINE DEPRECIATION
51       IF (CFC(I) - SV) 57,57,58
57       DEP(I) = 0.
         GO TO 50
58       IF (I-1) 10,10,11
10       IF (CFCZ.LE.SV) GO TO 50
         DEP(1) = (CFCZ-SV-CDEP)/(B+1.-A)
         GO TO 12
11       DEP(I) = (CFC(I-1)-SV-CDEP)/(B+1.-A)
12       CDEP = CDEP + DEP(I)
         GO TO 50
C        2  SUM-OF-YEARS DIGITS DEPRECIATION
52       IF (CFC(I)-SV) 59,59,60
59       DEP(I) = 0.
         GO TO 50
60       IF (I-1) 20,20,21
20       IF (CFCZ.LE.SV) GO TO 50
         DEP(1) = (2.*B)*(CFCZ-SV)/(B*B+B)
         GO TO 22
21       DEP(I) = (2.*B-2.*A+2.)*(CFC(I-1)-SV)/(B*B+B)
22       CDEP = CDEP + DEP(I)
         GO TO 50
```
```
         DO 501 I=1,K
         IF (ATCH(I)) 500,503,502
C        CASH FLOW IS NEGATIVE
500      IF (ATCH(I+1)) 503,503,504
C        CASH FLOW IS POSITIVE
502      IF (ATCH(I+1)) 504,503,503
504      L = L+1
503      FLOW = FLOW+ATCH(I+1)
501      CONTINUE
         IF (L-1) 506,509,510
C        ZERO SIGN CHANGES IN THE CASH FLOW
506      IF (FLOW) 507,507,508
507      TRR = 0.
         GO TO 610
508      TRR = 1.00
         GO TO 610
C        ONE SIGN CHANGE IN THE CASH FLOW
509      IF (FLOW) 507,507,510
C        MORE THAN ONE SIGN CHANGE IN THE CASH FLOW
510      CALL RORIE(IPR,ATCHZ,ATCH,N,EXT,ISWCH+1,TRR,MTH)
         IF (MTH.EQ.5) 615,705
615      TRR = -1.
         GO TO 610
C**********************************************************
C        SECTION 17 - PRINT SUMMARY STATISTICS FOR THE ITERATION
C**********************************************************
C        IN THE FIRST ITERATION PRINT THE HEADING
705      IF (IK.EQ.1) WRITE(IPR,1123)
1123     FORMAT(1H1,6H  ITER,7X,11HCAPITALIZED,61X,5HTOTAL,7X,
     1   5HTOTAL,3X,5HTOTAL,5X,7HSUM NET,2X,7HRATE OF,/1H ,
     2   6HYR  NO,8X,10HINVESTMENT,58X,20HBENEFITS     EXP COST
     3   3X,5HTAXES,3X,18HCASH FLOW  RETURN)
         WRITE(IPR,1122) N,IK,CI,TBENF,TEXPC,TTXS,FLOW,TRR
1122     FORMAT(1X,I2,I4,7X,F11.0,55X,4F11.0,F8.3)
C**********************************************************
C        SECTION 18 - KEEP STATISTICS ON TRIAL RATE OF RETURN (TRR)
C        FOR SUBROUTINE GRAPH.
C**********************************************************
         SUM1 = SUM1+TRR
         SUM2 = SUM2+TRR*TRR
         X = X+1.
         IF (TRR.LT.0.) GO TO 1200
         J = TRR*SCX*100.+3.5
         IF (J.GT.53) J=53
         GO TO 1201
1200     J = 1
1201     NR(J) = NR(J) +1
610      CONTINUE
```

```fortran
C       3   DOUBLE DECLINING BALANCE DEPRECIATION
   53   IF(CFC(I)-SV) 44,44,43
   44   DEP(I) = 0.
        GO TO 50
   43   IF(I-1) 30,31,31
   30   DEP(I) = 2.*CFCZ/B
        GO TO 63
   31   DEP(I) = 2.*(CFC(I-1)-CDEP)/B
        IF(DEP(I)-(CFC(I)-CDEP-SV)) 63,63,64
   64   DEP(I) = CFC(I-1)-CDEP-SV
   63   CDEP = CDEP+DEP(I)
        GO TO 50
C       4   DDB WITH CONVERSION TO STRAIGHT LINE DEPRECIATION
   54   IF (CFC(I)-SV) 81,81,80
   81   DEP(I) = 0.
        GO TO 50
   80   IF (I.LE.1) 82,83
   82   DEP(I) = 2.*CFCZ/B
        GO TO 84
   83   DEP(I) = 2.*(CFC(I-1)-CDEP)/B
        IF(DEP(I)-(CFC(I-1)-CDEP-SV)) 68,68,67
   67   DEP(I) = CFC(I-1)-CDEP-SV
   68   SL = (CFC(I-1)-SV-CDEP)/(B+1.-A)
        IF(DEP(I)-SL) 85,84,84
   85   DEP(I) = SL
   84   CDEP = CDEP + DEP(I)
        GO TO 50
C       5   150% DECLINING BALANCE DEPRECIATION
   55   IF (CFC(I)-SV) 90,90,91
   90   DEP(I) = 0.
        GO TO 50
   91   IF (I.LE.1) 92,93
   92   DEP(I) = 1.5*CFCZ/B
        GO TO 72
   93   DEP(I) = 1.5*(CFC(I-1)-CDEP)/B
        IF (DEP(I)-CFC(I-1)-CDEP-SV) 72,72,95
   95   DEP(I) = CFC(I-1)-CDEP-SV
   72   CDEP = CDEP + DEP(I)
   50   CONTINUE
        RETURN
        END
C**********************************************************
C**********************************************************
        SUBROUTINE RANG(SEED,R)
C       SUBROUTINE GENERATES A RANDOM NUMBER
C       BETWEEN 0.000 AND 0.999
        K = SEED
        DO 999 I=1,3
  999   K = MOD(5*K,8192)
        SEED = K
        R = SEED/8192.
        RETURN
        END
C**********************************************************
C**********************************************************
        FUNCTION MOD(I,J)
        MOD = I-(I/J)*J
        RETURN
        END
C**********************************************************
C**********************************************************
        SUBROUTINE NORML(SEED,RND)
C       PRODUCES A RANDOM SAMPLE FROM THE NORMAL DISTRIBUTION
C       WITH MEAN = 0 AND STANDARD DEVIATION = 1.
        CALL RANG(SEED,R)
        R1 = R
        CALL RANG(SEED,R)
        R2 = R
        RND=(2.*EXPF(R1))**0.5*COSF(6.28398*R2)*(-1.)
        RETURN
        END
C**********************************************************
C**********************************************************
        SUBROUTINE RLOGN(SEED,RLGN)
C       SAMPLE FROM LOGNORMAL DISTRIBUTION
        CALL NORML(SEED,RND)
        RLGN = EXPF(RND)
        RETURN
        END
C**********************************************************
C**********************************************************
        SUBROUTINE ERLAN(AVERL1,KERL1,SEED,ERLAN1)
C       SAMPLE FROM ERLANG DISTRIBUTION
        CALL RANG(SEED,R)
        IF (KERL1.GT.0) GO TO 20
        ERLAN1 = 9999.9
        RETURN
   20   IF (KERL1.GT.1) GO TO 30
        ERLAN1 = -AVERL1*ALOG(R)
        RETURN
   30   ERLAN1 = 0.
        DO 40 I=1,KERL1
   40   ERLAN1 = ERLAN1 -AVERL1*ALOG(R)
        RETURN
        END
C**********************************************************
```

```
C ***********************************************
C      SUBROUTINE GRAPH
C      RATE OF RETURN PLOT     RANGE       SCX
C                              0 - 25%      2.
C                              0 - 50%      1.
C
      DIMENSION ARRAY (26,53)
      COMMON NR(53),X,SUM1,SUM2,IPR,IRD,TRR,CFCZ,ATCH(50),SCX
      DATA (BLANK=1H ),(STAR=1H*),(DASH=1H-),(EDGE=1HI)
      DO 11 J=1,53
      DO 11 I=1,26
   11 ARRAY(I,J) =  BLANK
      DO 12 J=1,53
   12 ARRAY(26,J) = DASH
      DO 5 I=1,26
    5 ARRAY(I,2) = EDGE
      MXX = NR(1)
      DO 25 J=2,50
      IF (MXX-NR(J)) 26,26,25
   26 MXX = NR(J)
   25 CONTINUE
C      SCALING THE MAXIMUM VALUE
      FXM = MXX
      SCY = 25./FXM
      DO 27 J=1,53
      B = NR(J)
      I = 26.-B*SCY
   27 ARRAY(I,J) = STAR
      ARRAY(26,2) = EDGE
C      RESTORE PAGE TO PRINT GRAPH
      WRITE(IPR,100)
  100 FORMAT(1H1)
      DO 28 I=1,26
      KNT = I
      IF (KNT.NE.1) GO TO 65
      WRITE (IPR,72) (ARRAY(I,J),J=1,53)
   72 FORMAT(1H ,11X,3H100,53A1)
      GO TO 28
   65 IF (KNT.NE.13) GO TO 70
      WRITE(IPR,73) (ARRAY(I,J), J=1,53)
   73 FORMAT(1H ,8HRELATIVE,4X,2H50,53A1)
      GO TO 28
   70 IF (KNT.NE.14) GO TO 80
      WRITE(IPR,74) (ARRAY(I,J), J=1,53)
   74 FORMAT(1H ,9HFREQUENCY,5X,53A1)
      GO TO 28
   80 WRITE(IPR,75) (ARRAY(I,J),J=1,53)
   75 FORMAT(1H ,14X,53A1)
   28 CONTINUE

      IF(SCX.EQ.2.) WRITE(IPR,111)
  111 FORMAT(1H ,7X,9HNEGATIVE ,1H0,9X,1H5,8X,2H10,8X,2H15,,8X
     1 2H20,8X,2H25,11H OR GREATER,/8X,4HCELL, 54X,4HCELL)
      IF(SCX.EQ.1.) WRITE(IPR,110)
  110 FORMAT(1H ,7X,9HNEGATIVE ,21H0          10          20,
     1 30H          30          40          50,11H OR GREATER,/8X,
     2 4HCELL,54X,4HCELL)
      WRITE(IPR,86)
   86 FORMAT(1H0,36X,14HRATE  OF  RETURN)
      XBAR = SUM1/X
      XBAR = XBAR*100.
      VAR = (SUM2-(SUM1*SUM1)/X)/(X-1.)
      VAR = VAR*1000.
      IF (X.EQ.1.) VAR = 0.
      SDEV = SQRTF(VAR)
      WRITE(IPR,50) XBAR,SDEV
   50 FORMAT(/,17X,10HMEAN       =,F7.3,18X,9HSTD DEV  =,F6.3)
      WRITE(IPR,51) VAR,X
   51 FORMAT(17X,10HVARIANCE  =,F7.2,24X,3HN  =,F6.0)
      RETURN
      END
C ***********************************************
C ***********************************************
      SUBROUTINE RORIE(OUT,CASHO,CASH,N,EXT,ISW,RATE,MTH)
      DIMENSION CASH(50)
      INTEGER OUT
C      OUT = PRINTER DEVICE NUMBER (61)
C      CASHO = YEAR O CASH FLOW
C      CASH(I) = CASH FLOW AT END OF YEAR I
C      N = NO. OF INTEREST PERIODS (YEARS) CASH(I), I=1,N
C      EXT = EXTERNAL INTEREST RATE PER INTEREST PERIOD
C            STATED AS A DECIMAL
C      ISW = PRINTING SWITCH = 1 NO PRINTOUT IN SUBROUTINE
C                            = 2 PRINT TABULATION & RESULTS
C      RATE = COMPUTED INTERNAL RATE OF RETURN AS DECIMAL.
C      MTH = METHOD OF COMPUTATION
C            = 1 INPUT SIGNAL.    EXTERNAL I = INTERNAL ROR.
C            = 2 INPUT SIGNAL.    EXTERNAL I = EXT
C            = 3 OUTPUT SIGNAL.   EXTERNAL I NEEDED IN COMPUTATION
C            = 4 OUTPUT SIGNAL.   EXTERNAL I NOT NEEDED
C            = 5 OUTPUT SIGNAL.   NO RATE OF RETURN LOCATED.
C      COMPUTE ALGEBRAIC SUM OF CASH FLOW
      ASUM = CASHO
      DO 10 I=1,N
   10 ASUM = ASUM+CASH(I)
C      TRIAL RATE OF RETURN BY ORIGINAL BOOK METHOD
C      TRIAL RATE OF RETURN = (SUM/N)/CASHO + 0.10
      TRIAL = ASUM/(ABSF(CASHO)*FLOAT(N)) + 0.10
```

```
C     TLOW = LOWER BOUNDARY IN SEARCH
C     THI = UPPER BOUNDARY IN SEARCH
      THI = 1.00
      AMTH = 4
      ITER = 0
      IF (TRIAL.LT.0.) TRIAL = 0.05
      IF (TRIAL.GT.1.00) TRIAL = 0.80
      TLOW = 0.
15    IF (MTH.EQ.1) EXT = TRIAL
      ITER = ITER +1
      IF (ITER.GE.10) GO TO 75
      IF (ISW-2) 30,20,20
C     PRINT TABULATION
20    WRITE(OUT,21) TRIAL,EXT
21    FORMAT(//5X,31HTRIAL INTERNAL RATE OF RETURN = ,F7.4/
     1 12X,24HEXTERNAL INTEREST RATE =,F7.4//17X,4HCASH,66X,
     2 8HINTERNAL,14X,8HEXTERNAL/5X,4HYEAR,8X,4HFLOW,18X,
     3 11HCOMPUTATION,36X,10HINVESTMENT,12X,10HINVESTMENT/)
C     COMPUTE YEAR 0 LINE FOR TABLE
      PSUM = CASHO
30    IF (CASHO) 31,31,34
C     CASHO IS NEGATIVE OR ZERO
31    TSUM = CASHO*(1.+TRIAL)
      AEX = 0.
      AIN = CASHO
32    IF (ISW.EQ.1) GO TO 35
      WRITE(OUT,33) CASHO,AIN,AEX
33    FORMAT(7X,1H0,F12.2,65X,F12.2,10X,F12.2)
      GO TO 35
C     CASHO IS POSITIVE
34    TSUM = CASHO*(1.+EXT)
      AMTH = 3.
      AEX = CASHO
      AIN = 0.
      GO TO 32
35    DO 55 I=1,N
      IF (TSUM) 36,36,45
C     TSUM IS NEGATIVE OR ZERO
36    SUM = TSUM + CASH(I)
      IF (SUM) 37,37,38
37    AIN = CASH(I)
      AEX = 0.
      TSUM = SUM*(1.+TRIAL)
      GO TO 39
38    AIN = -TSUM
      AEX = CASH(I) + TSUM
      TSUM = SUM*(1.+EXT)

      WRITE(OUT,40) I,CASH(I),PSUM,TRIAL,CASH(I),SUM,AIN,AEX
40    FORMAT(6X,I2,F12.2,5X,F12.2,5H *(1+,F5.3,2H)+,F12.2,
     1 2H =,F12.2,2(10X,F12.2))
      GO TO 49
C     TSUM IS POSITIVE
45    SUM = TSUM + CASH(I)
      AMTH = 3.
      IF (SUM) 46,46,47
46    AIN = CASH(I) + TSUM
      AEX = -TSUM
      TSUM = SUM*(1.+TRIAL)
      GO TO 48
47    AIN = 0.
      AEX = CASH(I)
      TSUM = SUM*(1.+EXT)
48    IF (ISW.EQ.1) GO TO 49
      WRITE(OUT,40) I,CASH(I),PSUM,EXT,CASH(I),SUM,AIN,AEX
49    PSUM = SUM
55    CONTINUE
C     TEST VALUE OF SUM AT END OF YEAR N.  A NEGATIVE SUM
C     MEANS TRIAL TOO HIGH.  SUM = 0 INDICATES A PERFECT
C     TRIAL.  A POSITIVE SUM MEANS TRIAL TOO LOW.
      IF (SUM) 60,65,61
60    THI = TRIAL
      GO TO 62
61    TLOW = TRIAL
62    TRIAL = (TLOW + THI)/2.
      Y = TRIAL - TLOW
      IF (Y.LE.0.002) 63,15
63    IF (ISW.EQ.1) GO TO 70
      WRITE(OUT,64) TRIAL
64    FORMAT(//5X,64HCALCULATION INDICATES TRUE INTERNAL RATE
     1 OF RETURN VERY CLOSE TO,F7.4/5X,32HTHIS VALUE CONSIDE
     2RED THE ANSWER)
      GO TO 70
65    IF (ISW.EQ.1) GO TO 70
      WRITE(OUT,69) TRIAL
69    FORMAT(5X,32HINTERNAL RATE OF RETURN EQUAL TO,F7.4)
70    RATE = TRIAL
      MTH = AMTH
      RETURN
75    IF (ISW.EQ.1) GO TO 77
      WRITE(OUT,76)
76    FORMAT(//5X,25HNO RATE OF RETURN LOCATED)
77    RATE = TRIAL
      MTH = 5
      RETURN
```

428

# Appendix B
Compound
Interest Tables

**DISCRETE END-OF-PERIOD COMPOUNDING**

| | | | | | | |
|---|---|---|---|---|---|---|
| Tables for | $\frac{1}{4}\%$ | $\frac{1}{2}\%$ | $\frac{3}{4}\%$ | $1\%$ | $1\frac{1}{4}\%$ | $1\frac{1}{2}\%$ |
| | $1\frac{3}{4}\%$ | $2\%$ | $2\frac{1}{2}\%$ | $3\%$ | $3\frac{1}{2}\%$ | $4\%$ |
| | $4\frac{1}{2}\%$ | $5\%$ | $6\%$ | $7\%$ | $8\%$ | $9\%$ |
| | $10\%$ | $12\%$ | $15\%$ | $18\%$ | $20\%$ | $25\%$ |
| | $30\%$ | $35\%$ | $40\%$ | $45\%$ | $50\%$ | $60\%$ |

**CONTINUOUS COMPOUNDING—SINGLE PAYMENT FACTORS**

| | SINGLE PAYMENT | | UNIFORM PAYMENT SERIES | | | | GRADIENT SERIES | |
|---|---|---|---|---|---|---|---|---|
| | Compound Amount Factor | Present Worth Factor | Sinking Fund Factor | Capital Recovery Factor | Compound Amount Factor | Present Worth Factor | Gradient Uniform Series | Gradient Present Worth |
| n | Find F Given P F/P | Find P Given F P/F | Find A Given F A/F | Find A Given P A/P | Find F Given A F/A | Find P Given A P/A | Find A Given G A/G | Find P Given G P/G |
| 1 | 1.003 | .9975 | 1.0000 | 1.0025 | 1.000 | .998 | 0 | 0 |
| 2 | 1.005 | .9950 | .4994 | .5019 | 2.002 | 1.993 | .499 | .995 |
| 3 | 1.008 | .9925 | .3325 | .3350 | 3.008 | 2.985 | .998 | 2.980 |
| 4 | 1.010 | .9901 | .2491 | .2516 | 4.015 | 3.975 | 1.497 | 5.950 |
| 5 | 1.013 | .9876 | .1990 | .2015 | 5.025 | 4.963 | 1.995 | 9.901 |
| 6 | 1.015 | .9851 | .1656 | .1681 | 6.038 | 5.948 | 2.493 | 14.826 |
| 7 | 1.018 | .9827 | .1418 | .1443 | 7.053 | 6.931 | 2.990 | 20.722 |
| 8 | 1.020 | .9802 | .1239 | .1264 | 8.070 | 7.911 | 3.487 | 27.584 |
| 9 | 1.023 | .9778 | .1100 | .1125 | 9.091 | 8.889 | 3.983 | 35.406 |
| 10 | 1.025 | .9753 | .0989 | .1014 | 10.113 | 9.864 | 4.479 | 44.184 |
| 11 | 1.028 | .9729 | .0898 | .0923 | 11.139 | 10.837 | 4.975 | 53.913 |
| 12 | 1.030 | .9705 | .0822 | .0847 | 12.166 | 11.807 | 5.470 | 64.589 |
| 13 | 1.033 | .9681 | .0758 | .0783 | 13.197 | 12.775 | 5.965 | 76.205 |
| 14 | 1.036 | .9656 | .0703 | .0728 | 14.230 | 13.741 | 6.459 | 88.759 |
| 15 | 1.038 | .9632 | .0655 | .0680 | 15.265 | 14.704 | 6.953 | 102.244 |
| 16 | 1.041 | .9608 | .0613 | .0638 | 16.304 | 15.665 | 7.447 | 116.657 |
| 17 | 1.043 | .9584 | .0577 | .0602 | 17.344 | 16.623 | 7.940 | 131.992 |
| 18 | 1.046 | .9561 | .0544 | .0569 | 18.388 | 17.580 | 8.433 | 148.245 |
| 19 | 1.049 | .9537 | .0515 | .0540 | 19.434 | 18.533 | 8.925 | 165.411 |
| 20 | 1.051 | .9513 | .0488 | .0513 | 20.482 | 19.484 | 9.417 | 183.485 |
| 21 | 1.054 | .9489 | .0464 | .0489 | 21.533 | 20.433 | 9.908 | 202.463 |
| 22 | 1.056 | .9466 | .0443 | .0468 | 22.587 | 21.380 | 10.400 | 222.341 |
| 23 | 1.059 | .9442 | .0423 | .0448 | 23.644 | 22.324 | 10.890 | 243.113 |
| 24 | 1.062 | .9418 | .0405 | .0430 | 24.703 | 23.266 | 11.380 | 264.775 |
| 25 | 1.064 | .9395 | .0388 | .0413 | 25.765 | 24.205 | 11.870 | 287.323 |
| 26 | 1.067 | .9371 | .0373 | .0398 | 26.829 | 25.143 | 12.360 | 310.752 |
| 27 | 1.070 | .9348 | .0358 | .0383 | 27.896 | 26.077 | 12.849 | 335.057 |
| 28 | 1.072 | .9325 | .0345 | .0370 | 28.966 | 27.010 | 13.337 | 360.233 |
| 29 | 1.075 | .9301 | .0333 | .0358 | 30.038 | 27.940 | 13.825 | 386.278 |
| 30 | 1.078 | .9278 | .0321 | .0346 | 31.113 | 28.868 | 14.313 | 413.185 |
| 36 | 1.094 | .9140 | .0266 | .0291 | 37.621 | 34.386 | 17.231 | 592.499 |
| 40 | 1.105 | .9050 | .0238 | .0263 | 42.013 | 38.020 | 19.167 | 728.740 |
| 48 | 1.127 | .8871 | .0196 | .0221 | 50.931 | 45.179 | 23.021 | 1040.055 |
| 50 | 1.133 | .8826 | .0188 | .0213 | 53.189 | 46.946 | 23.980 | 1125.777 |
| 52 | 1.139 | .8782 | .0180 | .0205 | 55.457 | 48.705 | 24.938 | 1214.588 |
| 60 | 1.162 | .8609 | .0155 | .0180 | 64.647 | 55.652 | 28.751 | 1600.085 |
| 70 | 1.191 | .8396 | .0131 | .0156 | 76.394 | 64.144 | 33.481 | 2147.611 |
| 72 | 1.197 | .8355 | .0127 | .0152 | 78.779 | 65.817 | 34.422 | 2265.557 |
| 80 | 1.221 | .8189 | .0113 | .0138 | 88.439 | 72.426 | 38.169 | 2764.457 |
| 84 | 1.233 | .8108 | .0107 | .0132 | 93.342 | 75.681 | 40.033 | 3029.759 |
| 90 | 1.252 | .7987 | .0099 | .0124 | 100.788 | 80.504 | 42.816 | 3446.870 |
| 96 | 1.271 | .7869 | .0092 | .0117 | 108.347 | 85.255 | 45.584 | 3886.283 |
| 100 | 1.284 | .7790 | .0088 | .0113 | 113.450 | 88.382 | 47.422 | 4191.242 |
| 104 | 1.297 | .7713 | .0084 | .0109 | 118.604 | 91.479 | 49.252 | 4505.557 |
| 120 | 1.349 | .7411 | .0072 | .0097 | 139.741 | 103.562 | 56.508 | 5852.112 |
| 240 | 1.821 | .5492 | .0030 | .0055 | 328.302 | 180.311 | 107.586 | 19398.985 |
| 360 | 2.457 | .4070 | .0017 | .0042 | 582.737 | 237.189 | 152.890 | 36263.930 |
| 480 | 3.315 | .3016 | .0011 | .0036 | 926.059 | 279.342 | 192.670 | 53820.752 |

| | SINGLE PAYMENT | | UNIFORM PAYMENT SERIES | | | | GRADIENT SERIES | |
|---|---|---|---|---|---|---|---|---|
| | Compound Amount Factor | Present Worth Factor | Sinking Fund Factor | Capital Recovery Factor | Compound Amount Factor | Present Worth Factor | Gradient Uniform Series | Gradient Present Worth |
| *n* | Find *F* Given *P* *F/P* | Find *P* Given *F* *P/F* | Find *A* Given *F* *A/F* | Find *A* Given *P* *A/P* | Find *F* Given *A* *F/A* | Find *P* Given *A* *P/A* | Find *A* Given *G* *A/G* | Find *P* Given *G* *P/G* |
| 1 | 1.005 | .9950 | 1.0000 | 1.0050 | 1.000 | .995 | 0 | 0 |
| 2 | 1.010 | .9901 | .4988 | .5038 | 2.005 | 1.985 | .499 | .990 |
| 3 | 1.015 | .9851 | .3317 | .3367 | 3.015 | 2.970 | .997 | 2.960 |
| 4 | 1.020 | .9802 | .2481 | .2531 | 4.030 | 3.950 | 1.494 | 5.901 |
| 5 | 1.025 | .9754 | .1980 | .2030 | 5.050 | 4.926 | 1.990 | 9.803 |
| 6 | 1.030 | .9705 | .1646 | .1696 | 6.076 | 5.896 | 2.485 | 14.655 |
| 7 | 1.036 | .9657 | .1407 | .1457 | 7.106 | 6.862 | 2.980 | 20.449 |
| 8 | 1.041 | .9609 | .1228 | .1278 | 8.141 | 7.823 | 3.474 | 27.176 |
| 9 | 1.046 | .9561 | .1089 | .1139 | 9.182 | 8.779 | 3.967 | 34.824 |
| 10 | 1.051 | .9513 | .0978 | .1028 | 10.228 | 9.730 | 4.459 | 43.386 |
| 11 | 1.056 | .9466 | .0887 | .0937 | 11.279 | 10.677 | 4.950 | 52.853 |
| 12 | 1.062 | .9419 | .0811 | .0861 | 12.336 | 11.619 | 5.441 | 63.214 |
| 13 | 1.067 | .9372 | .0746 | .0796 | 13.397 | 12.556 | 5.930 | 74.460 |
| 14 | 1.072 | .9326 | .0691 | .0741 | 14.464 | 13.489 | 6.419 | 86.583 |
| 15 | 1.078 | .9279 | .0644 | .0694 | 15.537 | 14.417 | 6.907 | 99.574 |
| 16 | 1.083 | .9233 | .0602 | .0652 | 16.614 | 15.340 | 7.394 | 113.424 |
| 17 | 1.088 | .9187 | .0565 | .0615 | 17.697 | 16.259 | 7.880 | 128.123 |
| 18 | 1.094 | .9141 | .0532 | .0582 | 18.786 | 17.173 | 8.366 | 143.663 |
| 19 | 1.099 | .9096 | .0503 | .0553 | 19.880 | 18.082 | 8.850 | 160.036 |
| 20 | 1.105 | .9051 | .0477 | .0527 | 20.979 | 18.987 | 9.334 | 177.232 |
| 21 | 1.110 | .9006 | .0453 | .0503 | 22.084 | 19.888 | 9.817 | 195.243 |
| 22 | 1.116 | .8961 | .0431 | .0481 | 23.194 | 20.784 | 10.299 | 214.061 |
| 23 | 1.122 | .8916 | .0411 | .0461 | 24.310 | 21.676 | 10.781 | 233.677 |
| 24 | 1.127 | .8872 | .0393 | .0443 | 25.432 | 22.563 | 11.261 | 254.082 |
| 25 | 1.133 | .8828 | .0377 | .0427 | 26.559 | 23.446 | 11.741 | 275.269 |
| 26 | 1.138 | .8784 | .0361 | .0411 | 27.692 | 24.324 | 12.220 | 297.228 |
| 27 | 1.144 | .8740 | .0347 | .0397 | 28.830 | 25.198 | 12.698 | 319.952 |
| 28 | 1.150 | .8697 | .0334 | .0384 | 29.975 | 26.068 | 13.175 | 343.433 |
| 29 | 1.156 | .8653 | .0321 | .0371 | 31.124 | 26.933 | 13.651 | 367.663 |
| 30 | 1.161 | .8610 | .0310 | .0360 | 32.280 | 27.794 | 14.126 | 392.632 |
| 36 | 1.197 | .8356 | .0254 | .0304 | 39.336 | 32.871 | 16.962 | 557.560 |
| 40 | 1.221 | .8191 | .0226 | .0276 | 44.159 | 36.172 | 18.836 | 681.335 |
| 48 | 1.270 | .7871 | .0185 | .0235 | 54.098 | 42.580 | 22.544 | 959.919 |
| 50 | 1.283 | .7793 | .0177 | .0227 | 56.645 | 44.143 | 23.462 | 1035.697 |
| 52 | 1.296 | .7716 | .0169 | .0219 | 59.218 | 45.690 | 24.378 | 1113.816 |
| 60 | 1.349 | .7414 | .0143 | .0193 | 69.770 | 51.726 | 28.006 | 1448.646 |
| 70 | 1.418 | .7053 | .0120 | .0170 | 83.566 | 58.939 | 32.468 | 1913.643 |
| 72 | 1.432 | .6983 | .0116 | .0166 | 86.409 | 60.340 | 33.350 | 2012.348 |
| 80 | 1.490 | .6710 | .0102 | .0152 | 98.068 | 65.802 | 36.847 | 2424.646 |
| 84 | 1.520 | .6577 | .0096 | .0146 | 104.074 | 68.453 | 38.576 | 2640.664 |
| 90 | 1.567 | .6383 | .0088 | .0138 | 113.311 | 72.331 | 41.145 | 2976.077 |
| 96 | 1.614 | .6195 | .0081 | .0131 | 122.829 | 76.095 | 43.685 | 3324.185 |
| 100 | 1.647 | .6073 | .0077 | .0127 | 129.334 | 78.543 | 45.361 | 3562.793 |
| 104 | 1.680 | .5953 | .0074 | .0124 | 135.970 | 80.942 | 47.025 | 3806.286 |
| 120 | 1.819 | .5496 | .0061 | .0111 | 163.879 | 90.073 | 53.551 | 4823.505 |
| 240 | 3.310 | .3021 | .0022 | .0072 | 462.041 | 139.581 | 96.113 | 13415.540 |
| 360 | 6.023 | .1660 | .0010 | .0060 | 1004.515 | 166.792 | 128.324 | 21403.304 |
| 480 | 10.957 | .0913 | .0005 | .0055 | 1991.491 | 181.748 | 151.795 | 27588.357 |

| | SINGLE PAYMENT | | UNIFORM PAYMENT SERIES | | | | GRADIENT SERIES | |
|---|---|---|---|---|---|---|---|---|
| | Compound Amount Factor | Present Worth Factor | Sinking Fund Factor | Capital Recovery Factor | Compound Amount Factor | Present Worth Factor | Gradient Uniform Series | Gradient Present Worth |
| | Find $F$ Given $P$ | Find $P$ Given $F$ | Find $A$ Given $F$ | Find $A$ Given $P$ | Find $F$ Given $A$ | Find $P$ Given $A$ | Find $A$ Given $G$ | Find $P$ Given $G$ |
| $n$ | $F/P$ | $P/F$ | $A/F$ | $A/P$ | $F/A$ | $P/A$ | $A/G$ | $P/G$ |
| 1 | 1.008 | .9926 | 1.0000 | 1.0075 | 1.000 | .993 | 0 | 0 |
| 2 | 1.015 | .9852 | .4981 | .5056 | 2.008 | 1.978 | .498 | .985 |
| 3 | 1.023 | .9778 | .3308 | .3383 | 3.023 | 2.956 | .995 | 2.941 |
| 4 | 1.030 | .9706 | .2472 | .2547 | 4.045 | 3.926 | 1.491 | 5.852 |
| 5 | 1.038 | .9633 | .1970 | .2045 | 5.076 | 4.889 | 1.985 | 9.706 |
| 6 | 1.046 | .9562 | .1636 | .17-11 | 6.114 | 5.846 | 2.478 | 14.487 |
| 7 | 1.054 | .9490 | .1397 | .1472 | 7.159 | 6.795 | 2.970 | 20.181 |
| 8 | 1.062 | .9420 | .1218 | .1293 | 8.213 | 7.737 | 3.461 | 26.775 |
| 9 | 1.070 | .9350 | .1078 | .1153 | 9.275 | 8.672 | 3.950 | 34.254 |
| 10 | 1.078 | .9280 | .0967 | .1042 | 10.344 | 9.600 | 4.438 | 42.606 |
| 11 | 1.086 | .9211 | .0876 | .0951 | 11.422 | 10.521 | 4.925 | 51.817 |
| 12 | 1.094 | .9142 | .0800 | .0875 | 12.508 | 11.435 | 5.411 | 61.874 |
| 13 | 1.102 | .9074 | .0735 | .0810 | 13.601 | 12.342 | 5.895 | 72.763 |
| 14 | 1.110 | .9007 | .0680 | .0755 | 14.703 | 13.243 | 6.379 | 84.472 |
| 15 | 1.119 | .8940 | .0632 | .0707 | 15.814 | 14.137 | 6.861 | 96.988 |
| 16 | 1.127 | .8873 | .0591 | .0666 | 16.932 | 15.024 | 7.341 | 110.297 |
| 17 | 1.135 | .8807 | .0554 | .0629 | 18.059 | 15.905 | 7.821 | 124.389 |
| 18 | 1.144 | .8742 | .0521 | .0596 | 19.195 | 16.779 | 8.299 | 139.249 |
| 19 | 1.153 | .8676 | .0492 | .0567 | 20.339 | 17.647 | 8.776 | 154.867 |
| 20 | 1.161 | .8612 | .0465 | .0540 | 21.491 | 18.508 | 9.252 | 171.230 |
| 21 | 1.170 | .8548 | .0441 | .0516 | 22.652 | 19.363 | 9.726 | 188.325 |
| 22 | 1.179 | .8484 | .0420 | .0495 | 23.822 | 20.211 | 10.199 | 206.142 |
| 23 | 1.188 | .8421 | .0400 | .0475 | 25.001 | 21.053 | 10.671 | 224.668 |
| 24 | 1.196 | .8358 | .0382 | .0457 | 26.188 | 21.889 | 11.142 | 243.892 |
| 25 | 1.205 | .8296 | .0365 | .0440 | 27.385 | 22.719 | 11.612 | 263.803 |
| 26 | 1.214 | .8234 | .0350 | .0425 | 28.590 | 23.542 | 12.080 | 284.389 |
| 27 | 1.224 | .8173 | .0336 | .0411 | 29.805 | 24.359 | 12.547 | 305.639 |
| 28 | 1.233 | .8112 | .0322 | .0397 | 31.028 | 25.171 | 13.013 | 327.542 |
| 29 | 1.242 | .8052 | .0310 | .0385 | 32.261 | 25.976 | 13.477 | 350.087 |
| 30 | 1.251 | .7992 | .0298 | .0373 | 33.503 | 26.775 | 13.941 | 373.263 |
| 36 | 1.309 | .7641 | .0243 | .0318 | 41.153 | 31.447 | 16.695 | 524.992 |
| 40 | 1.348 | .7416 | .0215 | .0290 | 46.446 | 34.447 | 18.506 | 637.469 |
| 48 | 1.431 | .6986 | .0174 | .0249 | 57.521 | 40.185 | 22.069 | 886.840 |
| 50 | 1.453 | .6883 | .0166 | .0241 | 60.394 | 41.566 | 22.948 | 953.849 |
| 52 | 1.475 | .6780 | .0158 | .0233 | 63.311 | 42.928 | 23.821 | 1022.585 |
| 60 | 1.566 | .6387 | .0133 | .0208 | 75.424 | 48.173 | 27.266 | 1313.519 |
| 70 | 1.687 | .5927 | .0109 | .0184 | 91.620 | 54.305 | 31.463 | 1708.607 |
| 72 | 1.713 | .5839 | .0105 | .0180 | 95.007 | 55.477 | 32.288 | 1791.246 |
| 80 | 1.818 | .5500 | .0092 | .0167 | 109.073 | 59.994 | 35.539 | 2132.147 |
| 84 | 1.873 | .5338 | .0086 | .0161 | 116.427 | 62.154 | 37.136 | 2308.128 |
| 90 | 1.959 | .5104 | .0078 | .0153 | 127.879 | 65.275 | 39.495 | 2577.996 |
| 96 | 2.049 | .4881 | .0072 | .0147 | 139.856 | 68.258 | 41.811 | 2853.935 |
| 100 | 2.111 | .4737 | .0068 | .0143 | 148.145 | 70.175 | 43.331 | 3040.745 |
| 104 | 2.175 | .4597 | .0064 | .0139 | 156.684 | 72.034 | 44.833 | 3229.494 |
| 120 | 2.451 | .4079 | .0052 | .0127 | 193.514 | 78.942 | 50.652 | 3998.562 |
| 240 | 6.009 | .1664 | .0015 | .0090 | 667.887 | 111.145 | 85.421 | 9494.116 |
| 360 | 14.731 | .0679 | .0005 | .0080 | 1830.743 | 124.282 | 107.114 | 13312.387 |
| 480 | 36.110 | .0277 | .0002 | .0077 | 4681.320 | 129.641 | 119.662 | 15513.087 |

| | SINGLE PAYMENT | | UNIFORM PAYMENT SERIES | | | | GRADIENT SERIES | |
|---|---|---|---|---|---|---|---|---|
| | Compound Amount Factor | Present Worth Factor | Sinking Fund Factor | Capital Recovery Factor | Compound Amount Factor | Present Worth Factor | Gradient Uniform Series | Gradient Present Worth |
| $n$ | Find $F$ Given $P$ $F/P$ | Find $P$ Given $F$ $P/F$ | Find $A$ Given $F$ $A/F$ | Find $A$ Given $P$ $A/P$ | Find $F$ Given $A$ $F/A$ | Find $P$ Given $A$ $P/A$ | Find $A$ Given $G$ $A/G$ | Find $P$ Given $G$ $P/G$ |
| 1 | 1.010 | .9901 | 1.0000 | 1.0100 | 1.000 | .990 | 0 | 0 |
| 2 | 1.020 | .9803 | .4975 | .5075 | 2.010 | 1.970 | .498 | .980 |
| 3 | 1.030 | .9706 | .3300 | .3400 | 3.030 | 2.941 | .993 | 2.921 |
| 4 | 1.041 | .9610 | .2463 | .2563 | 4.060 | 3.902 | 1.488 | 5.804 |
| 5 | 1.051 | .9515 | .1960 | .2060 | 5.101 | 4.853 | 1.980 | 9.610 |
| 6 | 1.062 | .9420 | .1625 | .1725 | 6.152 | 5.795 | 2.471 | 14.321 |
| 7 | 1.072 | .9327 | .1386 | .1486 | 7.214 | 6.728 | 2.960 | 19.917 |
| 8 | 1.083 | .9235 | .1207 | .1307 | 8.286 | 7.652 | 3.448 | 26.381 |
| 9 | 1.094 | .9143 | .1067 | .1167 | 9.369 | 8.566 | 3.934 | 33.696 |
| 10 | 1.105 | .9053 | .0956 | .1056 | 10.462 | 9.471 | 4.418 | 41.843 |
| 11 | 1.116 | .8963 | .0865 | .0965 | 11.567 | 10.368 | 4.901 | 50.807 |
| 12 | 1.127 | .8874 | .0788 | .0888 | 12.683 | 11.255 | 5.381 | 60.569 |
| 13 | 1.138 | .8787 | .0724 | .0824 | 13.809 | 12.134 | 5.861 | 71.113 |
| 14 | 1.149 | .8700 | .0669 | .0769 | 14.947 | 13.004 | 6.338 | 82.422 |
| 15 | 1.161 | .8613 | .0621 | .0721 | 16.097 | 13.865 | 6.814 | 94.481 |
| 16 | 1.173 | .8528 | .0579 | .0679 | 17.258 | 14.718 | 7.289 | 107.273 |
| 17 | 1.184 | .8444 | .0543 | .0643 | 18.430 | 15.562 | 7.761 | 120.783 |
| 18 | 1.196 | .8360 | .0510 | .0610 | 19.615 | 16.398 | 8.232 | 134.996 |
| 19 | 1.208 | .8277 | .0481 | .0581 | 20.811 | 17.226 | 8.702 | 149.895 |
| 20 | 1.220 | .8195 | .0454 | .0554 | 22.019 | 18.046 | 9.169 | 165.466 |
| 21 | 1.232 | .8114 | .0430 | .0530 | 23.239 | 18.857 | 9.635 | 181.695 |
| 22 | 1.245 | .8034 | .0409 | .0509 | 24.472 | 19.660 | 10.100 | 198.566 |
| 23 | 1.257 | .7954 | .0389 | .0489 | 25.716 | 20.456 | 10.563 | 216.066 |
| 24 | 1.270 | .7876 | .0371 | .0471 | 26.973 | 21.243 | 11.024 | 234.180 |
| 25 | 1.282 | .7798 | .0354 | .0454 | 28.243 | 22.023 | 11.483 | 252.894 |
| 26 | 1.295 | .7720 | .0339 | .0439 | 29.526 | 22.795 | 11.941 | 272.196 |
| 27 | 1.308 | .7644 | .0324 | .0424 | 30.821 | 23.560 | 12.397 | 292.070 |
| 28 | 1.321 | .7568 | .0311 | .0411 | 32.129 | 24.316 | 12.852 | 312.505 |
| 29 | 1.335 | .7493 | .0299 | .0399 | 33.450 | 25.066 | 13.304 | 333.486 |
| 30 | 1.348 | .7419 | .0287 | .0387 | 34.785 | 25.808 | 13.756 | 355.002 |
| 36 | 1.431 | .6989 | .0232 | .0332 | 43.077 | 30.108 | 16.428 | 494.621 |
| 40 | 1.489 | .6717 | .0205 | .0305 | 48.886 | 32.835 | 18.178 | 596.856 |
| 48 | 1.612 | .6203 | .0163 | .0263 | 61.223 | 37.974 | 21.598 | 820.146 |
| 50 | 1.645 | .6080 | .0155 | .0255 | 64.463 | 39.196 | 22.436 | 879.418 |
| 52 | 1.678 | .5961 | .0148 | .0248 | 67.769 | 40.394 | 23.269 | 939.918 |
| 60 | 1.817 | .5504 | .0122 | .0222 | 81.670 | 44.955 | 26.533 | 1192.806 |
| 70 | 2.007 | .4983 | .0099 | .0199 | 100.676 | 50.169 | 30.470 | 1528.647 |
| 72 | 2.047 | .4885 | .0096 | .0196 | 104.710 | 51.150 | 31.239 | 1597.867 |
| 80 | 2.217 | .4511 | .0082 | .0182 | 121.672 | 54.888 | 34.249 | 1879.877 |
| 84 | 2.307 | .4335 | .0077 | .0177 | 130.672 | 56.648 | 35.717 | 2023.315 |
| 90 | 2.449 | .4084 | .0069 | .0169 | 144.863 | 59.161 | 37.872 | 2240.567 |
| 96 | 2.599 | .3847 | .0063 | .0163 | 159.927 | 61.528 | 39.973 | 2459.430 |
| 100 | 2.705 | .3697 | .0059 | .0159 | 170.481 | 63.029 | 41.343 | 2605.776 |
| 104 | 2.815 | .3553 | .0055 | .0155 | 181.464 | 64.471 | 42.688 | 2752.182 |
| 120 | 3.300 | .3030 | .0043 | .0143 | 230.039 | 69.701 | 47.835 | 3334.115 |
| 240 | 10.893 | .0918 | .0010 | .0110 | 989.255 | 90.819 | 75.739 | 6878.602 |
| 360 | 35.950 | .0278 | .0003 | .0103 | 3494.694 | 97.218 | 89.699 | 8720.432 |
| 480 | 118.648 | .0084 | .0001 | .0101 | 11764.773 | 99.157 | 95.920 | 9511.158 |

| | SINGLE PAYMENT | | UNIFORM PAYMENT SERIES | | | | GRADIENT SERIES | |
|---|---|---|---|---|---|---|---|---|
| | Compound Amount Factor | Present Worth Factor | Sinking Fund Factor | Capital Recovery Factor | Compound Amount Factor | Present Worth Factor | Gradient Uniform Series | Gradient Present Worth |
| $n$ | Find $F$ Given $P$ $F/P$ | Find $P$ Given $F$ $P/F$ | Find $A$ Given $F$ $A/F$ | Find $A$ Given $P$ $A/P$ | Find $F$ Given $A$ $F/A$ | Find $P$ Given $A$ $P/A$ | Find $A$ Given $G$ $A/G$ | Find $P$ Given $G$ $P/G$ |
| 1 | 1.013 | .9877 | 1.0000 | 1.0125 | 1.000 | .988 | 0 | 0 |
| 2 | 1.025 | .9755 | .4969 | .5094 | 2.013 | 1.963 | .497 | .975 |
| 3 | 1.038 | .9634 | .3292 | .3417 | 3.038 | 2.927 | .992 | 2.902 |
| 4 | 1.051 | .9515 | .2454 | .2579 | 4.076 | 3.878 | 1.484 | 5.757 |
| 5 | 1.064 | .9398 | .1951 | .2076 | 5.127 | 4.818 | 1.975 | 9.516 |
| 6 | 1.077 | .9282 | .1615 | .1740 | 6.191 | 5.746 | 2.464 | 14.157 |
| 7 | 1.091 | .9167 | .1376 | .1501 | 7.268 | 6.663 | 2.950 | 19.657 |
| 8 | 1.104 | .9054 | .1196 | .1321 | 8.359 | 7.568 | 3.435 | 25.995 |
| 9 | 1.118 | .8942 | .1057 | .1182 | 9.463 | 8.462 | 3.917 | 33.149 |
| 10 | 1.132 | .8832 | .0945 | .1070 | 10.582 | 9.346 | 4.398 | 41.097 |
| 11 | 1.146 | .8723 | .0854 | .0979 | 11.714 | 10.218 | 4.876 | 49.820 |
| 12 | 1.161 | .8615 | .0778 | .0903 | 12.860 | 11.079 | 5.352 | 59.297 |
| 13 | 1.175 | .8509 | .0713 | .0838 | 14.021 | 11.930 | 5.826 | 69.507 |
| 14 | 1.190 | .8404 | .0658 | .0783 | 15.196 | 12.771 | 6.298 | 80.432 |
| 15 | 1.205 | .8300 | .0610 | .0735 | 16.386 | 13.601 | 6.768 | 92.052 |
| 16 | 1.220 | .8197 | .0568 | .0693 | 17.591 | 14.420 | 7.236 | 104.348 |
| 17 | 1.235 | .8096 | .0532 | .0657 | 18.811 | 15.230 | 7.702 | 117.302 |
| 18 | 1.251 | .7996 | .0499 | .0624 | 20.046 | 16.030 | 8.166 | 130.896 |
| 19 | 1.266 | .7898 | .0470 | .0595 | 21.297 | 16.819 | 8.628 | 145.111 |
| 20 | 1.282 | .7800 | .0443 | .0568 | 22.563 | 17.599 | 9.087 | 159.932 |
| 21 | 1.298 | .7704 | .0419 | .0544 | 23.845 | 18.370 | 9.545 | 175.339 |
| 22 | 1.314 | .7609 | .0398 | .0523 | 25.143 | 19.131 | 10.001 | 191.317 |
| 23 | 1.331 | .7515 | .0378 | .0503 | 26.457 | 19.882 | 10.454 | 207.850 |
| 24 | 1.347 | .7422 | .0360 | .0485 | 27.788 | 20.624 | 10.906 | 224.920 |
| 25 | 1.364 | .7330 | .0343 | .0468 | 29.135 | 21.357 | 11.355 | 242.513 |
| 26 | 1.381 | .7240 | .0328 | .0453 | 30.500 | 22.081 | 11.802 | 260.613 |
| 27 | 1.399 | .7150 | .0314 | .0439 | 31.881 | 22.796 | 12.248 | 279.204 |
| 28 | 1.416 | .7062 | .0300 | .0425 | 33.279 | 23.503 | 12.691 | 298.272 |
| 29 | 1.434 | .6975 | .0288 | .0413 | 34.695 | 24.200 | 13.132 | 317.802 |
| 30 | 1.452 | .6889 | .0277 | .0402 | 36.129 | 24.889 | 13.571 | 337.780 |
| 36 | 1.564 | .6394 | .0222 | .0347 | 45.116 | 28.847 | 16.164 | 466.283 |
| 40 | 1.644 | .6084 | .0194 | .0319 | 51.490 | 31.327 | 17.851 | 559.232 |
| 48 | 1.815 | .5509 | .0153 | .0278 | 65.228 | 35.931 | 21.130 | 759.230 |
| 50 | 1.861 | .5373 | .0145 | .0270 | 68.882 | 37.013 | 21.929 | 811.674 |
| 52 | 1.908 | .5242 | .0138 | .0263 | 72.627 | 38.068 | 22.721 | 864.941 |
| 60 | 2.107 | .4746 | .0113 | .0238 | 88.575 | 42.035 | 25.808 | 1084.843 |
| 70 | 2.386 | .4191 | .0090 | .0215 | 110.872 | 46.470 | 29.491 | 1370.451 |
| 72 | 2.446 | .4088 | .0086 | .0211 | 115.674 | 47.292 | 30.205 | 1428.456 |
| 80 | 2.701 | .3702 | .0073 | .0198 | 136.119 | 50.387 | 32.982 | 1661.865 |
| 84 | 2.839 | .3522 | .0068 | .0193 | 147.129 | 51.822 | 34.326 | 1778.838 |
| 90 | 3.059 | .3269 | .0061 | .0186 | 164.705 | 53.846 | 36.285 | 1953.830 |
| 96 | 3.296 | .3034 | .0054 | .0179 | 183.641 | 55.725 | 38.179 | 2127.524 |
| 100 | 3.463 | .2887 | .0051 | .0176 | 197.072 | 56.901 | 39.406 | 2242.241 |
| 104 | 3.640 | .2747 | .0047 | .0172 | 211.188 | 58.021 | 40.604 | 2355.876 |
| 120 | 4.440 | .2252 | .0036 | .0161 | 275.217 | 61.983 | 45.118 | 2796.569 |
| 240 | 19.715 | .0507 | .0007 | .0132 | 1497.239 | 75.942 | 67.176 | 5101.529 |
| 360 | 87.541 | .0114 | .0001 | .0126 | 6923.280 | 79.086 | 75.840 | 5997.903 |
| 480 | 388.701 | .0026 | | .0125 | 31016.055 | 79.794 | 78.762 | 6284.744 |

*434*

| | SINGLE PAYMENT | | UNIFORM PAYMENT SERIES | | | | GRADIENT SERIES | |
|---|---|---|---|---|---|---|---|---|
| | Compound Amount Factor | Present Worth Factor | Sinking Fund Factor | Capital Recovery Factor | Compound Amount Factor | Present Worth Factor | Gradient Uniform Series | Gradient Present Worth |
| | Find $F$ Given $P$ | Find $P$ Given $F$ | Find $A$ Given $F$ | Find $A$ Given $P$ | Find $F$ Given $A$ | Find $P$ Given $A$ | Find $A$ Given $G$ | Find $P$ Given $G$ |
| $n$ | $F/P$ | $P/F$ | $A/F$ | $A/P$ | $F/A$ | $P/A$ | $A/G$ | $P/G$ |
| 1 | 1.015 | .9852 | 1.0000 | 1.0150 | 1.000 | .985 | 0 | 0 |
| 2 | 1.030 | .9707 | .4963 | .5113 | 2.015 | 1.956 | .496 | .971 |
| 3 | 1.046 | .9563 | .3284 | .3434 | 3.045 | 2.912 | .990 | 2.883 |
| 4 | 1.061 | .9422 | .2444 | .2594 | 4.091 | 3.854 | 1.481 | 5.710 |
| 5 | 1.077 | .9283 | .1941 | .2091 | 5.152 | 4.783 | 1.970 | 9.423 |
| 6 | 1.093 | .9145 | .1605 | .1755 | 6.230 | 5.697 | 2.457 | 13.996 |
| 7 | 1.110 | .9010 | .1366 | .1516 | 7.323 | 6.598 | 2.940 | 19.402 |
| 8 | 1.126 | .8877 | .1186 | .1336 | 8.433 | 7.486 | 3.422 | 25.616 |
| 9 | 1.143 | .8746 | .1046 | .1196 | 9.559 | 8.361 | 3.901 | 32.612 |
| 10 | 1.161 | .8617 | .0934 | .1084 | 10.703 | 9.222 | 4.377 | 40.367 |
| 11 | 1.178 | .8489 | .0843 | .0993 | 11.863 | 10.071 | 4.851 | 48.857 |
| 12 | 1.196 | .8364 | .0767 | .0917 | 13.041 | 10.908 | 5.323 | 58.057 |
| 13 | 1.214 | .8240 | .0702 | .0852 | 14.237 | 11.732 | 5.792 | 67.945 |
| 14 | 1.232 | .8118 | .0647 | .0797 | 15.450 | 12.543 | 6.258 | 78.499 |
| 15 | 1.250 | .7999 | .0599 | .0749 | 16.682 | 13.343 | 6.722 | 89.697 |
| 16 | 1.269 | .7880 | .0558 | .0708 | 17.932 | 14.131 | 7.184 | 101.518 |
| 17 | 1.288 | .7764 | .0521 | .0671 | 19.201 | 14.908 | 7.643 | 113.940 |
| 18 | 1.307 | .7649 | .0488 | .0638 | 20.489 | 15.673 | 8.100 | 126.943 |
| 19 | 1.327 | .7536 | .0459 | .0609 | 21.797 | 16.426 | 8.554 | 140.508 |
| 20 | 1.347 | .7425 | .0432 | .0582 | 23.124 | 17.169 | 9.006 | 154.615 |
| 21 | 1.367 | .7315 | .0409 | .0559 | 24.471 | 17.900 | 9.455 | 169.245 |
| 22 | 1.388 | .7207 | .0387 | .0537 | 25.838 | 18.621 | 9.902 | 184.380 |
| 23 | 1.408 | .7100 | .0367 | .0517 | 27.225 | 19.331 | 10.346 | 200.001 |
| 24 | 1.430 | .6995 | .0349 | .0499 | 28.634 | 20.030 | 10.788 | 216.090 |
| 25 | 1.451 | .6892 | .0333 | .0483 | 30.063 | 20.720 | 11.228 | 232.631 |
| 26 | 1.473 | .6790 | .0317 | .0467 | 31.514 | 21.399 | 11.665 | 249.607 |
| 27 | 1.495 | .6690 | .0303 | .0453 | 32.987 | 22.068 | 12.099 | 267.000 |
| 28 | 1.517 | .6591 | .0290 | .0440 | 34.481 | 22.727 | 12.531 | 284.796 |
| 29 | 1.540 | .6494 | .0278 | .0428 | 35.999 | 23.376 | 12.961 | 302.978 |
| 30 | 1.563 | .6398 | .0266 | .0416 | 37.539 | 24.016 | 13.388 | 321.531 |
| 36 | 1.709 | .5851 | .0212 | .0362 | 47.276 | 27.661 | 15.901 | 439.830 |
| 40 | 1.814 | .5513 | .0184 | .0334 | 54.268 | 29.916 | 17.528 | 524.357 |
| 48 | 2.043 | .4894 | .0144 | .0294 | 69.565 | 34.043 | 20.667 | 703.546 |
| 50 | 2.105 | .4750 | .0136 | .0286 | 73.683 | 35.000 | 21.428 | 749.964 |
| 52 | 2.169 | .4611 | .0128 | .0278 | 77.925 | 35.929 | 22.179 | 796.877 |
| 60 | 2.443 | .4093 | .0104 | .0254 | 96.215 | 39.380 | 25.093 | 988.167 |
| 70 | 2.835 | .3527 | .0082 | .0232 | 122.364 | 43.155 | 28.529 | 1231.166 |
| 72 | 2.921 | .3423 | .0078 | .0228 | 128.077 | 43.845 | 29.189 | 1279.794 |
| 80 | 3.291 | .3039 | .0065 | .0215 | 152.711 | 46.407 | 31.742 | 1473.074 |
| 84 | 3.493 | .2863 | .0060 | .0210 | 166.173 | 47.579 | 32.967 | 1568.514 |
| 90 | 3.819 | .2619 | .0053 | .0203 | 187.930 | 49.210 | 34.740 | 1709.544 |
| 96 | 4.176 | .2395 | .0047 | .0197 | 211.720 | 50.702 | 36.438 | 1847.473 |
| 100 | 4.432 | .2256 | .0044 | .0194 | 228.803 | 51.625 | 37.530 | 1937.451 |
| 104 | 4.704 | .2126 | .0040 | .0190 | 246.934 | 52.494 | 38.589 | 2025.705 |
| 120 | 5.969 | .1675 | .0030 | .0180 | 331.288 | 55.498 | 42.519 | 2359.711 |
| 240 | 35.633 | .0281 | .0004 | .0154 | 2308.854 | 64.796 | 59.737 | 3870.691 |
| 360 | 212.704 | .0047 | .0001 | .0151 | 14113.586 | 66.353 | 64.966 | 4310.716 |
| 480 | 1269.698 | .0008 | | .0150 | 84579.837 | 66.614 | 66.288 | 4415.741 |

| | SINGLE PAYMENT | | UNIFORM PAYMENT SERIES | | | | GRADIENT SERIES | |
|---|---|---|---|---|---|---|---|---|
| | Compound Amount Factor | Present Worth Factor | Sinking Fund Factor | Capital Recovery Factor | Compound Amount Factor | Present Worth Factor | Gradient Uniform Series | Gradient Present Worth |
| n | Find F Given P F/P | Find P Given F P/F | Find A Given F A/F | Find A Given P A/P | Find F Given A F/A | Find P Given A P/A | Find A Given G A/G | Find P Given G P/G |
| 1 | 1.017 | .9828 | 1.0000 | 1.0175 | 1.000 | .983 | 0 | 0 |
| 2 | 1.035 | .9659 | .4957 | .5132 | 2.017 | 1.949 | .496 | .966 |
| 3 | 1.053 | .9493 | .3276 | .3451 | 3.053 | 2.898 | .988 | 2.864 |
| 4 | 1.072 | .9330 | .2435 | .2610 | 4.106 | 3.831 | 1.478 | 5.663 |
| 5 | 1.091 | .9169 | .1931 | .2106 | 5.178 | 4.748 | 1.965 | 9.331 |
| 6 | 1.110 | .9011 | .1595 | .1770 | 6.269 | 5.649 | 2.449 | 13.837 |
| 7 | 1.129 | .8856 | .1355 | .1530 | 7.378 | 6.535 | 2.931 | 19.151 |
| 8 | 1.149 | .8704 | .1175 | .1350 | 8.508 | 7.405 | 3.409 | 25.243 |
| 9 | 1.169 | .8554 | .1036 | .1211 | 9.656 | 8.260 | 3.884 | 32.087 |
| 10 | 1.189 | .8407 | .0924 | .1099 | 10.825 | 9.101 | 4.357 | 39.654 |
| 11 | 1.210 | .8263 | .0832 | .1007 | 12.015 | 9.927 | 4.827 | 47.916 |
| 12 | 1.231 | .8121 | .0756 | .0931 | 13.225 | 10.740 | 5.293 | 56.849 |
| 13 | 1.253 | .7981 | .0692 | .0867 | 14.457 | 11.538 | 5.757 | 66.426 |
| 14 | 1.275 | .7844 | .0637 | .0812 | 15.710 | 12.322 | 6.218 | 76.623 |
| 15 | 1.297 | .7709 | .0589 | .0764 | 16.984 | 13.093 | 6.677 | 87.415 |
| 16 | 1.320 | .7576 | .0547 | .0722 | 18.282 | 13.850 | 7.132 | 98.779 |
| 17 | 1.343 | .7446 | .0510 | .0685 | 19.602 | 14.595 | 7.584 | 110.693 |
| 18 | 1.367 | .7318 | .0477 | .0652 | 20.945 | 15.327 | 8.034 | 123.133 |
| 19 | 1.390 | .7192 | .0448 | .0623 | 22.311 | 16.046 | 8.480 | 136.078 |
| 20 | 1.415 | .7068 | .0422 | .0597 | 23.702 | 16.753 | 8.924 | 149.508 |
| 21 | 1.440 | .6947 | .0398 | .0573 | 25.116 | 17.448 | 9.365 | 163.401 |
| 22 | 1.465 | .6827 | .0377 | .0552 | 26.556 | 18.130 | 9.803 | 177.738 |
| 23 | 1.490 | .6710 | .0357 | .0532 | 28.021 | 18.801 | 10.239 | 192.500 |
| 24 | 1.516 | .6594 | .0339 | .0514 | 29.511 | 19.461 | 10.671 | 207.667 |
| 25 | 1.543 | .6481 | .0322 | .0497 | 31.027 | 20.109 | 11.101 | 223.221 |
| 26 | 1.570 | .6369 | .0307 | .0482 | 32.570 | 20.746 | 11.527 | 239.145 |
| 27 | 1.597 | .6260 | .0293 | .0468 | 34.140 | 21.372 | 11.951 | 255.421 |
| 28 | 1.625 | .6152 | .0280 | .0455 | 35.738 | 21.987 | 12.372 | 272.032 |
| 29 | 1.654 | .6046 | .0268 | .0443 | 37.363 | 22.592 | 12.791 | 288.962 |
| 30 | 1.683 | .5942 | .0256 | .0431 | 39.017 | 23.186 | 13.206 | 306.195 |
| 36 | 1.867 | .5355 | .0202 | .0377 | 49.566 | 26.543 | 15.640 | 415.125 |
| 40 | 2.002 | .4996 | .0175 | .0350 | 57.234 | 28.594 | 17.207 | 492.011 |
| 48 | 2.300 | .4349 | .0135 | .0310 | 74.263 | 32.294 | 20.208 | 652.605 |
| 50 | 2.381 | .4200 | .0127 | .0302 | 78.902 | 33.141 | 20.932 | 693.701 |
| 52 | 2.465 | .4057 | .0119 | .0294 | 83.705 | 33.960 | 21.644 | 735.032 |
| 60 | 2.832 | .3531 | .0096 | .0271 | 104.675 | 36.964 | 24.388 | 901.495 |
| 70 | 3.368 | .2969 | .0074 | .0249 | 135.331 | 40.178 | 27.586 | 1108.333 |
| 72 | 3.487 | .2868 | .0070 | .0245 | 142.126 | 40.756 | 28.195 | 1149.118 |
| 80 | 4.006 | .2496 | .0058 | .0233 | 171.794 | 42.880 | 30.533 | 1309.248 |
| 84 | 4.294 | .2329 | .0053 | .0228 | 188.245 | 43.836 | 31.644 | 1387.158 |
| 90 | 4.765 | .2098 | .0046 | .0221 | 215.165 | 45.152 | 33.241 | 1500.880 |
| 96 | 5.288 | .1891 | .0041 | .0216 | 245.037 | 46.337 | 34.756 | 1610.472 |
| 100 | 5.668 | .1764 | .0037 | .0212 | 266.752 | 47.061 | 35.721 | 1681.089 |
| 104 | 6.075 | .1646 | .0034 | .0209 | 290.027 | 47.737 | 36.652 | 1749.675 |
| 120 | 8.019 | .1247 | .0025 | .0200 | 401.096 | 50.017 | 40.047 | 2003.027 |
| 240 | 64.307 | .0156 | .0003 | .0178 | 3617.560 | 56.254 | 53.352 | 3001.268 |
| 360 | 515.692 | .0019 | | .0175 | 29410.975 | 57.032 | 56.443 | 3219.083 |
| 480 | 4135.429 | .0002 | | .0175 | 236253.095 | 57.129 | 57.027 | 3257.884 |

| | SINGLE PAYMENT | | UNIFORM PAYMENT SERIES | | | | GRADIENT SERIES | |
|---|---|---|---|---|---|---|---|---|
| | Compound Amount Factor | Present Worth Factor | Sinking Fund Factor | Capital Recovery Factor | Compound Amount Factor | Present Worth Factor | Gradient Uniform Series | Gradient Present Worth |
| | Find F Given P | Find P Given F | Find A Given F | Find A Given P | Find F Given A | Find P Given A | Find A Given G | Find P Given G |
| n | F/P | P/F | A/F | A/P | F/A | P/A | A/G | P/G |
| 1 | 1.020 | .9804 | 1.0000 | 1.0200 | 1.000 | .980 | 0 | 0 |
| 2 | 1.040 | .9612 | .4950 | .5150 | 2.020 | 1.942 | .495 | .961 |
| 3 | 1.061 | .9423 | .3268 | .3468 | 3.060 | 2.884 | .987 | 2.846 |
| 4 | 1.082 | .9238 | .2426 | .2626 | 4.122 | 3.808 | 1.475 | 5.617 |
| 5 | 1.104 | .9057 | .1922 | .2122 | 5.204 | 4.713 | 1.960 | 9.240 |
| 6 | 1.126 | .8880 | .1585 | .1785 | 6.308 | 5.601 | 2.442 | 13.680 |
| 7 | 1.149 | .8706 | .1345 | .1545 | 7.434 | 6.472 | 2.921 | 18.903 |
| 8 | 1.172 | .8535 | .1165 | .1365 | 8.583 | 7.325 | 3.396 | 24.878 |
| 9 | 1.195 | .8368 | .1025 | .1225 | 9.755 | 8.162 | 3.868 | 31.572 |
| 10 | 1.219 | .8203 | .0913 | .1113 | 10.950 | 8.983 | 4.337 | 38.955 |
| 11 | 1.243 | .8043 | .0822 | .1022 | 12.169 | 9.787 | 4.802 | 46.998 |
| 12 | 1.268 | .7885 | .0746 | .0946 | 13.412 | 10.575 | 5.264 | 55.671 |
| 13 | 1.294 | .7730 | .0681 | .0881 | 14.680 | 11.348 | 5.723 | 64.948 |
| 14 | 1.319 | .7579 | .0626 | .0826 | 15.974 | 12.106 | 6.179 | 74.800 |
| 15 | 1.346 | .7430 | .0578 | .0778 | 17.293 | 12.849 | 6.631 | 85.202 |
| 16 | 1.373 | .7284 | .0537 | .0737 | 18.639 | 13.578 | 7.080 | 96.129 |
| 17 | 1.400 | .7142 | .0500 | .0700 | 20.012 | 14.292 | 7.526 | 107.555 |
| 18 | 1.428 | .7002 | .0467 | .0667 | 21.412 | 14.992 | 7.968 | 119.458 |
| 19 | 1.457 | .6864 | .0438 | .0638 | 22.841 | 15.678 | 8.407 | 131.814 |
| 20 | 1.486 | .6730 | .0412 | .0612 | 24.297 | 16.351 | 8.843 | 144.600 |
| 21 | 1.516 | .6598 | .0388 | .0588 | 25.783 | 17.011 | 9.276 | 157.796 |
| 22 | 1.546 | .6468 | .0366 | .0566 | 27.299 | 17.658 | 9.705 | 171.379 |
| 23 | 1.577 | .6342 | .0347 | .0547 | 28.845 | 18.292 | 10.132 | 185.331 |
| 24 | 1.608 | .6217 | .0329 | .0529 | 30.422 | 18.914 | 10.555 | 199.630 |
| 25 | 1.641 | .6095 | .0312 | .0512 | 32.030 | 19.523 | 10.974 | 214.259 |
| 26 | 1.673 | .5976 | .0297 | .0497 | 33.671 | 20.121 | 11.391 | 229.199 |
| 27 | 1.707 | .5859 | .0283 | .0483 | 35.344 | 20.707 | 11.804 | 244.431 |
| 28 | 1.741 | .5744 | .0270 | .0470 | 37.051 | 21.281 | 12.214 | 259.939 |
| 29 | 1.776 | .5631 | .0258 | .0458 | 38.792 | 21.844 | 12.621 | 275.706 |
| 30 | 1.811 | .5521 | .0246 | .0446 | 40.568 | 22.396 | 13.025 | 291.716 |
| 36 | 2.040 | .4902 | .0192 | .0392 | 51.994 | 25.489 | 15.381 | 392.040 |
| 40 | 2.208 | .4529 | .0166 | .0366 | 60.402 | 27.355 | 16.889 | 461.993 |
| 48 | 2.587 | .3865 | .0126 | .0326 | 79.354 | 30.673 | 19.756 | 605.966 |
| 50 | 2.692 | .3715 | .0118 | .0318 | 84.579 | 31.424 | 20.442 | 642.361 |
| 52 | 2.800 | .3571 | .0111 | .0311 | 90.016 | 32.145 | 21.116 | 678.785 |
| 60 | 3.281 | .3048 | .0088 | .0288 | 114.052 | 34.761 | 23.696 | 823.698 |
| 70 | 4.000 | .2500 | .0067 | .0267 | 149.978 | 37.499 | 26.663 | 999.834 |
| 72 | 4.161 | .2403 | .0063 | .0263 | 158.057 | 37.984 | 27.223 | 1034.056 |
| 80 | 4.875 | .2051 | .0052 | .0252 | 193.772 | 39.745 | 29.357 | 1166.787 |
| 84 | 5.277 | .1895 | .0047 | .0247 | 213.867 | 40.526 | 30.362 | 1230.419 |
| 90 | 5.943 | .1683 | .0040 | .0240 | 247.157 | 41.587 | 31.793 | 1322.170 |
| 96 | 6.693 | .1494 | .0035 | .0235 | 284.647 | 42.529 | 33.137 | 1409.297 |
| 100 | 7.245 | .1380 | .0032 | .0232 | 312.232 | 43.098 | 33.986 | 1464.753 |
| 104 | 7.842 | .1275 | .0029 | .0229 | 342.092 | 43.624 | 34.799 | 1518.087 |
| 120 | 10.765 | .0929 | .0020 | .0220 | 488.258 | 45.355 | 37.711 | 1710.416 |
| 240 | 115.889 | .0086 | .0002 | .0202 | 5744.437 | 49.569 | 47.911 | 2374.880 |
| 360 | 1247.561 | .0008 | | .0200 | 62328.056 | 49.960 | 49.711 | 2483.568 |
| 480 | 13430.199 | .0001 | | .0200 | 671459.945 | 49.996 | 49.964 | 2498.027 |

| | SINGLE PAYMENT | | UNIFORM PAYMENT SERIES | | | | GRADIENT SERIES | |
|---|---|---|---|---|---|---|---|---|
| | Compound Amount Factor | Present Worth Factor | Sinking Fund Factor | Capital Recovery Factor | Compound Amount Factor | Present Worth Factor | Gradient Uniform Series | Gradient Present Worth |
| n | Find F Given P F/P | Find P Given F P/F | Find A Given F A/F | Find A Given P A/P | Find F Given A F/A | Find P Given A P/A | Find A Given G A/G | Find P Given G P/G |
| 1 | 1.025 | .9756 | 1.0000 | 1.0250 | 1.000 | .976 | 0 | 0 |
| 2 | 1.051 | .9518 | .4938 | .5188 | 2.025 | 1.927 | .494 | .952 |
| 3 | 1.077 | .9286 | .3251 | .3501 | 3.076 | 2.856 | .984 | 2.809 |
| 4 | 1.104 | .9060 | .2408 | .2658 | 4.153 | 3.762 | 1.469 | 5.527 |
| 5 | 1.131 | .8839 | .1902 | .2152 | 5.256 | 4.646 | 1.951 | 9.062 |
| 6 | 1.160 | .8623 | .1565 | .1815 | 6.388 | 5.508 | 2.428 | 13.374 |
| 7 | 1.189 | .8413 | .1325 | .1575 | 7.547 | 6.349 | 2.901 | 18.421 |
| 8 | 1.218 | .8207 | .1145 | .1395 | 8.736 | 7.170 | 3.370 | 24.167 |
| 9 | 1.249 | .8007 | .1005 | .1255 | 9.955 | 7.971 | 3.836 | 30.572 |
| 10 | 1.280 | .7812 | .0893 | .1143 | 11.203 | 8.752 | 4.296 | 37.603 |
| 11 | 1.312 | .7621 | .0801 | .1051 | 12.483 | 9.514 | 4.753 | 45.225 |
| 12 | 1.345 | .7436 | .0725 | .0975 | 13.796 | 10.258 | 5.206 | 53.404 |
| 13 | 1.379 | .7254 | .0660 | .0910 | 15.140 | 10.983 | 5.655 | 62.109 |
| 14 | 1.413 | .7077 | .0605 | .0855 | 16.519 | 11.691 | 6.100 | 71.309 |
| 15 | 1.448 | .6905 | .0558 | .0808 | 17.932 | 12.381 | 6.540 | 80.976 |
| 16 | 1.485 | .6736 | .0516 | .0766 | 19.380 | 13.055 | 6.977 | 91.080 |
| 17 | 1.522 | .6572 | .0479 | .0729 | 20.865 | 13.712 | 7.409 | 101.595 |
| 18 | 1.560 | .6412 | .0447 | .0697 | 22.386 | 14.353 | 7.838 | 112.495 |
| 19 | 1.599 | .6255 | .0418 | .0668 | 23.946 | 14.979 | 8.262 | 123.755 |
| 20 | 1.639 | .6103 | .0391 | .0641 | 25.545 | 15.589 | 8.682 | 135.350 |
| 21 | 1.680 | .5954 | .0368 | .0618 | 27.183 | 16.185 | 9.099 | 147.257 |
| 22 | 1.722 | .5809 | .0346 | .0596 | 28.863 | 16.765 | 9.511 | 159.456 |
| 23 | 1.765 | .5667 | .0327 | .0577 | 30.584 | 17.332 | 9.919 | 171.923 |
| 24 | 1.809 | .5529 | .0309 | .0559 | 32.349 | 17.885 | 10.324 | 184.639 |
| 25 | 1.854 | .5394 | .0293 | .0543 | 34.158 | 18.424 | 10.724 | 197.584 |
| 26 | 1.900 | .5262 | .0278 | .0528 | 36.012 | 18.951 | 11.121 | 210.740 |
| 27 | 1.948 | .5134 | .0264 | .0514 | 37.912 | 19.464 | 11.513 | 224.089 |
| 28 | 1.996 | .5009 | .0251 | .0501 | 39.860 | 19.965 | 11.902 | 237.612 |
| 29 | 2.046 | .4887 | .0239 | .0489 | 41.856 | 20.454 | 12.286 | 251.295 |
| 30 | 2.098 | .4767 | .0228 | .0478 | 43.903 | 20.930 | 12.667 | 265.120 |
| 31 | 2.150 | .4651 | .0217 | .0467 | 46.000 | 21.395 | 13.044 | 279.074 |
| 32 | 2.204 | .4538 | .0208 | .0458 | 48.150 | 21.849 | 13.417 | 293.141 |
| 33 | 2.259 | .4427 | .0199 | .0449 | 50.354 | 22.292 | 13.786 | 307.307 |
| 34 | 2.315 | .4319 | .0190 | .0440 | 52.613 | 22.724 | 14.151 | 321.560 |
| 35 | 2.373 | .4214 | .0182 | .0432 | 54.928 | 23.145 | 14.512 | 335.887 |
| 40 | 2.685 | .3724 | .0148 | .0398 | 67.403 | 25.103 | 16.262 | 408.222 |
| 45 | 3.038 | .3292 | .0123 | .0373 | 81.516 | 26.833 | 17.918 | 480.807 |
| 50 | 3.437 | .2909 | .0103 | .0353 | 97.484 | 28.362 | 19.484 | 552.608 |
| 55 | 3.889 | .2572 | .0087 | .0337 | 115.551 | 29.714 | 20.961 | 622.828 |
| 60 | 4.400 | .2273 | .0074 | .0324 | 135.992 | 30.909 | 22.352 | 690.866 |
| 65 | 4.978 | .2009 | .0063 | .0313 | 159.118 | 31.965 | 23.660 | 756.281 |
| 70 | 5.632 | .1776 | .0054 | .0304 | 185.284 | 32.898 | 24.888 | 818.764 |
| 75 | 6.372 | .1569 | .0047 | .0297 | 214.888 | 33.723 | 26.039 | 878.115 |
| 80 | 7.210 | .1387 | .0040 | .0290 | 248.383 | 34.452 | 27.117 | 934.218 |
| 85 | 8.157 | .1226 | .0035 | .0285 | 286.279 | 35.096 | 28.123 | 987.027 |
| 90 | 9.229 | .1084 | .0030 | .0280 | 329.154 | 35.666 | 29.063 | 1036.550 |
| 95 | 10.442 | .0958 | .0026 | .0276 | 377.664 | 36.169 | 29.938 | 1082.838 |
| 100 | 11.814 | .0846 | .0023 | .0273 | 432.549 | 36.614 | 30.752 | 1125.975 |

| | SINGLE PAYMENT | | UNIFORM PAYMENT SERIES | | | | GRADIENT SERIES | |
|---|---|---|---|---|---|---|---|---|
| | Compound Amount Factor | Present Worth Factor | Sinking Fund Factor | Capital Recovery Factor | Compound Amount Factor | Present Worth Factor | Gradient Uniform Series | Gradient Present Worth |
| | Find $F$ Given $P$ | Find $P$ Given $F$ | Find $A$ Given $F$ | Find $A$ Given $P$ | Find $F$ Given $A$ | Find $P$ Given $A$ | Find $A$ Given $G$ | Find $P$ Given $G$ |
| $n$ | $F/P$ | $P/F$ | $A/F$ | $A/P$ | $F/A$ | $P/A$ | $A/G$ | $P/G$ |
| 1 | 1.030 | .9709 | 1.0000 | 1.0300 | 1.000 | .971 | 0 | 0 |
| 2 | 1.061 | .9426 | .4926 | .5226 | 2.030 | 1.913 | .493 | .943 |
| 3 | 1.093 | .9151 | .3235 | .3535 | 3.091 | 2.829 | .980 | 2.773 |
| 4 | 1.126 | .8885 | .2390 | .2690 | 4.184 | 3.717 | 1.463 | 5.438 |
| 5 | 1.159 | .8626 | .1884 | .2184 | 5.309 | 4.580 | 1.941 | 8.889 |
| 6 | 1.194 | .8375 | .1546 | .1846 | 6.468 | 5.417 | 2.414 | 13.076 |
| 7 | 1.230 | .8131 | .1305 | .1605 | 7.662 | 6.230 | 2.882 | 17.955 |
| 8 | 1.267 | .7894 | .1125 | .1425 | 8.892 | 7.020 | 3.345 | 23.481 |
| 9 | 1.305 | .7664 | .0984 | .1284 | 10.159 | 7.786 | 3.803 | 29.612 |
| 10 | 1.344 | .7441 | .0872 | .1172 | 11.464 | 8.530 | 4.256 | 36.309 |
| 11 | 1.384 | .7224 | .0781 | .1081 | 12.808 | 9.253 | 4.705 | 43.533 |
| 12 | 1.426 | .7014 | .0705 | .1005 | 14.192 | 9.954 | 5.148 | 51.248 |
| 13 | 1.469 | .6810 | .0640 | .0940 | 15.618 | 10.635 | 5.587 | 59.420 |
| 14 | 1.513 | .6611 | .0585 | .0885 | 17.086 | 11.296 | 6.021 | 68.014 |
| 15 | 1.558 | .6419 | .0538 | .0838 | 18.599 | 11.938 | 6.450 | 77.000 |
| 16 | 1.605 | .6232 | .0496 | .0796 | 20.157 | 12.561 | 6.874 | 86.348 |
| 17 | 1.653 | .6050 | .0460 | .0760 | 21.762 | 13.166 | 7.294 | 96.028 |
| 18 | 1.702 | .5874 | .0427 | .0727 | 23.414 | 13.754 | 7.708 | 106.014 |
| 19 | 1.754 | .5703 | .0398 | .0698 | 25.117 | 14.324 | 8.118 | 116.279 |
| 20 | 1.806 | .5537 | .0372 | .0672 | 26.870 | 14,877 | 8.523 | 126.799 |
| 21 | 1.860 | .5375 | .0349 | .0649 | 28.676 | 15.415 | 8.923 | 137.550 |
| 22 | 1.916 | .5219 | .0327 | .0627 | 30.537 | 15.937 | 9.319 | 148.509 |
| 23 | 1.974 | .5067 | .0308 | .0608 | 32.453 | 16.444 | 9.709 | 159.657 |
| 24 | 2.033 | .4919 | .0290 | .0590 | 34.426 | 16.936 | 10.095 | 170.971 |
| 25 | 2.094 | .4776 | .0274 | .0574 | 36.459 | 17.413 | 10.477 | 182.434 |
| 26 | 2.157 | .4637 | .0259 | .0559 | 38.553 | 17.877 | 10.853 | 194.026 |
| 27 | 2.221 | .4502 | .0246 | .0546 | 40.710 | 18.327 | 11.226 | 205.731 |
| 28 | 2.288 | .4371 | .0233 | .0533 | 42.931 | 18.764 | 11.593 | 217.532 |
| 29 | 2.357 | .4243 | .0221 | .0521 | 45.219 | 19.188 | 11.956 | 229.414 |
| 30 | 2.427 | .4120 | .0210 | .0510 | 47.575 | 19.600 | 12.314 | 241.361 |
| 31 | 2.500 | .4000 | .0200 | .0500 | 50.003 | 20.000 | 12.668 | 253.361 |
| 32 | 2.575 | .3883 | .0190 | .0490 | 52.503 | 20.389 | 13.017 | 265.399 |
| 33 | 2.652 | .3770 | .0182 | .0482 | 55.078 | 20.766 | 13.362 | 277.464 |
| 34 | 2.732 | .3660 | .0173 | .0473 | 57.730 | 21.132 | 13.702 | 289.544 |
| 35 | 2.814 | .3554 | .0165 | .0465 | 60.462 | 21.487 | 14.037 | 301.627 |
| 40 | 3.262 | .3066 | .0133 | .0433 | 75.401 | 23.115 | 15.650 | 361.750 |
| 45 | 3.782 | .2644 | .0108 | .0408 | 92.720 | 24.519 | 17.156 | 420.632 |
| 50 | 4.384 | .2281 | .0089 | .0389 | 112.797 | 25.730 | 18.558 | 477.480 |
| 55 | 5.082 | .1968 | .0073 | .0373 | 136.072 | 26.774 | 19.860 | 531.741 |
| 60 | 5.892 | .1697 | .0061 | .0361 | 163.053 | 27.676 | 21.067 | 583.053 |
| 65 | 6.830 | .1464 | .0051 | .0351 | 194.333 | 28.453 | 22.184 | 631.201 |
| 70 | 7.918 | .1263 | .0043 | .0343 | 230.594 | 29.123 | 23.215 | 676.087 |
| 75 | 9.179 | .1089 | .0037 | .0337 | 272.631 | 29.702 | 24.163 | 717.698 |
| 80 | 10.641 | .0940 | .0031 | .0331 | 321.363 | 30.201 | 25.035 | 756.087 |
| 85 | 12.336 | .0811 | .0026 | .0326 | 377.857 | 30.631 | 25.835 | 791.353 |
| 90 | 14.300 | .0699 | .0023 | .0323 | 443.349 | 31.002 | 26.567 | 823.630 |
| 95 | 16.578 | .0603 | .0019 | .0319 | 519.272 | 31.323 | 27.235 | 853.074 |
| 100 | 19.219 | .0520 | .0016 | .0316 | 607.288 | 31.599 | 27.844 | 879.854 |

| | SINGLE PAYMENT | | UNIFORM PAYMENT SERIES | | | | GRADIENT SERIES | |
|---|---|---|---|---|---|---|---|---|
| | Compound Amount Factor | Present Worth Factor | Sinking Fund Factor | Capital Recovery Factor | Compound Amount Factor | Present Worth Factor | Gradient Uniform Series | Gradient Present Worth |
| | Find F Given P | Find P Given F | Find A Given F | Find A Given P | Find F Given A | Find P Given A | Find A Given G | Find P Given G |
| n | F/P | P/F | A/F | A/P | F/A | P/A | A/G | P/G |
| 1 | 1.035 | .9662 | 1.0000 | 1.0350 | 1.000 | .966 | 0 | 0 |
| 2 | 1.071 | .9335 | .4914 | .5264 | 2.035 | 1.900 | .491 | .934 |
| 3 | 1.109 | .9019 | .3219 | .3569 | 3.106 | 2.802 | .977 | 2.737 |
| 4 | 1.148 | .8714 | .2373 | .2723 | 4.215 | 3.673 | 1.457 | 5.352 |
| 5 | 1.188 | .8420 | .1865 | .2215 | 5.362 | 4.515 | 1.931 | 8.720 |
| 6 | 1.229 | .8135 | .1527 | .1877 | 6.550 | 5.329 | 2.400 | 12.787 |
| 7 | 1.272 | .7860 | .1285 | .1635 | 7.779 | 6.115 | 2.863 | 17.503 |
| 8 | 1.317 | .7594 | .1105 | .1455 | 9.052 | 6.874 | 3.320 | 22.819 |
| 9 | 1.363 | .7337 | .0964 | .1314 | 10.368 | 7.608 | 3.771 | 28.689 |
| 10 | 1.411 | .7089 | .0852 | .1202 | 11.731 | 8.317 | 4.217 | 35.069 |
| 11 | 1.460 | .6849 | .0761 | .1111 | 13.142 | 9.002 | 4.657 | 41.919 |
| 12 | 1.511 | .6618 | .0685 | .1035 | 14.602 | 9.663 | 5.091 | 49.198 |
| 13 | 1.564 | .6394 | .0621 | .0971 | 16.113 | 10.303 | 5.520 | 56.871 |
| 14 | 1.619 | .6178 | .0566 | .0916 | 17.677 | 10.921 | 5.943 | 64.902 |
| 15 | 1.675 | .5969 | .0518 | .0868 | 19.296 | 11.517 | 6.361 | 73.259 |
| 16 | 1.734 | .5767 | .0477 | .0827 | 20.971 | 12.094 | 6.773 | 81.909 |
| 17 | 1.795 | .5572 | .0440 | .0790 | 22.705 | 12.651 | 7.179 | 90.824 |
| 18 | 1.857 | .5384 | .0408 | .0758 | 24.500 | 13.190 | 7.580 | 99.977 |
| 19 | 1.923 | .5202 | .0379 | .0729 | 26.357 | 13.710 | 7.975 | 109.339 |
| 20 | 1.990 | .5026 | .0354 | .0704 | 28.280 | 14.212 | 8.365 | 118.888 |
| 21 | 2.059 | .4856 | .0330 | .0680 | 30.269 | 14.698 | 8.749 | 128.600 |
| 22 | 2.132 | .4692 | .0309 | .0659 | 32.329 | 15.167 | 9.128 | 138.452 |
| 23 | 2.206 | .4533 | .0290 | .0640 | 34.460 | 15.620 | 9.502 | 148.424 |
| 24 | 2.283 | .4380 | .0273 | .0623 | 36.667 | 16.058 | 9.870 | 158.497 |
| 25 | 2.363 | .4231 | .0257 | .0607 | 38.950 | 16.482 | 10.233 | 168.653 |
| 26 | 2.446 | .4088 | .0242 | .0592 | 41.313 | 16.890 | 10.590 | 178.874 |
| 27 | 2.532 | .3950 | .0229 | .0579 | 43.759 | 17.285 | 10.942 | 189.144 |
| 28 | 2.620 | .3817 | .0216 | .0566 | 46.291 | 17.667 | 11.289 | 199.448 |
| 29 | 2.712 | .3687 | .0204 | .0554 | 48.911 | 18.036 | 11.631 | 209.773 |
| 30 | 2.807 | .3563 | .0194 | .0544 | 51.623 | 18.392 | 11.967 | 220.106 |
| 31 | 2.905 | .3442 | .0184 | .0534 | 54.429 | 18.736 | 12.299 | 230.432 |
| 32 | 3.007 | .3326 | .0174 | .0524 | 57.335 | 19.069 | 12.625 | 240.743 |
| 33 | 3.112 | .3213 | .0166 | .0516 | 60.341 | 19.390 | 12.946 | 251.026 |
| 34 | 3.221 | .3105 | .0158 | .0508 | 63.453 | 19.701 | 13.262 | 261.271 |
| 35 | 3.334 | .3000 | .0150 | .0500 | 66.674 | 20.001 | 13.573 | 271.471 |
| 40 | 3.959 | .2526 | .0118 | .0468 | 84.550 | 21.355 | 15.055 | 321.491 |
| 45 | 4.702 | .2127 | .0095 | .0445 | 105.782 | 22.495 | 16.417 | 369.308 |
| 50 | 5.585 | .1791 | .0076 | .0426 | 130.998 | 23.456 | 17.666 | 414.370 |
| 55 | 6.633 | .1508 | .0062 | .0412 | 160.947 | 24.264 | 18.808 | 456.353 |
| 60 | 7.878 | .1269 | .0051 | .0401 | 196.517 | 24.945 | 19.848 | 495.105 |
| 65 | 9.357 | .1069 | .0042 | .0392 | 238.763 | 25.518 | 20.793 | 530.599 |
| 70 | 11.113 | .0900 | .0035 | .0385 | 288.938 | 26.000 | 21.650 | 562.896 |
| 75 | 13.199 | .0758 | .0029 | .0379 | 348.530 | 26.407 | 22.423 | 592.121 |
| 80 | 15.676 | .0638 | .0024 | .0374 | 419.307 | 26.749 | 23.120 | 618.439 |
| 85 | 18.618 | .0537 | .0020 | .0370 | 503.367 | 27.037 | 23.747 | 642.037 |
| 90 | 22.112 | .0452 | .0017 | .0367 | 603.205 | 27.279 | 24.308 | 663.119 |
| 95 | 26.262 | .0381 | .0014 | .0364 | 721.781 | 27.484 | 24.811 | 681.890 |
| 100 | 31.191 | .0321 | .0012 | .0362 | 862.612 | 27.655 | 25.259 | 698.555 |

| | SINGLE PAYMENT | | UNIFORM PAYMENT SERIES | | | | GRADIENT SERIES | |
|---|---|---|---|---|---|---|---|---|
| | Compound Amount Factor | Present Worth Factor | Sinking Fund Factor | Capital Recovery Factor | Compound Amount Factor | Present Worth Factor | Gradient Uniform Series | Gradient Present Worth |
| | Find $F$ Given $P$ | Find $P$ Given $F$ | Find $A$ Given $F$ | Find $A$ Given $P$ | Find $F$ Given $A$ | Find $P$ Given $A$ | Find $A$ Given $G$ | Find $P$ Given $G$ |
| $n$ | $F/P$ | $P/F$ | $A/F$ | $A/P$ | $F/A$ | $P/A$ | $A/G$ | $P/G$ |
| 1 | 1.040 | .9615 | 1.0000 | 1.0400 | 1.000 | .962 | 0 | 0 |
| 2 | 1.082 | .9246 | .4902 | .5302 | 2.040 | 1.886 | .490 | .925 |
| 3 | 1.125 | .8890 | .3203 | .3603 | 3.122 | 2.775 | .974 | 2.703 |
| 4 | 1.170 | .8548 | .2355 | .2755 | 4.246 | 3.630 | 1.451 | 5.267 |
| 5 | 1.217 | .8219 | .1846 | .2246 | 5.416 | 4.452 | 1.922 | 8.555 |
| 6 | 1.265 | .7903 | .1508 | .1908 | 6.633 | 5.242 | 2.386 | 12.506 |
| 7 | 1.316 | .7599 | .1266 | .1666 | 7.898 | 6.002 | 2.843 | 17.066 |
| 8 | 1.369 | .7307 | .1085 | .1485 | 9.214 | 6.733 | 3.294 | 22.181 |
| 9 | 1.423 | .7026 | .0945 | .1345 | 10.583 | 7.435 | 3.739 | 27.801 |
| 10 | 1.480 | .6756 | .0833 | .1233 | 12.006 | 8.111 | 4.177 | 33.881 |
| 11 | 1.539 | .6496 | .0741 | .1141 | 13.486 | 8.760 | 4.609 | 40.377 |
| 12 | 1.601 | .6246 | .0666 | .1066 | 15.026 | 9.385 | 5.034 | 47.248 |
| 13 | 1.665 | .6006 | .0601 | .1001 | 16.627 | 9.986 | 5.453 | 54.455 |
| 14 | 1.732 | .5775 | .0547 | .0947 | 18.292 | 10.563 | 5.866 | 61.962 |
| 15 | 1.801 | .5553 | .0499 | .0899 | 20.024 | 11.118 | 6.272 | 69.735 |
| 16 | 1.873 | .5339 | .0458 | .0858 | 21.825 | 11.652 | 6.672 | 77.744 |
| 17 | 1.948 | .5134 | .0422 | .0822 | 23.698 | 12.166 | 7.066 | 85.958 |
| 18 | 2.026 | .4936 | .0390 | .0790 | 25.645 | 12.659 | 7.453 | 94.350 |
| 19 | 2.107 | .4746 | .0361 | .0761 | 27.671 | 13.134 | 7.834 | 102.893 |
| 20 | 2.191 | .4564 | .0336 | .0736 | 29.778 | 13.590 | 8.209 | 111.565 |
| 21 | 2.279 | .4388 | .0313 | .0713 | 31.969 | 14.029 | 8.578 | 120.341 |
| 22 | 2.370 | .4220 | .0292 | .0692 | 34.248 | 14.451 | 8.941 | 129.202 |
| 23 | 2.465 | .4057 | .0273 | .0673 | 36.618 | 14.857 | 9.297 | 138.128 |
| 24 | 2.563 | .3901 | .0256 | .0656 | 39.083 | 15.247 | 9.648 | 147.101 |
| 25 | 2.666 | .3751 | .0240 | .0640 | 41.646 | 15.622 | 9.993 | 156.104 |
| 26 | 2.772 | .3607 | .0226 | .0626 | 44.312 | 15.983 | 10.331 | 165.121 |
| 27 | 2.883 | .3468 | .0212 | .0612 | 47.084 | 16.330 | 10.664 | 174.138 |
| 28 | 2.999 | .3335 | .0200 | .0600 | 49.968 | 16.663 | 10.991 | 183.142 |
| 29 | 3.119 | .3207 | .0189 | .0589 | 52.966 | 16.984 | 11.312 | 192.121 |
| 30 | 3.243 | .3083 | .0178 | .0578 | 56.085 | 17.292 | 11.627 | 201.062 |
| 31 | 3.373 | .2965 | .0169 | .0569 | 59.328 | 17.588 | 11.937 | 209.956 |
| 32 | 3.508 | .2851 | .0159 | .0559 | 62.701 | 17.874 | 12.241 | 218.792 |
| 33 | 3.648 | .2741 | .0151 | .0551 | 66.210 | 18.148 | 12.540 | 227.563 |
| 34 | 3.794 | .2636 | .0143 | .0543 | 69.858 | 18.411 | 12.832 | 236.261 |
| 35 | 3.946 | .2534 | .0136 | .0536 | 73.652 | 18.665 | 13.120 | 244.877 |
| 40 | 4.801 | .2083 | .0105 | .0505 | 95.026 | 19.793 | 14.477 | 286.530 |
| 45 | 5.841 | .1712 | .0083 | .0483 | 121.029 | 20.720 | 15.705 | 325.403 |
| 50 | 7.107 | .1407 | .0066 | .0466 | 152.667 | 21.482 | 16.812 | 361.164 |
| 55 | 8.646 | .1157 | .0052 | .0452 | 191.159 | 22.109 | 17.807 | 393.689 |
| 60 | 10.520 | .0951 | .0042 | .0442 | 237.991 | 22.623 | 18.697 | 422.997 |
| 65 | 12.799 | .0781 | .0034 | .0434 | 294.968 | 23.047 | 19.491 | 449.201 |
| 70 | 15.572 | .0642 | .0027 | .0427 | 364.290 | 23.395 | 20.196 | 472.479 |
| 75 | 18.945 | .0528 | .0022 | .0422 | 448.631 | 23.680 | 20.821 | 493.041 |
| 80 | 23.050 | .0434 | .0018 | .0418 | 551.245 | 23.915 | 21.372 | 511.116 |
| 85 | 28.044 | .0357 | .0015 | .0415 | 676.090 | 24.109 | 21.857 | 526.938 |
| 90 | 34.119 | .0293 | .0012 | .0412 | 827.983 | 24.267 | 22.283 | 540.737 |
| 95 | 41.511 | .0241 | .0010 | .0410 | 1012.785 | 24.398 | 22.655 | 552.731 |
| 100 | 50.505 | .0198 | .0008 | .0408 | 1237.624 | 24.505 | 22.980 | 563.125 |

| | SINGLE PAYMENT | | UNIFORM PAYMENT SERIES | | | | GRADIENT SERIES | |
|---|---|---|---|---|---|---|---|---|
| | Compound Amount Factor | Present Worth Factor | Sinking Fund Factor | Capital Recovery Factor | Compound Amount Factor | Present Worth Factor | Gradient Uniform Series | Gradient Present Worth |
| | Find $F$ Given $P$ | Find $P$ Given $F$ | Find $A$ Given $F$ | Find $A$ Given $P$ | Find $F$ Given $A$ | Find $P$ Given $A$ | Find $A$ Given $G$ | Find $P$ Given $G$ |
| $n$ | $F/P$ | $P/F$ | $A/F$ | $A/P$ | $F/A$ | $P/A$ | $A/G$ | $P/G$ |
| 1 | 1.045 | .9569 | 1.0000 | 1.0450 | 1.000 | .957 | 0 | 0 |
| 2 | 1.092 | .9157 | .4890 | .5340 | 2.045 | 1.873 | .489 | .916 |
| 3 | 1.141 | .8763 | .3188 | .3638 | 3.137 | 2.749 | .971 | 2.668 |
| 4 | 1.193 | .8386 | .2337 | .2787 | 4.278 | 3.588 | 1.445 | 5.184 |
| 5 | 1.246 | .8025 | .1828 | .2278 | 5.471 | 4.390 | 1.912 | 8.394 |
| 6 | 1.302 | .7679 | .1489 | .1939 | 6.717 | 5.158 | 2.372 | 12.233 |
| 7 | 1.361 | .7348 | .1247 | .1697 | 8.019 | 5.893 | 2.824 | 16.642 |
| 8 | 1.422 | .7032 | .1066 | .1516 | 9.380 | 6.596 | 3.269 | 21.565 |
| 9 | 1.486 | .6729 | .0926 | .1376 | 10.802 | 7.269 | 3.707 | 26.948 |
| 10 | 1.553 | .6439 | .0814 | .1264 | 12.288 | 7.913 | 4.138 | 32.743 |
| 11 | 1.623 | .6162 | .0722 | .1172 | 13.841 | 8.529 | 4.562 | 38.905 |
| 12 | 1.696 | .5897 | .0647 | .1097 | 15.464 | 9.119 | 4.978 | 45.391 |
| 13 | 1.772 | .5643 | .0583 | .1033 | 17.160 | 9.683 | 5.387 | 52.163 |
| 14 | 1.852 | .5400 | .0528 | .0978 | 18.932 | 10.223 | 5.789 | 59.182 |
| 15 | 1.935 | .5167 | .0481 | .0931 | 20.784 | 10.740 | 6.184 | 66.416 |
| 16 | 2.022 | .4945 | .0440 | .0890 | 22.719 | 11.234 | 6.572 | 73.833 |
| 17 | 2.113 | .4732 | .0404 | .0854 | 24.742 | 11.707 | 6.953 | 81.404 |
| 18 | 2.208 | .4528 | .0372 | .0822 | 26.855 | 12.160 | 7.327 | 89.102 |
| 19 | 2.308 | .4333 | .0344 | .0794 | 29.064 | 12.593 | 7.695 | 96.901 |
| 20 | 2.412 | .4146 | .0319 | .0769 | 31.371 | 13.008 | 8.055 | 104.780 |
| 21 | 2.520 | .3968 | .0296 | .0746 | 33.783 | 13.405 | 8.409 | 112.715 |
| 22 | 2.634 | .3797 | .0275 | .0725 | 36.303 | 13.784 | 8.755 | 120.689 |
| 23 | 2.752 | .3634 | .0257 | .0707 | 38.937 | 14.148 | 9.096 | 128.683 |
| 24 | 2.876 | .3477 | .0240 | .0690 | 41.689 | 14.495 | 9.429 | 136.680 |
| 25 | 3.005 | .3327 | .0224 | .0674 | 44.565 | 14.828 | 9.756 | 144.665 |
| 26 | 3.141 | .3184 | .0210 | .0660 | 47.571 | 15.147 | 10.077 | 152.625 |
| 27 | 3.282 | .3047 | .0197 | .0647 | 50.711 | 15.451 | 10.391 | 160.547 |
| 28 | 3.430 | .2916 | .0185 | .0635 | 53.993 | 15.743 | 10.698 | 168.420 |
| 29 | 3.584 | .2790 | .0174 | .0624 | 57.423 | 16.022 | 10.999 | 176.232 |
| 30 | 3.745 | .2670 | .0164 | .0614 | 61.007 | 16.289 | 11.295 | 183.975 |
| 31 | 3.914 | .2555 | .0154 | .0604 | 64.752 | 16.544 | 11.583 | 191.640 |
| 32 | 4.090 | .2445 | .0146 | .0596 | 68.666 | 16.789 | 11.866 | 199.220 |
| 33 | 4.274 | .2340 | .0137 | .0587 | 72.756 | 17.023 | 12.143 | 206.707 |
| 34 | 4.466 | .2239 | .0130 | .0580 | 77.030 | 17.247 | 12.414 | 214.096 |
| 35 | 4.667 | .2143 | .0123 | .0573 | 81.497 | 17.461 | 12.679 | 221.380 |
| 40 | 5.816 | .1719 | .0093 | .0543 | 107.030 | 18.402 | 13.917 | 256.099 |
| 45 | 7.248 | .1380 | .0072 | .0522 | 138.850 | 19.156 | 15.020 | 287.732 |
| 50 | 9.033 | .1107 | .0056 | .0506 | 178.503 | 19.762 | 15.998 | 316.145 |
| 55 | 11.256 | .0888 | .0044 | .0494 | 227.918 | 20.248 | 16.860 | 341.375 |
| 60 | 14.027 | .0713 | .0035 | .0485 | 289.498 | 20.638 | 17.617 | 363.571 |
| 65 | 17.481 | .0572 | .0027 | .0477 | 366.238 | 20.951 | 18.278 | 382.947 |
| 70 | 21.784 | .0459 | .0022 | .0472 | 461.870 | 21.202 | 18.854 | 399.750 |
| 75 | 27.147 | .0368 | .0017 | .0467 | 581.044 | 21.404 | 19.354 | 414.242 |
| 80 | 33.830 | .0296 | .0014 | .0464 | 729.558 | 21.565 | 19.785 | 426.680 |
| 85 | 42.158 | .0237 | .0011 | .0461 | 914.632 | 21.695 | 20.157 | 437.309 |
| 90 | 52.537 | .0190 | .0009 | .0459 | 1145.269 | 21.799 | 20.476 | 446.359 |
| 95 | 65.471 | .0153 | .0007 | .0457 | 1432.684 | 21.883 | 20.749 | 454.039 |
| 100 | 81.589 | .0123 | .0006 | .0456 | 1790.856 | 21.950 | 20.981 | 460.538 |

*442*

| | SINGLE PAYMENT | | UNIFORM PAYMENT SERIES | | | | GRADIENT SERIES | |
|---|---|---|---|---|---|---|---|---|
| | Compound Amount Factor | Present Worth Factor | Sinking Fund Factor | Capital Recovery Factor | Compound Amount Factor | Present Worth Factor | Gradient Uniform Series | Gradient Present Worth |
| | Find F Given P | Find P Given F | Find A Given F | Find A Given P | Find F Given A | Find P Given A | Find A Given G | Find P Given G |
| n | F/P | P/F | A/F | A/P | F/A | P/A | A/G | P/G |
| 1 | 1.050 | .9524 | 1.0000 | 1.0500 | 1.000 | .952 | 0 | 0 |
| 2 | 1.102 | .9070 | .4878 | .5378 | 2.050 | 1.859 | .488 | .907 |
| 3 | 1.158 | .8638 | .3172 | .3672 | 3.152 | 2.723 | .967 | 2.635 |
| 4 | 1.216 | .8227 | .2320 | .2820 | 4.310 | 3.546 | 1.439 | 5.103 |
| 5 | 1.276 | .7835 | .1810 | .2310 | 5.526 | 4.329 | 1.903 | 8.237 |
| 6 | 1.340 | .7462 | .1470 | .1970 | 6.802 | 5.076 | 2.358 | 11.968 |
| 7 | 1.407 | .7107 | .1228 | .1728 | 8.142 | 5.786 | 2.805 | 16.232 |
| 8 | 1.477 | .6768 | .1047 | .1547 | 9.549 | 6.463 | 3.245 | 20.970 |
| 9 | 1.551 | .6446 | .0907 | .1407 | 11.027 | 7.108 | 3.676 | 26.127 |
| 10 | 1.629 | .6139 | .0795 | .1295 | 12.578 | 7.722 | 4.099 | 31.652 |
| 11 | 1.710 | .5847 | .0704 | .1204 | 14.207 | 8.306 | 4.514 | 37.499 |
| 12 | 1.796 | .5568 | .0628 | .1128 | 15.917 | 8.863 | 4.922 | 43.624 |
| 13 | 1.886 | .5303 | .0565 | .1065 | 17.713 | 9.394 | 5.322 | 49.988 |
| 14 | 1.980 | .5051 | .0510 | .1010 | 19.599 | 9.899 | 5.713 | 56.554 |
| 15 | 2.079 | .4810 | .0463 | .0963 | 21.579 | 10.380 | 6.097 | 63.288 |
| 16 | 2.183 | .4581 | .0423 | .0923 | 23.657 | 10.838 | 6.474 | 70.160 |
| 17 | 2.292 | .4363 | .0387 | .0887 | 25.840 | 11.274 | 6.842 | 77.140 |
| 18 | 2.407 | .4155 | .0355 | .0855 | 28.132 | 11.690 | 7.203 | 84.204 |
| 19 | 2.527 | .3957 | .0327 | .0827 | 30.539 | 12.085 | 7.557 | 91.328 |
| 20 | 2.653 | .3769 | .0302 | .0802 | 33.066 | 12.462 | 7.903 | 98.488 |
| 21 | 2.786 | .3589 | .0280 | .0780 | 35.719 | 12.821 | 8.242 | 105.667 |
| 22 | 2.925 | .3418 | .0260 | .0760 | 38.505 | 13.163 | 8.573 | 112.846 |
| 23 | 3.072 | .3256 | .0241 | .0741 | 41.430 | 13.489 | 8.897 | 120.009 |
| 24 | 3.225 | .3101 | .0225 | .0725 | 44.502 | 13.799 | 9.214 | 127.140 |
| 25 | 3.386 | .2953 | .0210 | .0710 | 47.727 | 14.094 | 9.524 | 134.228 |
| 26 | 3.556 | .2812 | .0196 | .0696 | 51.113 | 14.375 | 9.827 | 141.259 |
| 27 | 3.733 | .2678 | .0183 | .0683 | 54.669 | 14.643 | 10.122 | 148.223 |
| 28 | 3.920 | .2551 | .0171 | .0671 | 58.403 | 14.898 | 10.411 | 155.110 |
| 29 | 4.116 | .2429 | .0160 | .0660 | 62.323 | 15.141 | 10.694 | 161.913 |
| 30 | 4.322 | .2314 | .0151 | .0651 | 66.439 | 15.372 | 10.969 | 168.623 |
| 31 | 4.538 | .2204 | .0141 | .0641 | 70.761 | 15.593 | 11.238 | 175.233 |
| 32 | 4.765 | .2099 | .0133 | .0633 | 75.299 | 15.803 | 11.501 | 181.739 |
| 33 | 5.003 | .1999 | .0125 | .0625 | 80.064 | 16.003 | 11.757 | 188.135 |
| 34 | 5.253 | .1904 | .0118 | .0618 | 85.067 | 16.193 | 12.006 | 194.417 |
| 35 | 5.516 | .1813 | .0111 | .0611 | 90.320 | 16.374 | 12.250 | 200.581 |
| 40 | 7.040 | .1420 | .0083 | .0583 | 120.800 | 17.159 | 13.377 | 229.545 |
| 45 | 8.985 | .1113 | .0063 | .0563 | 159.700 | 17.774 | 14.364 | 255.315 |
| 50 | 11.467 | .0872 | .0048 | .0548 | 209.348 | 18.256 | 15.223 | 277.915 |
| 55 | 14.636 | .0683 | .0037 | .0537 | 272.713 | 18.633 | 15.966 | 297.510 |
| 60 | 18.679 | .0535 | .0028 | .0528 | 353.584 | 18.929 | 16.606 | 314.343 |
| 65 | 23.840 | .0419 | .0022 | .0522 | 456.798 | 19.161 | 17.154 | 328.691 |
| 70 | 30.426 | .0329 | .0017 | .0517 | 588.529 | 19.343 | 17.621 | 340.841 |
| 75 | 38.833 | .0258 | .0013 | .0513 | 756.654 | 19.485 | 18.018 | 351.072 |
| 80 | 49.561 | .0202 | .0010 | .0510 | 971.229 | 19.596 | 18.353 | 359.646 |
| 85 | 63.254 | .0158 | .0008 | .0508 | 1245.087 | 19.684 | 18.635 | 366.801 |
| 90 | 80.730 | .0124 | .0006 | .0506 | 1594.607 | 19.752 | 18.871 | 372.749 |
| 95 | 103.035 | .0097 | .0005 | .0505 | 2040.694 | 19.806 | 19.069 | 377.677 |
| 100 | 131.501 | .0076 | .0004 | .0504 | 2610.025 | 19.848 | 19.234 | 381.749 |

| | SINGLE PAYMENT | | UNIFORM PAYMENT SERIES | | | | GRADIENT SERIES | |
|---|---|---|---|---|---|---|---|---|
| | Compound Amount Factor | Present Worth Factor | Sinking Fund Factor | Capital Recovery Factor | Compound Amount Factor | Present Worth Factor | Gradient Uniform Series | Gradient Present Worth |
| | Find F Given P F/P | Find P Given F P/F | Find A Given F A/F | Find A Given P A/P | Find F Given A F/A | Find P Given A P/A | Find A Given G A/G | Find P Given G P/G |
| n | | | | | | | | |
| 1 | 1.060 | .9434 | 1.0000 | 1.0600 | 1.000 | .943 | 0 | 0 |
| 2 | 1.124 | .8900 | .4854 | .5454 | 2.060 | 1.833 | .485 | .890 |
| 3 | 1.191 | .8396 | .3141 | .3741 | 3.184 | 2.673 | .961 | 2.569 |
| 4 | 1.262 | .7921 | .2286 | .2886 | 4.375 | 3.465 | 1.427 | 4.946 |
| 5 | 1.338 | .7473 | .1774 | .2374 | 5.637 | 4.212 | 1.884 | 7.935 |
| 6 | 1.419 | .7050 | .1434 | .2034 | 6.975 | 4.917 | 2.330 | 11.459 |
| 7 | 1.504 | .6651 | .1191 | .1791 | 8.394 | 5.582 | 2.768 | 15.450 |
| 8 | 1.594 | .6274 | .1010 | .1610 | 9.897 | 6.210 | 3.195 | 19.842 |
| 9 | 1.689 | .5919 | .0870 | .1470 | 11.491 | 6.802 | 3.613 | 24.577 |
| 10 | 1.791 | .5584 | .0759 | .1359 | 13.181 | 7.360 | 4.022 | 29.602 |
| 11 | 1.898 | .5268 | .0668 | .1268 | 14.972 | 7.887 | 4.421 | 34.870 |
| 12 | 2.012 | .4970 | .0593 | .1193 | 16.870 | 8.384 | 4.811 | 40.337 |
| 13 | 2.133 | .4688 | .0530 | .1130 | 18.882 | 8.853 | 5.192 | 45.963 |
| 14 | 2.261 | .4423 | .0476 | .1076 | 21.015 | 9.295 | 5.564 | 51.713 |
| 15 | 2.397 | .4173 | .0430 | .1030 | 23.276 | 9.712 | 5.926 | 57.555 |
| 16 | 2.540 | .3936 | .0390 | .0990 | 25.673 | 10.106 | 6.279 | 63.459 |
| 17 | 2.693 | .3714 | .0354 | .0954 | 28.213 | 10.477 | 6.624 | 69.401 |
| 18 | 2.854 | .3503 | .0324 | .0924 | 30.906 | 10.828 | 6.960 | 75.357 |
| 19 | 3.026 | .3305 | .0296 | .0896 | 33.760 | 11.158 | 7.287 | 81.306 |
| 20 | 3.207 | .3118 | .0272 | .0872 | 36.786 | 11.470 | 7.605 | 87.230 |
| 21 | 3.400 | .2942 | .0250 | .0850 | 39.993 | 11.764 | 7.915 | 93.114 |
| 22 | 3.604 | .2775 | .0230 | .0830 | 43.392 | 12.042 | 8.217 | 98.941 |
| 23 | 3.820 | .2618 | .0213 | .0813 | 46.996 | 12.303 | 8.510 | 104.701 |
| 24 | 4.049 | .2470 | .0197 | .0797 | 50.816 | 12.550 | 8.795 | 110.381 |
| 25 | 4.292 | .2330 | .0182 | .0782 | 54.865 | 12.783 | 9.072 | 115.973 |
| 26 | 4.549 | .2198 | .0169 | .0769 | 59.156 | 13.003 | 9.341 | 121.468 |
| 27 | 4.822 | .2074 | .0157 | .0757 | 63.706 | 13.211 | 9.603 | 126.860 |
| 28 | 5.112 | .1956 | .0146 | .0746 | 68.528 | 13.406 | 9.857 | 132.142 |
| 29 | 5.418 | .1846 | .0136 | .0736 | 73.640 | 13.591 | 10.103 | 137.310. |
| 30 | 5.743 | .1741 | .0126 | .0726 | 79.058 | 13.765 | 10.342 | 142.359 |
| 31 | 6.088 | .1643 | .0118 | .0718 | 84.802 | 13.929 | 10.547 | 147.286 |
| 32 | 6.453 | .1550 | .0110 | .0710 | 90.890 | 14.084 | 10.799 | 152.090 |
| 33 | 6.841 | .1462 | .0103 | .0703 | 97.343 | 14.230 | 11.017 | 156.768 |
| 34 | 7.251 | .1379 | .0096 | .0696 | 104.184 | 14.368 | 11.228 | 161.319 |
| 35 | 7.686 | .1301 | .0090 | .0690 | 111.435 | 14.498 | 11.432 | 165.743 |
| 40 | 10.286 | .0972 | .0065 | .0665 | 154.762 | 15.046 | 12.359 | 185.957 |
| 45 | 13.765 | .0727 | .0047 | .0647 | 212.744 | 15.456 | 13.141 | 203.110 |
| 50 | 18.420 | .0543 | .0034 | .0634 | 290.336 | 15.762 | 13.796 | 217.457 |
| 55 | 24.650 | .0406 | .0025 | .0625 | 394.172 | 15.991 | 14.341 | 229.322 |
| 60 | 32.988 | .0303 | .0019 | .0619 | 533.128 | 16.161 | 14.791 | 239.043 |
| 65 | 44.145 | .0227 | .0014 | .0614 | 719.083 | 16.289 | 15.160 | 246.945 |
| 70 | 59.076 | .0169 | .0010 | .0610 | 967.932 | 16.385 | 15.461 | 253.327 |
| 75 | 79.057 | .0126 | .0008 | .0608 | 1300.949 | 16.456 | 15.706 | 258.453 |
| 80 | 105.796 | .0095 | .0006 | .0606 | 1746.600 | 16.509 | 15.903 | 262.549 |
| 85 | 141.579 | .0071 | .0004 | .0604 | 2342.982 | 16.549 | 16.062 | 265.810 |
| 90 | 189.465 | .0053 | .0003 | .0603 | 3141.075 | 16.579 | 16.189 | 268.395 |
| 95 | 253.546 | .0039 | .0002 | .0602 | 4209.104 | 16.601 | 16.290 | 270.437 |
| 100 | 339.302 | .0029 | .0002 | .0602 | 5638.368 | 16.618 | 16.371 | 272.047 |

| | SINGLE PAYMENT | | UNIFORM PAYMENT SERIES | | | | GRADIENT SERIES | |
|---|---|---|---|---|---|---|---|---|
| | Compound Amount Factor | Present Worth Factor | Sinking Fund Factor | Capital Recovery Factor | Compound Amount Factor | Present Worth Factor | Gradient Uniform Series | Gradient Present Worth |
| | Find $F$ Given $P$ $F/P$ | Find $P$ Given $F$ $P/F$ | Find $A$ Given $F$ $A/F$ | Find $A$ Given $P$ $A/P$ | Find $F$ Given $A$ $F/A$ | Find $P$ Given $A$ $P/A$ | Find $A$ Given $G$ $A/G$ | Find $P$ Given $G$ $P/G$ |
| $n$ | | | | | | | | |
| 1 | 1.070 | .9346 | 1.0000 | 1.0700 | 1.000 | .935 | 0 | 0 |
| 2 | 1.145 | .8734 | .4831 | .5531 | 2.070 | 1.808 | .483 | .873 |
| 3 | 1.225 | .8163 | .3111 | .3811 | 3.215 | 2.624 | .955 | 2.506 |
| 4 | 1.311 | .7629 | .2252 | .2952 | 4.440 | 3.387 | 1.416 | 4.795 |
| 5 | 1.403 | .7130 | .1739 | .2439 | 5.751 | 4.100 | 1.865 | 7.647 |
| 6 | 1.501 | .6663 | .1398 | .2098 | 7.153 | 4.767 | 2.303 | 10.978 |
| 7 | 1.606 | .6227 | .1156 | .1856 | 8.654 | 5.389 | 2.730 | 14.715 |
| 8 | 1.718 | .5820 | .0975 | .1675 | 10.260 | 5.971 | 3.147 | 18.789 |
| 9 | 1.838 | .5439 | .0835 | .1535 | 11.978 | 6.515 | 3.552 | 23.140 |
| 10 | 1.967 | .5083 | .0724 | .1424 | 13.816 | 7.024 | 3.946 | 27.716 |
| 11 | 2.105 | .4751 | .0634 | .1334 | 15.784 | 7.499 | 4.330 | 32.466 |
| 12 | 2.252 | .4440 | .0559 | .1259 | 17.888 | 7.943 | 4.703 | 37.351 |
| 13 | 2.410 | .4150 | .0497 | .1197 | 20.141 | 8.358 | 5.065 | 42.330 |
| 14 | 2.579 | .3878 | .0443 | .1143 | 22.550 | 8.745 | 5.417 | 47.372 |
| 15 | 2.759 | .3624 | .0398 | .1098 | 25.129 | 9.108 | 5.758 | 52.446 |
| 16 | 2.952 | .3387 | .0359 | .1059 | 27.888 | 9.447 | 6.090 | 57.527 |
| 17 | 3.159 | .3166 | .0324 | .1024 | 30.840 | 9.763 | 6.411 | 62.592 |
| 18 | 3.380 | .2959 | .0294 | .0994 | 33.999 | 10.059 | 6.722 | 67.622 |
| 19 | 3.617 | .2765 | .0268 | .0968 | 37.379 | 10.336 | 7.024 | 72.599 |
| 20 | 3.870 | .2584 | .0244 | .0944 | 40.995 | 10.594 | 7.316 | 77.509 |
| 21 | 4.141 | .2415 | .0223 | .0923 | 44.865 | 10.836 | 7.599 | 82.339 |
| 22 | 4.430 | .2257 | .0204 | .0904 | 49.006 | 11.061 | 7.872 | 87.079 |
| 23 | 4.741 | .2109 | .0187 | .0887 | 53.436 | 11.272 | 8.137 | 91.720 |
| 24 | 5.072 | .1971 | .0172 | .0872 | 58.177 | 11.469 | 8.392 | 96.255 |
| 25 | 5.427 | .1842 | .0158 | .0858 | 63.249 | 11.654 | 8.639 | 100.676 |
| 26 | 5.807 | .1722 | .0146 | .0846 | 68.676 | 11.826 | 8.877 | 104.981 |
| 27 | 6.214 | .1609 | .0134 | .0834 | 74.484 | 11.987 | 9.107 | 109.166 |
| 28 | 6.649 | .1504 | .0124 | .0824 | 80.698 | 12.137 | 9.329 | 113.226 |
| 29 | 7.114 | .1406 | .0114 | .0814 | 87.347 | 12.278 | 9.543 | 117.162 |
| 30 | 7.612 | .1314 | .0106 | .0806 | 94.461 | 12.409 | 9.749 | 120.972 |
| 31 | 8.145 | .1228 | .0098 | .0798 | 102.073 | 12.532 | 9.947 | 124.655 |
| 32 | 8.715 | .1147 | .0091 | .0791 | 110.218 | 12.647 | 10.138 | 128.212 |
| 33 | 9.325 | .1072 | .0084 | .0784 | 118.933 | 12.754 | 10.322 | 131.643 |
| 34 | 9.978 | .1002 | .0078 | .0778 | 128.259 | 12.854 | 10.499 | 134.951 |
| 35 | 10.677 | .0937 | .0072 | .0772 | 138.237 | 12.948 | 10.669 | 138.135 |
| 40 | 14.974 | .0668 | .0050 | .0750 | 199.635 | 13.332 | 11.423 | 152.293 |
| 45 | 21.002 | .0476 | .0035 | .0735 | 285.749 | 13.606 | 12.036 | 163.756 |
| 50 | 29.457 | .0339 | .0025 | .0725 | 406.529 | 13.801 | 12.529 | 172.905 |
| 55 | 41.315 | .0242 | .0017 | .0717 | 575.929 | 13.940 | 12.921 | 180.124 |
| 60 | 57.946 | .0173 | .0012 | .0712 | 813.520 | 14.039 | 13.232 | 185.768 |
| 65 | 81.273 | .0123 | .0009 | .0709 | 1146.755 | 14.110 | 13.476 | 190.145 |
| 70 | 113.989 | .0088 | .0006 | .0706 | 1614.134 | 14.160 | 13.666 | 193.519 |
| 75 | 159.876 | .0063 | .0004 | .0704 | 2269.657 | 14.196 | 13.814 | 196.104 |
| 80 | 224.234 | .0045 | .0003 | .0703 | 3189.063 | 14.222 | 13.927 | 198.075 |
| 85 | 314.500 | .0032 | .0002 | .0702 | 4478.576 | 14.240 | 14.015 | 199.572 |
| 90 | 441.103 | .0023 | .0002 | .0702 | 6287.185 | 14.253 | 14.081 | 200.704 |
| 95 | 618.670 | .0016 | .0001 | .0701 | 8823.854 | 14.263 | 14.132 | 201.558 |
| 100 | 867.716 | .0012 | .0001 | .0701 | 12381.662 | 14.269 | 14.170 | 202.200 |

| | SINGLE PAYMENT | | UNIFORM PAYMENT SERIES | | | | GRADIENT SERIES | |
|---|---|---|---|---|---|---|---|---|
| | Compound Amount Factor | Present Worth Factor | Sinking Fund Factor | Capital Recovery Factor | Compound Amount Factor | Present Worth Factor | Gradient Uniform Series | Gradient Present Worth |
| | Find $F$ Given $P$ | Find $P$ Given $F$ | Find $A$ Given $F$ | Find $A$ Given $P$ | Find $F$ Given $A$ | Find $P$ Given $A$ | Find $A$ Given $G$ | Find $P$ Given $G$ |
| $n$ | $F/P$ | $P/F$ | $A/F$ | $A/P$ | $F/A$ | $P/A$ | $A/G$ | $P/G$ |
| 1 | 1.080 | .9259 | 1.0000 | 1.0800 | 1.000 | .926 | 0 | 0 |
| 2 | 1.166 | .8573 | .4808 | .5608 | 2.080 | 1.783 | .481 | .857 |
| 3 | 1.260 | .7938 | .3080 | .3880 | 3.246 | 2.577 | .949 | 2.445 |
| 4 | 1.360 | .7350 | .2219 | .3019 | 4.506 | 3.312 | 1.404 | 4.650 |
| 5 | 1.469 | .6806 | .1705 | .2505 | 5.867 | 3.993 | 1.846 | 7.372 |
| 6 | 1.587 | .6302 | .1363 | .2163 | 7.336 | 4.623 | 2.276 | 10.523 |
| 7 | 1.714 | .5835 | .1121 | .1921 | 8.923 | 5.206 | 2.694 | 14.024 |
| 8 | 1.851 | .5403 | .0940 | .1740 | 10.637 | 5.747 | 3.099 | 17.806 |
| 9 | 1.999 | .5002 | .0801 | .1601 | 12.488 | 6.247 | 3.491 | 21.808 |
| 10 | 2.159 | .4632 | .0690 | .1490 | 14.487 | 6.710 | 3.871 | 25.977 |
| 11 | 2.332 | .4289 | .0601 | .1401 | 16.645 | 7.139 | 4.240 | 30.266 |
| 12 | 2.518 | .3971 | .0527 | .1327 | 18.977 | 7.536 | 4.596 | 34.634 |
| 13 | 2.720 | .3677 | .0465 | .1265 | 21.495 | 7.904 | 4.940 | 39.046 |
| 14 | 2.937 | .3405 | .0413 | .1213 | 24.215 | 8.244 | 5.273 | 43.472 |
| 15 | 3.172 | .3152 | .0368 | .1168 | 27.152 | 8.559 | 5.594 | 47.886 |
| 16 | 3.426 | .2919 | .0330 | .1130 | 30.324 | 8.851 | 5.905 | 52.264 |
| 17 | 3.700 | .2703 | .0296 | .1096 | 33.750 | 9.122 | 6.204 | 56.588 |
| 18 | 3.996 | .2502 | .0267 | .1067 | 37.450 | 9.372 | 6.492 | 60.843 |
| 19 | 4.316 | .2317 | .0241 | .1041 | 41.446 | 9.604 | 6.770 | 65.013 |
| 20 | 4.661 | .2145 | .0219 | .1019 | 45.762 | 9.818 | 7.037 | 69.090 |
| 21 | 5.034 | .1987 | .0198 | .0998 | 50.423 | 10.017 | 7.294 | 73.063 |
| 22 | 5.437 | .1839 | .0180 | .0980 | 55.457 | 10.201 | 7.541 | 76.926 |
| 23 | 5.871 | .1703 | .0164 | .0964 | 60.893 | 10.371 | 7.779 | 80.673 |
| 24 | 6.341 | .1577 | .0150 | .0950 | 66.765 | 10.529 | 8.007 | 84.300 |
| 25 | 6.848 | .1460 | .0137 | .0937 | 73.106 | 10.675 | 8.225 | 87.804 |
| 26 | 7.396 | .1352 | .0125 | .0925 | 79.954 | 10.810 | 8.435 | 91.184 |
| 27 | 7.988 | .1252 | .0114 | .0914 | 87.351 | 10.935 | 8.636 | 94.439 |
| 28 | 8.627 | .1159 | .0105 | .0905 | 95.339 | 11.051 | 8.829 | 97.569 |
| 29 | 9.317 | .1073 | .0096 | .0896 | 103.966 | 11.158 | 9.013 | 100.574 |
| 30 | 10.063 | .0994 | .0088 | .0888 | 113.283 | 11.258 | 9.190 | 103.456 |
| 31 | 10.868 | .0920 | .0081 | .0881 | 123.346 | 11.350 | 9.358 | 106.216 |
| 32 | 11.737 | .0852 | .0075 | .0875 | 134.214 | 11.435 | 9.520 | 108.857 |
| 33 | 12.676 | .0789 | .0069 | .0869 | 145.951 | 11.514 | 9.674 | 111.382 |
| 34 | 13.690 | .0730 | .0063 | .0863 | 158.627 | 11.587 | 9.821 | 113.792 |
| 35 | 14.785 | .0676 | .0058 | .0858 | 172.317 | 11.655 | 9.961 | 116.092 |
| 40 | 21.725 | .0460 | .0039 | .0839 | 259.057 | 11.925 | 10.570 | 126.042 |
| 45 | 31.920 | .0313 | .0026 | .0826 | 386.506 | 12.108 | 11.045 | 133.733 |
| 50 | 46.902 | .0213 | .0017 | .0817 | 573.770 | 12.233 | 11.411 | 139.593 |
| 55 | 68.914 | .0145 | .0012 | .0812 | 848.923 | 12.319 | 11.690 | 144.006 |
| 60 | 101.257 | .0099 | .0008 | .0808 | 1253.213 | 12.377 | 11.902 | 147.300 |
| 65 | 148.780 | .0067 | .0005 | .0805 | 1847.248 | 12.416 | 12.060 | 149.739 |
| 70 | 218.606 | .0046 | .0004 | .0804 | 2720.080 | 12.443 | 12.178 | 151.533 |
| 75 | 321.205 | .0031 | .0002 | .0802 | 4002.557 | 12.461 | 12.266 | 152.845 |
| 80 | 471.955 | .0021 | .0002 | .0802 | 5886.935 | 12.474 | 12.330 | 153.800 |
| 85 | 693.456 | .0014 | .0001 | .0801 | 8655.706 | 12.482 | 12.377 | 154.492 |
| 90 | 1018.915 | .0010 | .0001 | .0801 | 12723.939 | 12.488 | 12.412 | 154.993 |
| 95 | 1497.121 | .0007 | .0001 | .0801 | 18701.507 | 12.492 | 12.437 | 155.352 |
| 100 | 2199.761 | .0005 | | .0800 | 27484.516 | 12.494 | 12.455 | 155.611 |

| | SINGLE PAYMENT | | UNIFORM PAYMENT SERIES | | | | GRADIENT SERIES | |
|---|---|---|---|---|---|---|---|---|
| | Compound Amount Factor | Present Worth Factor | Sinking Fund Factor | Capital Recovery Factor | Compound Amount Factor | Present Worth Factor | Gradient Uniform Series | Gradient Present Worth |
| | Find F Given P | Find P Given F | Find A Given F | Find A Given P | Find F Given A | Find P Given A | Find A Given G | Find P Given G |
| n | F/P | P/F | A/F | A/P | F/A | P/A | A/G | P/G |
| 1 | 1.090 | .9174 | 1.0000 | 1.0900 | 1.000 | .917 | 0 | 0 |
| 2 | 1.188 | .8417 | .4785 | .5685 | 2.090 | 1.759 | .478 | .842 |
| 3 | 1.295 | .7722 | .3051 | .3951 | 3.278 | 2.531 | .943 | 2.386 |
| 4 | 1.412 | .7084 | .2187 | .3087 | 4.573 | 3.240 | 1.393 | 4.511 |
| 5 | 1.539 | .6499 | .1671 | .2571 | 5.985 | 3.890 | 1.828 | 7.111 |
| 6 | 1.677 | .5963 | .1329 | .2229 | 7.523 | 4.486 | 2.250 | 10.092 |
| 7 | 1.828 | .5470 | .1087 | .1987 | 9.200 | 5.033 | 2.657 | 13.375 |
| 8 | 1.993 | .5019 | .0907 | .1807 | 11.028 | 5.535 | 3.051 | 16.888 |
| 9 | 2.172 | .4604 | .0768 | .1668 | 13.021 | 5.995 | 3.431 | 20.571 |
| 10 | 2.367 | .4224 | .0658 | .1558 | 15.193 | 6.418 | 3.798 | 24.373 |
| 11 | 2.580 | .3875 | .0569 | .1469 | 17.560 | 6.805 | 4.151 | 28.248 |
| 12 | 2.813 | .3555 | .0497 | .1397 | 20.141 | 7.161 | 4.491 | 32.159 |
| 13 | 3.066 | .3262 | .0436 | .1336 | 22.953 | 7.487 | 4.818 | 36.073 |
| 14 | 3.342 | .2992 | .0384 | .1284 | 26.019 | 7.786 | 5.133 | 39.963 |
| 15 | 3.642 | .2745 | .0341 | .1241 | 29.361 | 8.061 | 5.435 | 43.807 |
| 16 | 3.970 | .2519 | .0303 | .1203 | 33.003 | 8.313 | 5.724 | 47.585 |
| 17 | 4.328 | .2311 | .0270 | .1170 | 36.974 | 8.544 | 6.002 | 51.282 |
| 18 | 4.717 | .2120 | .0242 | .1142 | 41.301 | 8.756 | 6.269 | 54.886 |
| 19 | 5.142 | .1945 | .0217 | .1117 | 46.018 | 8.950 | 6.524 | 58.387 |
| 20 | 5.604 | .1784 | .0195 | .1095 | 51.160 | 9.129 | 6.767 | 61.777 |
| 21 | 6.109 | .1637 | .0176 | .1076 | 56.765 | 9.292 | 7.001 | 65.051 |
| 22 | 6.659 | .1502 | .0159 | .1059 | 62.873 | 9.442 | 7.223 | 68.205 |
| 23 | 7.258 | .1378 | .0144 | .1044 | 69.532 | 9.580 | 7.436 | 71.236 |
| 24 | 7.911 | .1264 | .0130 | .1030 | 76.790 | 9.707 | 7.638 | 74.143 |
| 25 | 8.623 | .1160 | .0118 | .1018 | 84.701 | 9.823 | 7.832 | 76.926 |
| 26 | 9.399 | .1064 | .0107 | .1007 | 93.324 | 9.929 | 8.016 | 79.586 |
| 27 | 10.245 | .0976 | .0097 | .0997 | 102.723 | 10.027 | 8.191 | 82.124 |
| 28 | 11.167 | .0895 | .0089 | .0989 | 112.968 | 10.116 | 8.357 | 84.542 |
| 29 | 12.172 | .0822 | .0081 | .0981 | 124.135 | 10.198 | 8.515 | 86.842 |
| 30 | 13.268 | .0754 | .0073 | .0973 | 136.308 | 10.274 | 8.666 | 89.028 |
| 31 | 14.462 | .0691 | .0067 | .0967 | 149.575 | 10.343 | 8.808 | 91.102 |
| 32 | 15.763 | .0634 | .0061 | .0961 | 164.037 | 10.406 | 8.944 | 93.069 |
| 33 | 17.182 | .0582 | .0056 | .0956 | 179.800 | 10.464 | 9.072 | 94.931 |
| 34 | 18.728 | .0534 | .0051 | .0951 | 196.982 | 10.518 | 9.193 | 96.693 |
| 35 | 20.414 | .0490 | .0046 | .0946 | 215.711 | 10.567 | 9.308 | 98.359 |
| 40 | 31.409 | .0318 | .0030 | .0930 | 337.882 | 10.757 | 9.796 | 105.376 |
| 45 | 48.327 | .0207 | .0019 | .0919 | 525.859 | 10.881 | 10.160 | 110.556 |
| 50 | 74.358 | .0134 | .0012 | .0912 | 815.084 | 10.962 | 10.430 | 114.325 |
| 55 | 114.408 | .0087 | .0008 | .0908 | 1260.092 | 11.014 | 10.626 | 117.036 |
| 60 | 176.031 | .0057 | .0005 | .0905 | 1944.792 | 11.048 | 10.768 | 118.968 |
| 65 | 270.846 | .0037 | .0003 | .0903 | 2998.288 | 11.070 | 10.870 | 120.334 |
| 70 | 416.730 | .0024 | .0002 | .0902 | 4619.223 | 11.084 | 10.943 | 121.294 |
| 75 | 641.191 | .0016 | .0001 | .0901 | 7113.232 | 11.094 | 10.994 | 121.965 |
| 80 | 986.552 | .0010 | .0001 | .0901 | 10950.574 | 11.100 | 11.030 | 122.431 |
| 85 | 1517.932 | .0007 | .0001 | .0901 | 16854.800 | 11.104 | 11.055 | 122.753 |
| 90 | 2335.527 | .0004 | | .0900 | 25939.184 | 11.106 | 11.073 | 122.976 |
| 95 | 3593.497 | .0003 | | .0900 | 39916.635 | 11.108 | 11.085 | 123.129 |
| 100 | 5529.041 | .0002 | | .0900 | 61422.675 | 11.109 | 11.093 | 123.234 |

| | SINGLE PAYMENT | | UNIFORM PAYMENT SERIES | | | | GRADIENT SERIES | |
|---|---|---|---|---|---|---|---|---|
| | Compound Amount Factor | Present Worth Factor | Sinking Fund Factor | Capital Recovery Factor | Compound Amount Factor | Present Worth Factor | Gradient Uniform Series | Gradient Present Worth |
| | Find F Given P | Find P Given F | Find A Given F | Find A Given P | Find F Given A | Find P Given A | Find A Given G | Find P Given G |
| n | F/P | P/F | A/F | A/P | F/A | P/A | A/G | P/G |
| 1 | 1.100 | .9091 | 1.0000 | 1.1000 | 1.000 | .909 | 0 | 0 |
| 2 | 1.210 | .8264 | .4762 | .5762 | 2.100 | 1.736 | .476 | .826 |
| 3 | 1.331 | .7513 | .3021 | .4021 | 3.310 | 2.487 | .937 | 2.329 |
| 4 | 1.464 | .6830 | .2155 | .3155 | 4.641 | 3.170 | 1.381 | 4.378 |
| 5 | 1.611 | .6209 | .1638 | .2638 | 6.105 | 3.791 | 1.810 | 6.862 |
| 6 | 1.772 | .5645 | .1296 | .2296 | 7.716 | 4.355 | 2.224 | 9.684 |
| 7 | 1.949 | .5132 | .1054 | .2054 | 9.487 | 4.868 | 2.622 | 12.763 |
| 8 | 2.144 | .4665 | .0874 | .1874 | 11.436 | 5.335 | 3.004 | 16.029 |
| 9 | 2.358 | .4241 | .0736 | .1736 | 13.579 | 5.759 | 3.372 | 19.421 |
| 10 | 2.594 | .3855 | .0627 | .1627 | 15.937 | 6.145 | 3.725 | 22.891 |
| 11 | 2.853 | .3505 | .0540 | .1540 | 18.531 | 6.495 | 4.064 | 26.396 |
| 12 | 3.138 | .3186 | .0468 | .1468 | 21.384 | 6.814 | 4.388 | 29.901 |
| 13 | 3.452 | .2897 | .0408 | .1408 | 24.523 | 7.103 | 4.699 | 33.377 |
| 14 | 3.797 | .2633 | .0357 | .1357 | 27.975 | 7.367 | 4.996 | 36.800 |
| 15 | 4.177 | .2394 | .0315 | .1315 | 31.772 | 7.606 | 5.279 | 40.152 |
| 16 | 4.595 | .2176 | .0278 | .1278 | 35.950 | 7.824 | 5.549 | 43.416 |
| 17 | 5.054 | .1978 | .0247 | .1247 | 40.545 | 8.022 | 5.807 | 46.582 |
| 18 | 5.560 | .1799 | .0219 | .1219 | 45.599 | 8.201 | 6.053 | 49.640 |
| 19 | 6.116 | .1635 | .0195 | .1195 | 51.159 | 8.365 | 6.286 | 52.583 |
| 20 | 6.727 | .1486 | .0175 | .1175 | 57.275 | 8.514 | 6.508 | 55.407 |
| 21 | 7.400 | .1351 | .0156 | .1156 | 64.002 | 8.649 | 6.719 | 58.110 |
| 22 | 8.140 | .1228 | .0140 | .1140 | 71.403 | 8.772 | 6.919 | 60.689 |
| 23 | 8.954 | .1117 | .0126 | .1126 | 79.543 | 8.883 | 7.108 | 63.146 |
| 24 | 9.850 | .1015 | .0113 | .1113 | 88.497 | 8.985 | 7.288 | 65.481 |
| 25 | 10.835 | .0923 | .0102 | .1102 | 98.347 | 9.077 | 7.458 | 67.696 |
| 26 | 11.918 | .0839 | .0092 | .1092 | 109.182 | 9.161 | 7.619 | 69.794 |
| 27 | 13.110 | .0763 | .0083 | .1083 | 121.100 | 9.237 | 7.770 | 71.777 |
| 28 | 14.421 | .0693 | .0075 | .1075 | 134.210 | 9.307 | 7.914 | 73.650 |
| 29 | 15.863 | .0630 | .0067 | .1067 | 148.631 | 9.370 | 8.049 | 75.415 |
| 30 | 17.449 | .0573 | .0061 | .1061 | 164.494 | 9.427 | 8.176 | 77.077 |
| 31 | 19.194 | .0521 | .0055 | .1055 | 181.943 | 9.479 | 8.296 | 78.640 |
| 32 | 21.114 | .0474 | .0050 | .1050 | 201.138 | 9.526 | 8.409 | 80.108 |
| 33 | 23.225 | .0431 | .0045 | .1045 | 222.252 | 9.569 | 8.515 | 81.486 |
| 34 | 25.548 | .0391 | .0041 | .1041 | 245.477 | 9.609 | 8.615 | 82.777 |
| 35 | 28.102 | .0356 | .0037 | .1037 | 271.024 | 9.644 | 8.709 | 83.987 |
| 40 | 45.259 | .0221 | .0023 | .1023 | 442.593 | 9.779 | 9.096 | 88.953 |
| 45 | 72.890 | .0137 | .0014 | .1014 | 718.905 | 9.863 | 9.374 | 92.454 |
| 50 | 117.391 | .0085 | .0009 | .1009 | 1163.909 | 9.915 | 9.570 | 94.889 |
| 55 | 189.059 | .0053 | .0005 | .1005 | 1880.591 | 9.947 | 9.708 | 96.562 |
| 60 | 304.482 | .0033 | .0003 | .1003 | 3034.816 | 9.967 | 9.802 | 97.701 |
| 65 | 490.371 | .0020 | .0002 | .1002 | 4893.707 | 9.980 | 9.867 | 98.471 |
| 70 | 789.747 | .0013 | .0001 | .1001 | 7887.470 | 9.987 | 9.911 | 98.987 |
| 75 | 1271.895 | .0008 | .0001 | .1001 | 12708.954 | 9.992 | 9.941 | 99.332 |
| 80 | 2048.400 | .0005 | | .1000 | 20474.002 | 9.995 | 9.961 | 99.561 |
| 85 | 3298.969 | .0003 | | .1000 | 32979.690 | 9.997 | 9.974 | 99.712 |
| 90 | 5313.023 | .0002 | | .1000 | 53120.226 | 9.998 | 9.983 | 99.812 |
| 95 | 8556.676 | .0001 | | .1000 | 85556.761 | 9.999 | 9.989 | 99.877 |
| 100 | 13780.612 | .0001 | | .1000 | 137796.123 | 9.999 | 9.993 | 99.920 |

| | SINGLE PAYMENT | | UNIFORM PAYMENT SERIES | | | | GRADIENT SERIES | |
|---|---|---|---|---|---|---|---|---|
| | Compound Amount Factor | Present Worth Factor | Sinking Fund Factor | Capital Recovery Factor | Compound Amount Factor | Present Worth Factor | Gradient Uniform Series | Gradient Present Worth |
| | Find F Given P | Find P Given F | Find A Given F | Find A Given P | Find F Given A | Find P Given A | Find A Given G | Find P Given G |
| n | F/P | P/F | A/F | A/P | F/A | P/A | A/G | P/G |
| 1 | 1.120 | .8929 | 1.0000 | 1.1200 | 1.000 | .893 | 0 | 0 |
| 2 | 1.254 | .7972 | .4717 | .5917 | 2.120 | 1.690 | .472 | .797 |
| 3 | 1.405 | .7118 | .2963 | .4163 | 3.374 | 2.402 | .925 | 2.221 |
| 4 | 1.574 | .6355 | .2092 | .3292 | 4.779 | 3.037 | 1.359 | 4.127 |
| 5 | 1.762 | .5674 | .1574 | .2774 | 6.353 | 3.605 | 1.775 | 6.397 |
| 6 | 1.974 | .5066 | .1232 | .2432 | 8.115 | 4.111 | 2.172 | 8.930 |
| 7 | 2.211 | .4523 | .0991 | .2191 | 10.089 | 4.564 | 2.551 | 11.644 |
| 8 | 2.476 | .4039 | .0813 | .2013 | 12.300 | 4.968 | 2.913 | 14.471 |
| 9 | 2.773 | .3606 | .0677 | .1877 | 14.776 | 5.328 | 3.257 | 17.356 |
| 10 | 3.106 | .3220 | .0570 | .1770 | 17.549 | 5.650 | 3.585 | 20.254 |
| 11 | 3.479 | .2875 | .0484 | .1684 | 20.655 | 5.938 | 3.895 | 23.129 |
| 12 | 3.896 | .2567 | .0414 | .1614 | 24.133 | 6.194 | 4.190 | 25.952 |
| 13 | 4.363 | .2292 | .0357 | .1557 | 28.029 | 6.424 | 4.468 | 28.702 |
| 14 | 4.887 | .2046 | .0309 | .1509 | 32.393 | 6.628 | 4.732 | 31.362 |
| 15 | 5.474 | .1827 | .0268 | .1468 | 37.280 | 6.811 | 4.980 | 33.920 |
| 16 | 6.130 | .1631 | .0234 | .1434 | 42.753 | 6.974 | 5.215 | 36.367 |
| 17 | 6.866 | .1456 | .0205 | .1405 | 48.884 | 7.120 | 5.435 | 38.697 |
| 18 | 7.690 | .1300 | .0179 | .1379 | 55.750 | 7.250 | 5.643 | 40.908 |
| 19 | 8.613 | .1161 | .0158 | .1358 | 63.440 | 7.366 | 5.838 | 42.998 |
| 20 | 9.646 | .1037 | .0139 | .1339 | 72.052 | 7.469 | 6.020 | 44.968 |
| 21 | 10.804 | .0926 | .0122 | .1322 | 81.699 | 7.562 | 6.191 | 46.819 |
| 22 | 12.100 | .0826 | .0108 | .1308 | 92.503 | 7.645 | 6.351 | 48.554 |
| 23 | 13.552 | .0738 | .0096 | .1296 | 104.603 | 7.718 | 6.501 | 50.178 |
| 24 | 15.179 | .0659 | .0085 | .1285 | 118.155 | 7.784 | 6.641 | 51.693 |
| 25 | 17.000 | .0588 | .0075 | .1275 | 133.334 | 7.843 | 6.771 | 53.105 |
| 26 | 19.040 | .0525 | .0067 | .1267 | 150.334 | 7.896 | 6.892 | 54.418 |
| 27 | 21.325 | .0469 | .0059 | .1259 | 169.374 | 7.943 | 7.005 | 55.637 |
| 28 | 23.884 | .0419 | .0052 | .1252 | 190.699 | 7.984 | 7.110 | 56.767 |
| 29 | 26.750 | .0374 | .0047 | .1247 | 214.583 | 8.022 | 7.207 | 57.814 |
| 30 | 29.960 | .0334 | .0041 | .1241 | 241.333 | 8.055 | 7.297 | 58.782 |
| 31 | 33.555 | .0298 | .0037 | .1237 | 271.293 | 8.085 | 7.381 | 59.676 |
| 32 | 37.582 | .0266 | .0033 | .1233 | 304.848 | 8.112 | 7.459 | 60.501 |
| 33 | 42.092 | .0238 | .0029 | .1229 | 342.429 | 8.135 | 7.530 | 61.261 |
| 34 | 47.143 | .0212 | .0026 | .1226 | 384.521 | 8.157 | 7.596 | 61.961 |
| 35 | 52.800 | .0189 | .0023 | .1223 | 431.663 | 8.176 | 7.658 | 62.605 |
| 40 | 93.051 | .0107 | .0013 | .1213 | 767.091 | 8.244 | 7.899 | 65.116 |
| 45 | 163.988 | .0061 | .0007 | .1207 | 1358.230 | 8.283 | 8.057 | 66.734 |
| 50 | 289.002 | .0035 | .0004 | .1204 | 2400.018 | 8.304 | 8.160 | 67.762 |
| 55 | 509.321 | .0020 | .0002 | .1202 | 4236.005 | 8.317 | 8.225 | 68.408 |
| 60 | 897.597 | .0011 | .0001 | .1201 | 7471.641 | 8.324 | 8.266 | 68.810 |
| 65 | 1581.872 | .0006 | .0001 | .1201 | 13173.937 | 8.328 | 8.292 | 69.058 |
| 70 | 2787.800 | .0004 | | .1200 | 23223.332 | 8.330 | 8.308 | 69.210 |
| 75 | 4913.056 | .0002 | | .1200 | 40933.799 | 8.332 | 8.318 | 69.303 |
| 80 | 8658.483 | .0001 | | .1200 | 72145.692 | 8.332 | 8.324 | 69.359 |
| 85 | 15259.206 | .0001 | | .1200 | 127151.714 | 8.333 | 8.328 | 69.393 |
| 90 | 26891.934 | | | .1200 | 224091.118 | 8.333 | 8.330 | 69.414 |
| 95 | 47392.777 | | | .1200 | 394931.471 | 8.333 | 8.331 | 69.426 |
| 100 | 83522.266 | | | .1200 | 696010.547 | 8.333 | 8.332 | 69.434 |

| | SINGLE PAYMENT | | UNIFORM PAYMENT SERIES | | | | GRADIENT SERIES | |
|---|---|---|---|---|---|---|---|---|
| | Compound Amount Factor | Present Worth Factor | Sinking Fund Factor | Capital Recovery Factor | Compound Amount Factor | Present Worth Factor | Gradient Uniform Series | Gradient Present Worth |
| | Find F Given P | Find P Given F | Find A Given F | Find A Given P | Find F Given A | Find P Given A | Find A Given G | Find P Given G |
| n | F/P | P/F | A/F | A/P | F/A | P/A | A/G | P/G |
| 1 | 1.150 | .8696 | 1.0000 | 1.1500 | 1.000 | .870 | 0 | 0 |
| 2 | 1.323 | .7561 | .4651 | .6151 | 2.150 | 1.626 | .465 | .756 |
| 3 | 1.521 | .6575 | .2880 | .4380 | 3.472 | 2.283 | .907 | 2.071 |
| 4 | 1.749 | .5718 | .2003 | .3503 | 4.993 | 2.855 | 1.326 | 3.786 |
| 5 | 2.011 | .4972 | .1483 | .2983 | 6.742 | 3.352 | 1.723 | 5.775 |
| 6 | 2.313 | .4323 | .1142 | .2642 | 8.754 | 3.784 | 2.097 | 7.937 |
| 7 | 2.660 | .3759 | .0904 | .2404 | 11.067 | 4.160 | 2.450 | 10.192 |
| 8 | 3.059 | .3269 | .0729 | .2229 | 13.727 | 4.487 | 2.781 | 12.481 |
| 9 | 3.518 | .2843 | .0596 | .2096 | 16.786 | 4.772 | 3.092 | 14.755 |
| 10 | 4.046 | .2472 | .0493 | .1993 | 20.304 | 5.019 | 3.383 | 16.979 |
| 11 | 4.652 | .2149 | .0411 | .1911 | 24.349 | 5.234 | 3.655 | 19.129 |
| 12 | 5.350 | .1869 | .0345 | .1845 | 29.002 | 5.421 | 3.908 | 21.185 |
| 13 | 6.153 | .1625 | .0291 | .1791 | 34.352 | 5.583 | 4.144 | 23.135 |
| 14 | 7.076 | .1413 | .0247 | .1747 | 40.505 | 5.724 | 4.362 | 24.972 |
| 15 | 8.137 | .1229 | .0210 | .1710 | 47.580 | 5.847 | 4.565 | 26.693 |
| 16 | 9.358 | .1069 | .0179 | .1679 | 55.717 | 5.954 | 4.752 | 28.296 |
| 17 | 10.761 | .0929 | .0154 | .1654 | 65.075 | 6.047 | 4.925 | 29.783 |
| 18 | 12.375 | .0808 | .0132 | .1632 | 75.836 | 6.128 | 5.084 | 31.156 |
| 19 | 14.232 | .0703 | .0113 | .1613 | 88.212 | 6.198 | 5.231 | 32.421 |
| 20 | 16.367 | .0611 | .0098 | .1598 | 102.444 | 6.259 | 5.365 | 33.582 |
| 21 | 18.822 | .0531 | .0084 | .1584 | 118.810 | 6.312 | 5.488 | 34.645 |
| 22 | 21.645 | .0462 | .0073 | .1573 | 137.632 | 6.359 | 5.601 | 35.615 |
| 23 | 24.891 | .0402 | .0063 | .1563 | 159.276 | 6.399 | 5.704 | 36.499 |
| 24 | 28.625 | .0349 | .0054 | .1554 | 184.168 | 6.434 | 5.798 | 37.302 |
| 25 | 32.919 | .0304 | .0047 | .1547 | 212.793 | 6.464 | 5.883 | 38.031 |
| 26 | 37.857 | .0264 | .0041 | .1541 | 245.712 | 6.491 | 5.961 | 38.692 |
| 27 | 43.535 | .0230 | .0035 | .1535 | 283.569 | 6.514 | 6.032 | 39.289 |
| 28 | 50.066 | .0200 | .0031 | .1531 | 327.104 | 6.534 | 6.096 | 39.828 |
| 29 | 57.575 | .0174 | .0027 | .1527 | 377.170 | 6.551 | 6.154 | 40.315 |
| 30 | 66.212 | .0151 | .0023 | .1523 | 434.745 | 6.566 | 6.207 | 40.753 |
| 31 | 76.144 | .0131 | .0020 | .1520 | 500.957 | 6.579 | 6.254 | 41.147 |
| 32 | 87.565 | .0114 | .0017 | .1517 | 577.100 | 6.591 | 6.297 | 41.501 |
| 33 | 100.700 | .0099 | .0015 | .1515 | 664.666 | 6.600 | 6.336 | 41.818 |
| 34 | 115.805 | .0086 | .0013 | .1513 | 765.365 | 6.609 | 6.371 | 42.103 |
| 35 | 133.176 | .0075 | .0011 | .1511 | 881.170 | 6.617 | 6.402 | 42.359 |
| 40 | 267.864 | .0037 | .0006 | .1506 | 1779.090 | 6.642 | 6.517 | 43.283 |
| 45 | 538.769 | .0019 | .0003 | .1503 | 3585.128 | 6.654 | 6.583 | 43.805 |
| 50 | 1083.657 | .0009 | .0001 | .1501 | 7217.716 | 6.661 | 6.620 | 44.096 |
| 55 | 2179.622 | .0005 | .0001 | .1501 | 14524.148 | 6.664 | 6.641 | 44.256 |
| 60 | 4383.999 | .0002 | | .1500 | 29219.992 | 6.665 | 6.653 | 44.343 |
| 65 | 8817.787 | .0001 | | .1500 | 58778.583 | 6.666 | 6.659 | 44.390 |
| 70 | 17735.720 | .0001 | | .1500 | 118231.467 | 6.666 | 6.663 | 44.416 |
| 75 | 35672.868 | | | .1500 | 237812.453 | 6.666 | 6.665 | 44.429 |
| 80 | 71750.879 | | | .1500 | 478332.529 | 6.667 | 6.666 | 44.436 |
| 85 | 144316.647 | | | .1500 | 962104.313 | 6.667 | 6.666 | 44.440 |
| 90 | 290272.325 | | | .1500 | 1935142.168 | 6.667 | 6.666 | 44.442 |
| 95 | 583841.328 | | | .1500 | 3892268.851 | 6.667 | 6.667 | 44.443 |
| 100 | 1174313.451 | | | .1500 | 7828749.671 | 6.667 | 6.667 | 44.444 |

| | SINGLE PAYMENT | | UNIFORM PAYMENT SERIES | | | | GRADIENT SERIES | |
|---|---|---|---|---|---|---|---|---|
| | Compound Amount Factor | Present Worth Factor | Sinking Fund Factor | Capital Recovery Factor | Compound Amount Factor | Present Worth Factor | Gradient Uniform Series | Gradient Present Worth |
| | Find F Given P F/P | Find P Given F P/F | Find A Given F A/F | Find A Given P A/P | Find F Given A F/A | Find P Given A P/A | Find A Given G A/G | Find P Given G P/G |
| n | | | | | | | | |
| 1 | 1.180 | .8475 | 1.0000 | 1.1800 | 1.000 | .847 | 0 | 0 |
| 2 | 1.392 | .7182 | .4587 | .6387 | 2.180 | 1.566 | .459 | .718 |
| 3 | 1.643 | .6086 | .2799 | .4599 | 3.572 | 2.174 | .890 | 1.935 |
| 4 | 1.939 | .5158 | .1917 | .3717 | 5.215 | 2.690 | 1.295 | 3.483 |
| 5 | 2.288 | .4371 | .1398 | .3198 | 7.154 | 3.127 | 1.673 | 5.231 |
| 6 | 2.700 | .3704 | .1059 | .2859 | 9.442 | 3.498 | 2.025 | 7.083 |
| 7 | 3.185 | .3139 | .0824 | .2624 | 12.142 | 3.812 | 2.353 | 8.967 |
| 8 | 3.759 | .2660 | .0652 | .2452 | 15.327 | 4.078 | 2.656 | 10.829 |
| 9 | 4.435 | .2255 | .0524 | .2324 | 19.086 | 4.303 | 2.936 | 12.633 |
| 10 | 5.234 | .1911 | .0425 | .2225 | 23.521 | 4.494 | 3.194 | 14.352 |
| 11 | 6.176 | .1619 | .0348 | .2148 | 28.755 | 4.656 | 3.430 | 15.972 |
| 12 | 7.288 | .1372 | .0286 | .2086 | 34.931 | 4.793 | 3.647 | 17.481 |
| 13 | 8.599 | .1163 | .0237 | .2037 | 42.219 | 4.910 | 3.845 | 18.877 |
| 14 | 10.147 | .0985 | .0197 | .1997 | 50.818 | 5.008 | 4.025 | 20.158 |
| 15 | 11.974 | .0835 | .0164 | .1964 | 60.965 | 5.092 | 4.189 | 21.327 |
| 16 | 14.129 | .0708 | .0137 | .1937 | 72.939 | 5.162 | 4.337 | 22.389 |
| 17 | 16.672 | .0600 | .0115 | .1915 | 87.068 | 5.222 | 4.471 | 23.348 |
| 18 | 19.673 | .0508 | .0096 | .1896 | 103.740 | 5.273 | 4.592 | 24.212 |
| 19 | 23.214 | .0431 | .0081 | .1881 | 123.414 | 5.316 | 4.700 | 24.988 |
| 20 | 27.393 | .0365 | .0068 | .1868 | 146.628 | 5.353 | 4.798 | 25.681 |
| 21 | 32.324 | .0309 | .0057 | .1857 | 174.021 | 5.384 | 4.885 | 26.300 |
| 22 | 38.142 | .0262 | .0048 | .1848 | 206.345 | 5.410 | 4.963 | 26.851 |
| 23 | 45.008 | .0222 | .0041 | .1841 | 244.487 | 5.432 | 5.033 | 27.339 |
| 24 | 53.109 | .0188 | .0035 | .1835 | 289.494 | 5.451 | 5.095 | 27.772 |
| 25 | 62.669 | .0160 | .0029 | .1829 | 342.603 | 5.467 | 5.150 | 28.155 |
| 26 | 73.949 | .0135 | .0025 | .1825 | 405.272 | 5.480 | 5.199 | 28.494 |
| 27 | 87.260 | .0115 | .0021 | .1821 | 479.221 | 5.492 | 5.243 | 28.791 |
| 28 | 102.967 | .0097 | .0018 | .1818 | 566.481 | 5.502 | 5.281 | 29.054 |
| 29 | 121.501 | .0082 | .0015 | .1815 | 669.447 | 5.510 | 5.315 | 29.284 |
| 30 | 143.371 | .0070 | .0013 | .1813 | 790.948 | 5.517 | 5.345 | 29.486 |
| 31 | 169.177 | .0059 | .0011 | .1811 | 934.319 | 5.523 | 5.371 | 29.664 |
| 32 | 199.629 | .0050 | .0009 | .1809 | 1103.496 | 5.528 | 5.394 | 29.819 |
| 33 | 235.563 | .0042 | .0008 | .1808 | 1303.125 | 5.532 | 5.415 | 29.955 |
| 34 | 277.964 | .0036 | .0006 | .1806 | 1538.688 | 5.536 | 5.433 | 30.074 |
| 35 | 327.997 | .0030 | .0006 | .1806 | 1816.652 | 5.539 | 5.449 | 30.177 |
| 40 | 750.378 | .0013 | .0002 | .1802 | 4163.213 | 5.548 | 5.502 | 30.527 |
| 45 | 1716.684 | .0006 | .0001 | .1801 | 9531.577 | 5.552 | 5.529 | 30.701 |
| 50 | 3927.357 | .0003 | | .1800 | 21813.094 | 5.554 | 5.543 | 30.786 |
| 55 | 8984.841 | .0001 | | .1800 | 49910.228 | 5.555 | 5.549 | 30.827 |
| 60 | 20555.140 | | | .1800 | 114189.666 | 5.555 | 5.553 | 30.846 |
| 65 | 47025.181 | | | .1800 | 261245.449 | 5.555 | 5.554 | 30.856 |
| 70 | 107582.222 | | | .1800 | 597673.458 | 5.556 | 5.555 | 30.860 |
| 75 | 246122.064 | | | .1800 | 1367339.243 | 5.556 | 5.555 | 30.862 |
| 80 | 563067.660 | | | .1800 | 3128148.114 | 5.556 | 5.555 | 30.863 |
| 85 | 1288162.408 | | | .1800 | 7156452.266 | 5.556 | 5.555 | 30.864 |

| | SINGLE PAYMENT | | UNIFORM PAYMENT SERIES | | | | GRADIENT SERIES | |
|---|---|---|---|---|---|---|---|---|
| | Compound Amount Factor | Present Worth Factor | Sinking Fund Factor | Capital Recovery Factor | Compound Amount Factor | Present Worth Factor | Gradient Uniform Series | Gradient Present Worth |
| | Find $F$ Given $P$ | Find $P$ Given $F$ | Find $A$ Given $F$ | Find $A$ Given $P$ | Find $F$ Given $A$ | Find $P$ Given $A$ | Find $A$ Given $G$ | Find $P$ Given $G$ |
| $n$ | $F/P$ | $P/F$ | $A/F$ | $A/P$ | $F/A$ | $P/A$ | $A/G$ | $P/G$ |
| 1 | 1.200 | .8333 | 1.0000 | 1.2000 | 1.000 | .833 | 0 | 0 |
| 2 | 1.440 | .6944 | .4545 | .6545 | 2.200 | 1.528 | .455 | .694 |
| 3 | 1.728 | .5787 | .2747 | .4747 | 3.640 | 2.106 | .879 | 1.852 |
| 4 | 2.074 | .4823 | .1863 | .3863 | 5.368 | 2.589 | 1.274 | 3.299 |
| 5 | 2.488 | .4019 | .1344 | .3344 | 7.442 | 2.991 | 1.641 | 4.906 |
| 6 | 2.986 | .3349 | .1007 | .3007 | 9.930 | 3.326 | 1.979 | 6.581 |
| 7 | 3.583 | .2791 | .0774 | .2774 | 12.916 | 3.605 | 2.290 | 8.255 |
| 8 | 4.300 | .2326 | .0606 | .2606 | 16.499 | 3.837 | 2.576 | 9.883 |
| 9 | 5.160 | .1938 | .0481 | .2481 | 20.799 | 4.031 | 2.836 | 11.434 |
| 10 | 6.192 | .1615 | .0385 | .2385 | 25.959 | 4.192 | 3.074 | 12.887 |
| 11 | 7.430 | .1346 | .0311 | .2311 | 32.150 | 4.327 | 3.289 | 14.233 |
| 12 | 8.916 | .1122 | .0253 | .2253 | 39.581 | 4.439 | 3.484 | 15.467 |
| 13 | 10.699 | .0935 | .0206 | .2206 | 48.497 | 4.533 | 3.660 | 16.588 |
| 14 | 12.839 | .0779 | .0169 | .2169 | 59.196 | 4.611 | 3.817 | 17.601 |
| 15 | 15.407 | .0649 | .0139 | .2139 | 72.035 | 4.675 | 3.959 | 18.509 |
| 16 | 18.488 | .0541 | .0114 | .2114 | 87.442 | 4.730 | 4.085 | 19.321 |
| 17 | 22.186 | .0451 | .0094 | .2094 | 105.931 | 4.775 | 4.198 | 20.042 |
| 18 | 26.623 | .0376 | .0078 | .2078 | 128.117 | 4.812 | 4.298 | 20.680 |
| 19 | 31.948 | .0313 | .0065 | .2065 | 154.740 | 4.843 | 4.386 | 21.244 |
| 20 | 38.338 | .0261 | .0054 | .2054 | 186.688 | 4.870 | 4.464 | 21.739 |
| 21 | 46.005 | .0217 | .0044 | .2044 | 225.026 | 4.891 | 4.533 | 22.174 |
| 22 | 55.206 | .0181 | .0037 | .2037 | 271.031 | 4.909 | 4.594 | 22.555 |
| 23 | 66.247 | .0151 | .0031 | .2031 | 326.237 | 4.925 | 4.647 | 22.887 |
| 24 | 79.497 | .0126 | .0025 | .2025 | 392.484 | 4.937 | 4.694 | 23.176 |
| 25 | 95.396 | .0105 | .0021 | .2021 | 471.981 | 4.948 | 4.735 | 23.428 |
| 26 | 114.475 | .0087 | .0018 | .2018 | 567.377 | 4.956 | 4.771 | 23.646 |
| 27 | 137.371 | .0073 | .0015 | .2015 | 681.853 | 4.964 | 4.802 | 23.835 |
| 28 | 164.845 | .0061 | .0012 | .2012 | 819.223 | 4.970 | 4.829 | 23.999 |
| 29 | 197.814 | .0051 | .0010 | .2010 | 984.068 | 4.975 | 4.853 | 24.141 |
| 30 | 237.376 | .0042 | .0008 | .2008 | 1181.882 | 4.979 | 4.873 | 24.263 |
| 31 | 284.852 | .0035 | .0007 | .2007 | 1419.258 | 4.982 | 4.891 | 24.368 |
| 32 | 341.822 | .0029 | .0006 | .2006 | 1704.109 | 4.985 | 4.906 | 24.459 |
| 33 | 410.186 | .0024 | .0005 | .2005 | 2045.931 | 4.988 | 4.919 | 24.537 |
| 34 | 492.224 | .0020 | .0004 | .2004 | 2456.118 | 4.990 | 4.931 | 24.604 |
| 35 | 590.668 | .0017 | .0003 | .2003 | 2948.341 | 4.992 | 4.941 | 24.661 |
| 40 | 1469.772 | .0007 | .0001 | .2001 | 7343.858 | 4.997 | 4.973 | 24.847 |
| 45 | 3657.262 | .0003 | .0001 | .2001 | 18281.310 | 4.999 | 4.988 | 24.932 |
| 50 | 9100.438 | .0001 | | .2000 | 45497.191 | 4.999 | 4.995 | 24.970 |
| 55 | 22644.802 | | | .2000 | 113219.011 | 5.000 | 4.998 | 24.987 |
| 60 | 56347.514 | | | .2000 | 281732.572 | 5.000 | 4.999 | 24.994 |
| 65 | 140210.647 | | | .2000 | 701048.235 | 5.000 | 5.000 | 24.998 |
| 70 | 348888.957 | | | .2000 | 1744439.785 | 5.000 | 5.000 | 24.999 |
| 75 | 868147.369 | | | .2000 | 4340731.847 | 5.000 | 5.000 | 25.000 |

# 25% Compound Interest Factors 25%

| | SINGLE PAYMENT | | UNIFORM PAYMENT SERIES | | | | GRADIENT SERIES | |
|---|---|---|---|---|---|---|---|---|
| | Compound Amount Factor | Present Worth Factor | Sinking Fund Factor | Capital Recovery Factor | Compound Amount Factor | Present Worth Factor | Gradient Uniform Series | Gradient Present Worth |
| n | Find F Given P F/P | Find P Given F P/F | Find A Given F A/F | Find A Given P A/P | Find F Given A F/A | Find P Given A P/A | Find A Given G A/G | Find P Given G P/G |
| 1 | 1.250 | .8000 | 1.0000 | 1.2500 | 1.000 | .800 | 0 | 0 |
| 2 | 1.563 | .6400 | .4444 | .6944 | 2.250 | 1.440 | .444 | .640 |
| 3 | 1.953 | .5120 | .2623 | .5123 | 3.813 | 1.952 | .852 | 1.664 |
| 4 | 2.441 | .4096 | .1734 | .4234 | 5.766 | 2.362 | 1.225 | 2.893 |
| 5 | 3.052 | .3277 | .1218 | .3718 | 8.207 | 2.689 | 1.563 | 4.204, |
| 6 | 3.815 | .2621 | .0888 | .3388 | 11.259 | 2.951 | 1.868 | 5.514 |
| 7 | 4.768 | .2097 | .0663 | .3163 | 15.073 | 3.161 | 2.142 | 6.773 |
| 8 | 5.960 | .1678 | .0504 | .3004 | 19.842 | 3.329 | 2.387 | 7.947 |
| 9 | 7.451 | .1342 | .0388 | .2888 | 25.802 | 3.463 | 2.605 | 9.021 |
| 10 | 9.313 | .1074 | .0301 | .2801 | 33.253 | 3.571 | 2.797 | 9.987 |
| 11 | 11.642 | .0859 | .0235 | .2735 | 42.566 | 3.656 | 2.966 | 10.864 |
| 12 | 14.552 | .0687 | .0184 | .2684 | 54.208 | 3.725 | 3.115 | 11.602 |
| 13 | 18.190 | .0550 | .0145 | .2645 | 68.760 | 3.780 | 3.244 | 12.262 |
| 14 | 22.737 | .0440 | .0115 | .2615 | 86.949 | 3.824 | 3.356 | 12.833 |
| 15 | 28.422 | .0352 | .0091 | .2591 | 109.687 | 3.859 | 3.453 | 13.326 |
| 16 | 35.527 | .0281 | .0072 | .2572 | 138.109 | 3.887 | 3.537 | 13.748 |
| 17 | 44.409 | .0225 | .0058 | .2558 | 173.636 | 3.910 | 3.608 | 14.108 |
| 18 | 55.511 | .0180 | .0046 | .2546 | 218.045 | 3.928 | 3.670 | 14.415 |
| 19 | 69.389 | .0144 | .0037 | .2537 | 273.556 | 3.942 | 3.722 | 14.674 |
| 20 | 86.736 | .0115 | .0029 | .2529 | 342.945 | 3.954 | 3.767 | 14.893 |
| 21 | 108.420 | .0092 | .0023 | .2523 | 429.681 | 3.963 | 3.805 | 15.078 |
| 22 | 135.525 | .0074 | .0019 | .2519 | 538.101 | 3.970 | 3.836 | 15.233 |
| 23 | 169.407 | .0059 | .0015 | .2515 | 673.626 | 3.976 | 3.863 | 15.362 |
| 24 | 211.758 | .0047 | .0012 | .2512 | 843.033 | 3.981 | 3.886 | 15.471 |
| 25 | 264.698 | .0038 | .0009 | .2509 | 1054.791 | 3.985 | 3.905 | 15.562 |
| 26 | 330.872 | .0030 | .0008 | .2508 | 1319.489 | 3.988 | 3.921 | 15.637 |
| 27 | 413.590 | .0024 | .0006 | .2506 | 1650.361 | 3.990 | 3.935 | 15.700 |
| 28 | 516.988 | .0019 | .0005 | .2505 | 2063.952 | 3.992 | 3.946 | 15.752 |
| 29 | 646.235 | .0015 | .0004 | .2504 | 2580.939 | 3.994 | 3.955 | 15.796 |
| 30 | 807.794 | .0012 | .0003 | .2503 | 3227.174 | 3.995 | 3.963 | 15.832 |
| 31 | 1009.742 | .0010 | .0002 | .2502 | 4034.968 | 3.996 | 3.969 | 15.861 |
| 32 | 1262.177 | .0008 | .0002 | .2502 | 5044.710 | 3.997 | 3.975 | 15.886 |
| 33 | 1577.722 | .0006 | .0002 | .2502 | 6306.887 | 3.997 | 3.979 | 15.906 |
| 34 | 1972.152 | .0005 | .0001 | .2501 | 7884.609 | 3.998 | 3.983 | 15.923 |
| 35 | 2465.190 | .0004 | .0001 | .2501 | 9856.761 | 3.998 | 3.986 | 15.937 |
| 40 | 7523.164 | .0001 | | .2500 | 30088.655 | 3.999 | 3.995 | 15.977 |
| 45 | 22958.874 | | | .2500 | 91831.496 | 4.000 | 3.998 | 15.991 |
| 50 | 70064.923 | | | .2500 | 280255.693 | 4.000 | 3.999 | 15.997 |
| 55 | 213821.177 | | | .2500 | 855280.707 | 4.000 | 4.000 | 15.999 |
| 60 | 652530.447 | | | .2500 | 2610117.787 | 4.000 | 4.000 | 16.000 |

| | SINGLE PAYMENT | | UNIFORM PAYMENT SERIES | | | | GRADIENT SERIES | |
|---|---|---|---|---|---|---|---|---|
| | Compound Amount Factor | Present Worth Factor | Sinking Fund Factor | Capital Recovery Factor | Compound Amount Factor | Present Worth Factor | Gradient Uniform Series | Gradient Present Worth |
| $n$ | Find $F$ Given $P$ $F/P$ | Find $P$ Given $F$ $P/F$ | Find $A$ Given $F$ $A/F$ | Find $A$ Given $P$ $A/P$ | Find $F$ Given $A$ $F/A$ | Find $P$ Given $A$ $P/A$ | Find $A$ Given $G$ $A/G$ | Find $P$ Given $G$ $P/G$ |
| 1 | 1.300 | .7692 | 1.0000 | 1.3000 | 1.000 | .769 | 0 | 0 |
| 2 | 1.690 | .5917 | .4348 | .7348 | 2.300 | 1.361 | .435 | .592 |
| 3 | 2.197 | .4552 | .2506 | .5506 | 3.990 | 1.816 | .827 | 1.502 |
| 4 | 2.856 | .3501 | .1616 | .4616 | 6.187 | 2.166 | 1.178 | 2.552 |
| 5 | 3.713 | .2693 | .1106 | .4106 | 9.043 | 2.436 | 1.490 | 3.630 |
| 6 | 4.827 | .2072 | .0784 | .3784 | 12.756 | 2.643 | 1.765 | 4.666 |
| 7 | 6.275 | .1594 | .0569 | .3569 | 17.583 | 2.802 | 2.006 | 5.622 |
| 8 | 8.157 | .1226 | .0419 | .3419 | 23.858 | 2.925 | 2.216 | 6.480 |
| 9 | 10.604 | .0943 | .0312 | .3312 | 32.015 | 3.019 | 2.396 | 7.234 |
| 10 | 13.786 | .0725 | .0235 | .3235 | 42.619 | 3.092 | 2.551 | 7.887 |
| 11 | 17.922 | .0558 | .0177 | .3177 | 56.405 | 3.147 | 2.683 | 8.445 |
| 12 | 23.298 | .0429 | .0135 | .3135 | 74.327 | 3.190 | 2.795 | 8.917 |
| 13 | 30.288 | .0330 | .0102 | .3102 | 97.625 | 3.223 | 2.889 | 9.314 |
| 14 | 39.374 | .0254 | .0078 | .3078 | 127.913 | 3.249 | 2.969 | 9.644 |
| 15 | 51.186 | .0195 | .0060 | .3060 | 167.286 | 3.268 | 3.034 | 9.917 |
| 16 | 66.542 | .0150 | .0046 | .3046 | 218.472 | 3.283 | 3.089 | 10.143 |
| 17 | 86.504 | .0116 | .0035 | .3035 | 285.014 | 3.295 | 3.135 | 10.328 |
| 18 | 112.455 | .0089 | .0027 | .3027 | 371.518 | 3.304 | 3.172 | 10.479 |
| 19 | 146.192 | .0068 | .0021 | .3021 | 483.973 | 3.311 | 3.202 | 10.602 |
| 20 | 190.050 | .0053 | .0016 | .3016 | 630.165 | 3.316 | 3.228 | 10.702 |
| 21 | 247.065 | .0040 | .0012 | .3012 | 820.215 | 3.320 | 3.248 | 10.783 |
| 22 | 321.184 | .0031 | .0009 | .3009 | 1067.280 | 3.323 | 3.265 | 10.848 |
| 23 | 417.539 | .0024 | .0007 | .3007 | 1388.464 | 3.325 | 3.278 | 10.901 |
| 24 | 542.801 | .0018 | .0006 | .3006 | 1806.003 | 3.327 | 3.289 | 10.943 |
| 25 | 705.641 | .0014 | .0004 | .3004 | 2348.803 | 3.329 | 3.298 | 10.977 |
| 26 | 917.333 | .0011 | .0003 | .3003 | 3054.444 | 3.330 | 3.305 | 11.005 |
| 27 | 1192.533 | .0008 | .0003 | .3003 | 3971.778 | 3.331 | 3.311 | 11.026 |
| 28 | 1550.293 | .0006 | .0002 | .3002 | 5164.311 | 3.331 | 3.315 | 11.044 |
| 29 | 2015.381 | .0005 | .0001 | .3001 | 6714.604 | 3.332 | 3.319 | 11.058 |
| 30 | 2619.996 | .0004 | .0001 | .3001 | 8729.985 | 3.332 | 3.322 | 11.069 |
| 31 | 3405.994 | .0003 | .0001 | .3001 | 11349.981 | 3.332 | 3.324 | 11.078 |
| 32 | 4427.793 | .0002 | .0001 | .3001 | 14755.975 | 3.333 | 3.326 | 11.085 |
| 33 | 5756.130 | .0002 | .0001 | .3001 | 19183.768 | 3.333 | 3.328 | 11.090 |
| 34 | 7482.970 | .0001 | | .3000 | 24939.899 | 3.333 | 3.329 | 11.094 |
| 35 | 9727.860 | .0001 | | .3000 | 32422.868 | 3.333 | 3.330 | 11.098 |
| 40 | 36118.865 | | | .3000 | 120392.883 | 3.333 | 3.332 | 11.107 |
| 45 | 134106.817 | | | .3000 | 447019.389 | 3.333 | 3.333 | 11.110 |
| 50 | 497929.223 | | | .3000 | 1659760.745 | 3.333 | 3.333 | 11.111 |
| 55 | 1848776.352 | | | .3000 | 6162584.505 | 3.333 | 3.333 | 11.111 |
| 60 | 6864377.179 | | | .3000 | 22881253.930 | 3.333 | 3.333 | 11.111 |

| | SINGLE PAYMENT | | UNIFORM PAYMENT SERIES | | | | GRADIENT SERIES | |
|---|---|---|---|---|---|---|---|---|
| | Compound Amount Factor | Present Worth Factor | Sinking Fund Factor | Capital Recovery Factor | Compound Amount Factor | Present Worth Factor | Gradient Uniform Series | Gradient Present Worth |
| n | Find F Given P F/P | Find P Given F P/F | Find A Given F A/F | Find A Given P A/P | Find F Given A F/A | Find P Given A P/A | Find A Given G A/G | Find P Given G P/G |
| 1 | 1.350 | .7407 | 1.000 | 1.3500 | 1.000 | .741 | 0 | 0 |
| 2 | 1.823 | .5487 | .4255 | .7755 | 2.350 | 1.289 | .426 | .549 |
| 3 | 2.460 | .4064 | .2397 | .5897 | 4.172 | 1.696 | .803 | 1.362 |
| 4 | 3.322 | .3011 | .1508 | .5008 | 6.633 | 1.997 | 1.134 | 2.265 |
| 5 | 4.484 | .2230 | .1005 | .4505 | 9.954 | 2.220 | 1.422 | 3.157 |
| 6 | 6.053 | .1652 | .0693 | .4193 | 14.438 | 2.385 | 1.670 | 3.983 |
| 7 | 8.172 | .1224 | .0488 | .3988 | 20.492 | 2.508 | 1.881 | 4.717 |
| 8 | 11.032 | .0906 | .0349 | .3849 | 28.664 | 2.598 | 2.060 | 5.352 |
| 9 | 14.894 | .0671 | .0252 | .3752 | 39.696 | 2.665 | 2.209 | 5.889 |
| 10 | 20.107 | .0497 | .0183 | .3683 | 54.590 | 2.715 | 2.334 | 6.336 |
| 11 | 27.144 | .0368 | .0134 | .3634 | 74.697 | 2.752 | 2.436 | 6.705 |
| 12 | 36.644 | .0273 | .0098 | .3598 | 101.841 | 2.779 | 2.520 | 7.005 |
| 13 | 49.470 | .0202 | .0072 | .3572 | 138.485 | 2.799 | 2.589 | 7.247 |
| 14 | 66.784 | .0150 | .0053 | .3553 | 187.954 | 2.814 | 2.644 | 7.442 |
| 15 | 90.158 | .0111 | .0039 | .3539 | 254.738 | 2.825 | 2.689 | 7.597 |
| 16 | 121.714 | .0082 | .0029 | .3529 | 344.897 | 2.834 | 2.725 | 7.721 |
| 17 | 164.314 | .0061 | .0021 | .3521 | 466.611 | 2.840 | 2.753 | 7.818 |
| 18 | 221.824 | .0045 | .0016 | .3516 | 630.925 | 2.844 | 2.776 | 7.895 |
| 19 | 299.462 | .0033 | .0012 | .3512 | 852.748 | 2.848 | 2.793 | 7.955 |
| 20 | 404.274 | .0025 | .0009 | .3509 | 1152.210 | 2.850 | 2.808 | 8.002 |
| 21 | 545.769 | .0018 | .0006 | .3506 | 1556.484 | 2.852 | 2.819 | 8.038 |
| 22 | 736.789 | .0014 | .0005 | .3505 | 2102.253 | 2.853 | 2.827 | 8.067 |
| 23 | 994.665 | .0010 | .0004 | .3504 | 2839.042 | 2.854 | 2.834 | 8.089 |
| 24 | 1342.797 | .0007 | .0003 | .3503 | 3833.706 | 2.855 | 2.839 | 8.106 |
| 25 | 1812.776 | .0006 | .0002 | .3502 | 5176.504 | 2.856 | 2.843 | 8.119 |
| 26 | 2447.248 | .0004 | .0001 | .3501 | 6989.280 | 2.856 | 2.847 | 8.130 |
| 27 | 3303.785 | .0003 | .0001 | .3501 | 9436.528 | 2.856 | 2.849 | 8.137 |
| 28 | 4460.109 | .0002 | .0001 | .3501 | 12740.313 | 2.857 | 2.851 | 8.143 |
| 29 | 6021.148 | .0002 | .0001 | .3501 | 17200.422 | 2.857 | 2.852 | 8.148 |
| 30 | 8128.550 | .0001 | | .3500 | 23221.570 | 2.857 | 2.853 | 8.152 |
| 31 | 10973.542 | .0001 | | .3500 | 31350.120 | 2.857 | 2.854 | 8.154 |
| 32 | 14814.281 | .0001 | | .3500 | 42323.661 | 2.857 | 2.855 | 8.157 |
| 33 | 19999.280 | .0001 | | .3500 | 57137.943 | 2.857 | 2.855 | 8.158 |
| 34 | 26999.028 | | | .3500 | 77137.223 | 2.857 | 2.856 | 8.159 |
| 35 | 36448.688 | | | .3500 | 104136.251 | 2.857 | 2.856 | 8.160 |
| 40 | 163437.135 | | | .3500 | 466960.385 | 2.857 | 2.857 | 8.163 |
| 45 | 732857.577 | | | .3500 | 2093875.934 | 2.857 | 2.857 | 8.163 |
| 50 | 3286157.879 | | | .3500 | 9389019.655 | 2.857 | 2.857 | 8.163 |

| | SINGLE PAYMENT | | UNIFORM PAYMENT SERIES | | | | GRADIENT SERIES | |
|---|---|---|---|---|---|---|---|---|
| | Compound Amount Factor | Present Worth Factor | Sinking Fund Factor | Capital Recovery Factor | Compound Amount Factor | Present Worth Factor | Gradient Uniform Series | Gradient Present Worth |
| | Find $F$ Given $P$ | Find $P$ Given $F$ | Find $A$ Given $F$ | Find $A$ Given $P$ | Find $F$ Given $A$ | Find $P$ Given $A$ | Find $A$ Given $G$ | Find $P$ Given $G$ |
| $n$ | $F/P$ | $P/F$ | $A/F$ | $A/P$ | $F/A$ | $P/A$ | $A/G$ | $P/G$ |
| 1 | 1.400 | .7143 | 1.0000 | 1.4000 | 1.000 | .714 | 0 | 0 |
| 2 | 1.960 | .5102 | .4167 | .8167 | 2.400 | 1.224 | .417 | .510 |
| 3 | 2.744 | .3644 | .2294 | .6294 | 4.360 | 1.589 | .780 | 1.239 |
| 4 | 3.842 | .2603 | .1408 | .5408 | 7.104 | 1.849 | 1.092 | 2.020 |
| 5 | 5.378 | .1859 | .0914 | .4914 | 10.946 | 2.035 | 1.358 | 2.764 |
| 6 | 7.530 | .1328 | .0613 | .4613 | 16.324 | 2.168 | 1.581 | 3.428 |
| 7 | 10.541 | .0949 | .0419 | .4419 | 23.853 | 2.263 | 1.766 | 3.997 |
| 8 | 14.758 | .0678 | .0291 | .4291 | 34.395 | 2.331 | 1.919 | 4.471 |
| 9 | 20.661 | .0484 | .0203 | .4203 | 49.153 | 2.379 | 2.042 | 4.858 |
| 10 | 28.925 | .0346 | .0143 | .4143 | 69.814 | 2.414 | 2.142 | 5.170 |
| 11 | 40.496 | .0247 | .0101 | .4101 | 98.739 | 2.438 | 2.221 | 5.417 |
| 12 | 56.694 | .0176 | .0072 | .4072 | 139.235 | 2.456 | 2.285 | 5.611 |
| 12 | 79.371 | .0126 | .0051 | .4051 | 195.929 | 2.469 | 2.334 | 5.762 |
| 14 | 111.120 | .0090 | .0036 | .4036 | 275.300 | 2.478 | 2.373 | 5.879 |
| 15 | 155.568 | .0064 | .0026 | .4026 | 386.420 | 2.484 | 2.403 | 5.969 |
| 16 | 217.795 | .0046 | .0018 | .4018 | 541.988 | 2.489 | 2.426 | 6.038 |
| 17 | 304.913 | .0033 | .0013 | .4013 | 759.784 | 2.492 | 2.444 | 6.090 |
| 18 | 426.879 | .0023 | .0009 | .4009 | 1064.697 | 2.494 | 2.458 | 6.130 |
| 19 | 597.630 | .0017 | .0007 | .4007 | 1491.576 | 2.496 | 2.468 | 6.160 |
| 20 | 836.683 | .0012 | .0005 | .4005 | 2089.206 | 2.497 | 2.476 | 6.183 |
| 21 | 1171.356 | .0009 | .0003 | .4003 | 2925.889 | 2.498 | 2.482 | 6.200 |
| 22 | 1639.898 | .0006 | .0002 | .4002 | 4097.245 | 2.498 | 2.487 | 6.213 |
| 23 | 2295.857 | .0004 | .0002 | .4002 | 5737.142 | 2.499 | 2.490 | 6.222 |
| 24 | 3214.200 | .0003 | .0001 | .4001 | 8032.999 | 2.499 | 2.493 | 6.229 |
| 25 | 4499.880 | .0002 | .0001 | .4001 | 11247.199 | 2.499 | 2.494 | 6.235 |
| 26 | 6299.831 | .0002 | .0001 | .4001 | 15747.079 | 2.500 | 2.496 | 6.239 |
| 27 | 8819.764 | .0001 | | .4000 | 22046.910 | 2.500 | 2.497 | 6.242 |
| 28 | 12347.670 | .0001 | | .4000 | 30866.674 | 2.500 | 2.498 | 6.244 |
| 29 | 17286.737 | .0001 | | .4000 | 43214.344 | 2.500 | 2.498 | 6.245 |
| 30 | 24201.432 | | | .4000 | 60501.081 | 2.500 | 2.499 | 6.247 |
| 31 | 33882.005 | | | .4000 | 84702.513 | 2.500 | 2.499 | 6.248 |
| 32 | 47434.807 | | | .4000 | 118584.519 | 2.500 | 2.499 | 6.248 |
| 33 | 66408.730 | | | .4000 | 166019.326 | 2.500 | 2.500 | 6.249 |
| 34 | 92972.223 | | | .4000 | 232428.057 | 2.500 | 2.500 | 6.249 |
| 35 | 130161.112 | | | .4000 | 325400.279 | 2.500 | 2.500 | 6.249 |
| 40 | 700037.697 | | | .4000 | 1750091.743 | 2.500 | 2.500 | 6.250 |
| 45 | 3764970.745 | | | .4000 | 9412424.362 | 2.500 | 2.500 | 6.250 |
| 50 | 20248916.262 | | | .4000 | 50622288.153 | 2.500 | 2.500 | 6.250 |

| | SINGLE PAYMENT | | UNIFORM PAYMENT SERIES | | | | GRADIENT SERIES | |
|---|---|---|---|---|---|---|---|---|
| | Compound Amount Factor | Present Worth Factor | Sinking Fund Factor | Capital Recovery Factor | Compound Amount Factor | Present Worth Factor | Gradient Uniform Series | Gradient Present Worth |
| | Find $F$ Given $P$ | Find $P$ Given $F$ | Find $A$ Given $F$ | Find $A$ Given $P$ | Find $F$ Given $A$ | Find $P$ Given $A$ | Find $A$ Given $G$ | Find $P$ Given $G$ |
| $n$ | $F/P$ | $P/F$ | $A/F$ | $A/P$ | $F/A$ | $P/A$ | $A/G$ | $P/G$ |
| 1 | 1.450 | .6897 | 1.0000 | 1.4500 | 1.000 | .690 | 0 | 0 |
| 2 | 2.103 | .4756 | .4082 | .8582 | 2.450 | 1.165 | .408 | .476 |
| 3 | 3.049 | .3280 | .2197 | .6697 | 4.553 | 1.493 | .758 | 1.132 |
| 4 | 4.421 | .2262 | .1316 | .5816 | 7.601 | 1.720 | 1.053 | 1.810 |
| 5 | 6.410 | .1560 | .0832 | .5332 | 12.022 | 1.876 | 1.298 | 2.434 |
| 6 | 9.294 | .1076 | .0543 | .5043 | 18.431 | 1.983 | 1.499 | 2.972 |
| 7 | 13.476 | .0742 | .0361 | .4861 | 27.725 | 2.057 | 1.661 | 3.418 |
| 8 | 19.541 | .0512 | .0243 | .4743 | 41.202 | 2.109 | 1.791 | 3.776 |
| 9 | 28.334 | .0353 | .0165 | .4665 | 60.743 | 2.144 | 1.893 | 4.058 |
| 10 | 41.085 | .0243 | .0112 | .4612 | 89.077 | 2.168 | 1.973 | 4.277 |
| 11 | 59.573 | .0168 | .0077 | .4577 | 130.162 | 2.185 | 2.034 | 4.445 |
| 12 | 86.381 | .0116 | .0053 | .4553 | 189.735 | 2.196 | 2.082 | 4.572 |
| 13 | 125.252 | .0080 | .0036 | .4536 | 276.115 | 2.204 | 2.118 | 4.668 |
| 14 | 181.615 | .0055 | .0025 | .4525 | 401.367 | 2.210 | 2.145 | 4.740 |
| 15 | 263.342 | .0038 | .0017 | .4517 | 582.982 | 2.214 | 2.165 | 4.793 |
| 16 | 381.846 | .0026 | .0012 | .4512 | 846.324 | 2.216 | 2.180 | 4.832 |
| 17 | 553.676 | .0018 | .0008 | .4508 | 1228.170 | 2.218 | 2.191 | 4.861 |
| 18 | 802.831 | .0012 | .0006 | .4506 | 1781.846 | 2.219 | 2.200 | 4.882 |
| 19 | 1164.105 | .0009 | .0004 | .4504 | 2584.677 | 2.220 | 2.206 | 4.898 |
| 20 | 1687.952 | .0006 | .0003 | .4503 | 3748.782 | 2.221 | 2.210 | 4.909 |
| 21 | 2447.530 | .0004 | .0002 | .4502 | 5436.734 | 2.221 | 2.214 | 4.917 |
| 22 | 3548.919 | .0003 | .0001 | .4501 | 7884.264 | 2.222 | 2.216 | 4.923 |
| 23 | 5145.932 | .0002 | .0001 | .4501 | 11433.182 | 2.222 | 2.218 | 4.927 |
| 24 | 7461.602 | .0001 | .0001 | .4501 | 16579.115 | 2.222 | 2.219 | 4.930 |
| 25 | 10819.322 | .0001 | | .4500 | 24040.716 | 2.222 | 2.220 | 4.933 |
| 26 | 15688.017 | .0001 | | .4500 | 34860.038 | 2.222 | 2.221 | 4.934 |
| 27 | 22747.625 | | | .4500 | 50548.056 | 2.222 | 2.221 | 4.935 |
| 28 | 32984.056 | | | .4500 | 73295.681 | 2.222 | 2.221 | 4.936 |
| 29 | 47826.882 | | | .4500 | 106279.737 | 2.222 | 2.222 | 4.937 |
| 30 | 69348.978 | | | .4500 | 154106.618 | 2.222 | 2.222 | 4.937 |
| 31 | 100556.019 | | | .4500 | 223455.597 | 2.222 | 2.222 | 4.938 |
| 32 | 145806.227 | | | .4500 | 324011.615 | 2.222 | 2.222 | 4.938 |
| 33 | 211419.029 | | | .4500 | 469817.842 | 2.222 | 2.222 | 4.938 |
| 34 | 306557.592 | | | .4500 | 681236.871 | 2.222 | 2.222 | 4.938 |
| 35 | 444508.508 | | | .4500 | 987794.463 | 2.222 | 2.222 | 4.938 |

| | SINGLE PAYMENT | | UNIFORM PAYMENT SERIES | | | | GRADIENT SERIES | |
|---|---|---|---|---|---|---|---|---|
| | Compound Amount Factor | Present Worth Factor | Sinking Fund Factor | Capital Recovery Factor | Compound Amount Factor | Present Worth Factor | Gradient Uniform Series | Gradient Present Worth |
| *n* | Find *F* Given *P* F/P | Find *P* Given *F* P/F | Find *A* Given *F* A/F | Find *A* Given *P* A/P | Find *F* Given *A* F/A | Find *P* Given *A* P/A | Find *A* Given *G* A/G | Find *P* Given *G* P/G |
| 1 | 1.500 | .6667 | 1.0000 | 1.5000 | 1.000 | .667 | 0 | 0 |
| 2 | 2.250 | .4444 | .4000 | .9000 | 2.500 | 1.111 | .400 | .444 |
| 3 | 3.375 | .2963 | .2105 | .7105 | 4.750 | 1.407 | .737 | 1.037 |
| 4 | 5.063 | .1975 | .1231 | .6231 | 8.125 | 1.605 | 1.015 | 1.630 |
| 5 | 7.594 | .1317 | .0758 | .5758 | 13.188 | 1.737 | 1.242 | 2.156 |
| 6 | 11.391 | .0878 | .0481 | .5481 | 20.781 | 1.824 | 1.423 | 2.595 |
| 7 | 17.086 | .0585 | .0311 | .5311 | 32.172 | 1.883 | 1.565 | 2.947 |
| 8 | 25.629 | .0390 | .0203 | .5203 | 49.258 | 1.922 | 1.675 | 3.220 |
| 9 | 38.443 | .0260 | .0134 | .5134 | 74.887 | 1.948 | 1.760 | 3.428 |
| 10 | 57.665 | .0173 | .0088 | .5088 | 113.330 | 1.965 | 1.824 | 3.584 |
| 11 | 86.498 | .0116 | .0058 | .5058 | 170.995 | 1.977 | 1.871 | 3.699 |
| 12 | 129.746 | .0077 | .0039 | .5039 | 257.493 | 1.985 | 1.907 | 3.784 |
| 13 | 194.620 | .0051 | .0026 | .5026 | 387.239 | 1.990 | 1.933 | 3.846 |
| 14 | 291.929 | .0034 | .0017 | .5017 | 581.859 | 1.993 | 1.952 | 3.890 |
| 15 | 437.894 | .0023 | .0011 | .5011 | 873.788 | 1.995 | 1.966 | 3.922 |
| 16 | 656.841 | .0015 | .0008 | .5008 | 1311.682 | 1.997 | 1.976 | 3.945 |
| 17 | 985.261 | .0010 | .0005 | .5005 | 1968.523 | 1.998 | 1.983 | 3.961 |
| 18 | 1477.892 | .0007 | .0003 | .5003 | 2953.784 | 1.999 | 1.988 | 3.973 |
| 19 | 2216.838 | .0005 | .0002 | .5002 | 4431.676 | 1.999 | 1.991 | 3.981 |
| 20 | 3325.257 | .0003 | .0002 | .5002 | 6648.513 | 1.999 | 1.994 | 3.987 |
| 21 | 4987.885 | .0002 | .0001 | .5001 | 9973.770 | 2.000 | 1.996 | 3.991 |
| 22 | 7481.828 | .0001 | .0001 | .5001 | 14961.655 | 2.000 | 1.997 | 3.994 |
| 23 | 11222.741 | .0001 | | .5000 | 22443.483 | 2.000 | 1.998 | 3.996 |
| 24 | 16834.112 | .0001 | | .5000 | 33666.224 | 2.000 | 1.999 | 3.997 |
| 25 | 25251.168 | | | .5000 | 50500.337 | 2.000 | 1.999 | 3.998 |
| 26 | 37876.752 | | | .5000 | 75751.505 | 2.000 | 1.999 | 3.999 |
| 27 | 56815.129 | | | .5000 | 113628.257 | 2.000 | 2.000 | 3.999 |
| 28 | 85222.693 | | | .5000 | 170443.386 | 2.000 | 2.000 | 3.999 |
| 29 | 127834.039 | | | .5000 | 255666.079 | 2.000 | 2.000 | 4.000 |
| 30 | 191751.059 | | | .5000 | 383500.118 | 2.000 | 2.000 | 4.000 |
| 31 | 287626.589 | | | .5000 | 575251.178 | 2.000 | 2.000 | 4.000 |
| 32 | 431439.883 | | | .5000 | 862877.767 | 2.000 | 2.000 | 4.000 |
| 33 | 647159.825 | | | .5000 | 1294317.650 | 2.000 | 2.000 | 4.000 |
| 34 | 970739.737 | | | .5000 | 1941477.475 | 2.000 | 2.000 | 4.000 |
| 35 | 1456109.606 | | | .5000 | 2912217.212 | 2.000 | 2.000 | 4.000 |

| | SINGLE PAYMENT | | UNIFORM PAYMENT SERIES | | | | GRADIENT SERIES | |
|---|---|---|---|---|---|---|---|---|
| | Compound Amount Factor | Present Worth Factor | Sinking Fund Factor | Capital Recovery Factor | Compound Amount Factor | Present Worth Factor | Gradient Uniform Series | Gradient Present Worth |
| n | Find F Given P F/P | Find P Given F P/F | Find A Given F A/F | Find A Given P A/P | Find F Given A F/A | Find P Given A P/A | Find A Given G A/G | Find P Given G P/G |
| 1 | 1.600 | .6250 | 1.0000 | 1.6000 | 1.000 | .625 | 0 | 0 |
| 2 | 2.560 | .3906 | .3846 | .9846 | 2.600 | 1.016 | .385 | .391 |
| 3 | 4.096 | .2441 | .1938 | .7938 | 5.160 | 1.260 | .698 | .879 |
| 4 | 6.554 | .1526 | .1080 | .7080 | 9.256 | 1.412 | .946 | 1.337 |
| 5 | 10.486· | .0954 | .0633 | .6633 | 15.810 | 1.508 | 1.140 | 1.718 |
| 6 | 16.777 | .0596 | .0380 | .6380 | 26.295 | 1.567 | 1.286 | 2.016 |
| 7 | 26.844 | .0373 | .0232 | .6232 | 43.073 | 1.605 | 1.396 | 2.240 |
| 8 | 42.950 | .0233 | .0143 | .6143 | 69.916 | 1.628 | 1.476 | 2.403 |
| 9 | 68.719 | .0146 | .0089 | .6089 | 112.866 | 1.642 | 1.534 | 2.519 |
| 10 | 109.951 | .0091 | .0055 | .6055 | 181.585 | 1.652 | 1.575 | 2.601 |
| 11 | 175.922 | .0057 | .0034 | .6034 | 291.536 | 1.657 | 1.604 | 2.658 |
| 12 | 281.475 | .0036 | .0021 | .6021 | 467.458 | 1.661 | 1.624 | 2.697 |
| 13 | 450.360 | .0022 | .0013 | .6013 | 748.933 | 1.663 | 1.638 | 2.724 |
| 14 | 720.576 | .0014 | .0008 | .6008 | 1199.293 | 1.664 | 1.647 | 2.742 |
| 15 | 1152.922 | .0009 | .0005 | .6005 | 1919.869 | 1.665 | 1.654 | 2.754 |
| 16 | 1844.674 | .0005 | .0003 | .6003 | 3072.791 | 1.666 | 1.658 | 2.762 |
| 17 | 2951.479 | .0003 | .0002 | .6002 | 4917.465 | 1.666 | 1.661 | 2.767 |
| 18 | 4722.366 | .0002 | .0001 | .6001 | 7868.944 | 1.666 | 1.663 | 2.771 |
| 19 | 7555.786 | .0001 | .0001 | .6001 | 12591.311 | 1.666 | 1.664 | 2.773 |
| 20 | 12089.258 | .0001 | | .6000 | 20147.097 | 1.667 | 1.665 | 2.775 |
| 21 | 19342.813 | .0001 | | .6000 | 32236.355 | 1.667 | 1.666 | 2.776 |
| 22 | 30948.501 | | | .6000 | 51579.168 | 1.667 | 1.666 | 2.777 |
| 23 | 49517.602 | | | .6000 | 82527.669 | 1.667 | 1.666 | 2.777 |
| 24 | 79228.163 | | | .6000 | 132045.271 | 1.667 | 1.666 | 2.777 |
| 25 | 126765.060 | | | .6000 | 211273.433 | 1.667 | 1.666 | 2.777 |
| 26 | 202824.096 | | | .6000 | 338038.493 | 1.667 | 1.667 | 2.778 |
| 27 | 324518.554 | | | .6000 | 540862.589 | 1.667 | 1.667 | 2.778 |
| 28 | 519229.686 | | | .6000 | 865381.143 | 1.667 | 1.667 | 2.778 |
| 29 | 830767.497 | | | .6000 | 1384610.829 | 1.667 | 1.667 | 2.778 |
| 30 | 1329227.996 | | | .6000 | 2215378.326 | 1.667 | 1.667 | 2.778 |

# Continuous Compounding—Single Payment Factors

| rn | Compound Amount Factor $e^{rn}$<br>Find F Given P F/P | Present Worth Factor $e^{-rn}$<br>Find P Given F P/F | rn | Compound Amount Factor $e^{rn}$<br>Find F Given P F/P | Present Worth Factor $e^{-rn}$<br>Find P Given F P/F |
|---|---|---|---|---|---|
| .01 | 1.0101 | .9900 | .51 | 1.6653 | .6005 |
| .02 | 1.0202 | .9802 | .52 | 1.6820 | .5945 |
| .03 | 1.0305 | .9704 | .53 | 1.6989 | .5886 |
| .04 | 1.0408 | .9608 | .54 | 1.7160 | .5827 |
| .05 | 1.0513 | .9512 | .55 | 1.7333 | .5769 |
| .06 | 1.0618 | .9418 | .56 | 1.7507 | .5712 |
| .07 | 1.0725 | .9324 | .57 | 1.7683 | .5655 |
| .08 | 1.0833 | .9231 | .58 | 1.7860 | .5599 |
| .09 | 1.0942 | .9139 | .59 | 1.8040 | .5543 |
| .10 | 1.1052 | .9048 | .60 | 1.8221 | .5488 |
| .11 | 1.1163 | .8958 | .61 | 1.8404 | .5434 |
| .12 | 1.1275 | .8869 | .62 | 1.8589 | .5379 |
| .13 | 1.1388 | .8781 | .63 | 1.8776 | .5326 |
| .14 | 1.1503 | .8694 | .64 | 1.8965 | .5273 |
| .15 | 1.1618 | .8607 | .65 | 1.9155 | .5220 |
| .16 | 1.1735 | .8521 | .66 | 1.9348 | .5169 |
| .17 | 1.1853 | .8437 | .67 | 1.9542 | .5117 |
| .18 | 1.1972 | .8353 | .68 | 1.9739 | .5066 |
| .19 | 1.2092 | .8270 | .69 | 1.9937 | .5016 |
| .20 | 1.2214 | .8187 | .70 | 2.0138 | .4966 |
| .21 | 1.2337 | .8106 | .71 | 2.0340 | .4916 |
| .22 | 1.2461 | .8025 | .72 | 2.0544 | .4868 |
| .23 | 1.2586 | .7945 | .73 | 2.0751 | .4819 |
| .24 | 1.2712 | .7866 | .74 | 2.0959 | .4771 |
| .25 | 1.2840 | .7788 | .75 | 2.1170 | .4724 |
| .26 | 1.2969 | .7711 | .76 | 2.1383 | .4677 |
| .27 | 1.3100 | .7634 | .77 | 2.1598 | .4630 |
| .28 | 1.3231 | .7558 | .78 | 2.1815 | .4584 |
| .29 | 1.3364 | .7483 | .79 | 2.2034 | .4538 |
| .30 | 1.3499 | .7408 | .80 | 2.2255 | .4493 |
| .31 | 1.3634 | .7334 | .81 | 2.2479 | .4449 |
| .32 | 1.3771 | .7261 | .82 | 2.2705 | .4404 |
| .33 | 1.3910 | .7189 | .83 | 2.2933 | .4360 |
| .34 | 1.4049 | .7118 | .84 | 2.3164 | .4317 |
| .35 | 1.4191 | .7047 | .85 | 2.3396 | .4274 |
| .36 | 1.4333 | .6977 | .86 | 2.3632 | .4232 |
| .37 | 1.4477 | .6907 | .87 | 2.3869 | .4190 |
| .38 | 1.4623 | .6839 | .88 | 2.4109 | .4148 |
| .39 | 1.4770 | .6771 | .89 | 2.4351 | .4107 |
| .40 | 1.4918 | .6703 | .90 | 2.4596 | .4066 |
| .41 | 1.5068 | .6637 | .91 | 2.4843 | .4025 |
| .42 | 1.5220 | .6570 | .92 | 2.5093 | .3985 |
| .43 | 1.5373 | .6505 | .93 | 2.5345 | .3946 |
| .44 | 1.5527 | .6440 | .94 | 2.5600 | .3906 |
| .45 | 1.5683 | .6376 | .95 | 2.5857 | .3867 |
| .46 | 1.5841 | .6313 | .96 | 2.6117 | .3829 |
| .47 | 1.6000 | .6250 | .97 | 2.6379 | .3791 |
| .48 | 1.6161 | .6188 | .98 | 2.6645 | .3753 |
| .49 | 1.6323 | .6126 | .99 | 2.6912 | .3716 |
| .50 | 1.6487 | .6065 | 1.00 | 2.7183 | .3679 |

# Appendix C
## Selected
## References

AASHO. *Road User Benefit Analyses for Highway Improvements.* Washington, D.C.: American Association of State Highway Officials, 1960.

Alfred, A. M., and Evans, J. B. *Appraisal of Investment Projects by Discounted Cash Flow: Principles and Some Short Cut Techniques,* 3d ed. London: Chapman and Hall Ltd., 1971.

American Telephone and Telegraph Co. *Engineering Economy,* 3d ed. New York: McGraw-Hill, 1977.

Barish, N. N., and Kaplan, S. *Economic Analysis for Engineering and Managerial Decision Making,* 2d ed. New York: McGraw-Hill, 1978.

Baumol, W. J. *Economic Theory and Operations Analysis,* 2d ed. Englewood Cliffs, N.J.: Prentice-Hall, Inc., 1965.

Bernhard, R. H. "A Comprehensive Comparison and Critique of Discounting Indices Proposed for Capital Investment Evaluation," *The Engineering Economist.* Vol. 16, No. 3 (Spring, 1971), 157–186.

Bierman, H., and Smidt, S. *The Capital Budgeting Decision,* 3d ed. New York: The Macmillan Co., 1971.

Bower, J. L. *Managing the Resource Allocation Process: A Study of Corporate Planning and Investment.* Boston: Harvard University, 1970.

Canada, J. R. *Intermediate Economic Analysis for Management and Engineering.* Englewood Cliffs, N.J.: Prentice-Hall, Inc. 1971.

DeGarmo, E. P., and Canada, J. R. *Engineering Economy,* 5th ed. New York: The Macmillan Co., 1973.

de Neufville, R., and Stafford, J. H. *Systems Analysis for Engineers and Managers.* New York: McGraw Hill, 1971.

Eilon, S. "What is a Decision?," *Management Science,* Vol. 16, No. 4 (December, 1969), B172–B189.

Emerson, C. R., and Taylor, W. R.   *An Introduction to Engineering Economy*. Bozeman, Montana: Cardinal Publishers, 1973.

*Engineering Economist, The.*   A quarterly journal of the Engineering Economy Divisions of ASEE and AIIE. For subscription information write to American Institute of Industrial Engineers, 25 Technology Park/Atlanta, Norcross, GA 30092.

English, J. M., ed.   *Cost Effectiveness: The Economic Evaluation of Engineered Systems*. New York: John Wiley & Sons, 1968.

Fabrycky, W. J., and Thuesen, G. J.   *Economic Decision Analysis*. Englewood Cliffs, N.J.: Prentice-Hall, Inc., 1974.

Fleischer, G. A.   *Capital Allocation Theory: The Study of Investment Decisions*. New York: Appleton-Century-Crofts, 1969.

Grant, E. L., Ireson, W. G., and Leavenworth, R. S.   *Principles of Engineering Economy*, 6th ed. New York: The Ronald Press Co., 1976.

Hanssmann, F.   *Operations Research Techniques for Capital Investment*. New York: John Wiley & Sons, 1968.

Hillier, F. S.   *The Evaluation of Risky Interrelated Investments*. Amsterdam: North-Holland Publishing Co., 1970.

ICE.   *An Introduction to Engineering Economics*. London: The Institution of Civil Engineers, 1969.

James, L. D., and Lee, R. R.   *Economics of Water Resources Planning*. New York: McGraw-Hill, 1971.

Jean, W. H.   "On Multiple Rates of Return," *The Journal of Finance*. Vol. 23 (March, 1968), 187–191.

Jelen, F. C.   *Cost and Optimization Engineering*. New York: McGraw-Hill, 1970.

Jeynes, P. H.   *Profitability and Economic Choice*. Ames, Iowa: The Iowa State University Press, 1968.

Kaplan, S.   "A Note on a Method for Precisely Determining the Uniqueness or Nonuniqueness of the Internal Rate of Return for a Proposed Investment," *The Journal of Industrial Engineering*. January-February, 1965, 70–71.

Lorie, J. H., and Savage, L. J.   "Three Problems in Rationing Capital," *The Journal of Business*. Vol. 28, No. 4, 229–239.

Mao, J. C. T.   *Quantative Analysis of Financial Decisions*. New York: The Macmillan Co., 1969.

Marglin, S. A.   *Approaches to Dynamic Investment Planning*. Amsterdam: North-Holland Publishing Co., 1963.

Masse, P.   *Optimal Investment Decisions: Rules for Action and Criteria for Choice*. Englewood Cliffs, N.J.: Prentice-Hall, Inc., 1962.

McKean, R. N.   *Efficiency in Government Through Systems Analysis*. New York: John Wiley & Sons, 1958.

Merrett, A. J., and Sykes, A.   *The Finance and Analysis of Capital Projects*, 2d ed. London: Longman, 1973.

Morris, W. T. *Engineering Economic Analysis.* Reston, Va.: Reston Publishing, 1976.

Newnan, D. G. *An Economic Analysis of Railway Grade Crossings on the California State Highway System, Report EEP-16.* Stanford, California: Stanford University, 1965.

——. "Determining Rate of Return by Means of Payback Period and Useful Life," *The Engineering Economist.* Vol. 15, No. 1 (Fall, 1969), 29–39.

Norstrøm, C. J. "A Sufficient Condition for a Unique Nonnegative Internal Rate of Return," *Journal of Financial and Quantitative Analysis.* June, 1972, 1835–1839.

Oakford, R. V. *Capital Budgeting.* New York: The Ronald Press Co., 1970.

Oglesby, C. H. *Highway Engineering,* 3d ed. New York: John Wiley & Sons, 1975.

Ostwald, P. F. *Cost Estimating for Engineering and Management.* Englewood Cliffs, N.J.: Prentice-Hall, Inc., 1974.

Park, W. R. *Cost Engineering Analysis.* New York: John Wiley & Sons, 1973.

Pearce, D. W. *Cost-Benefit Analysis.* London: The Macmillan Press, Ltd., 1971.

Reisman, A. *Managerial and Engineering Economics.* Boston: Allyn and Bacon, 1971.

Riggs, J. L. *Engineering Economics.* New York: McGraw-Hill, 1977.

Schweyer, H. E. *Analytic Models for Managerial and Engineering Economics.* New York: Reinhold Publishing Corp., 1964.

Smith, G. W. *Engineering Economy: Analysis of Capital Expenditures,* 2d ed. Ames, Iowa: The Iowa State University Press, 1973.

Soper, C. S. "The Marginal Efficiency of Capital: A Further Note," *The Economic Journal.* Vol. 69, 174–177.

Swalm, R. "Utility Theory—Insights into Risk Taking," *Harvard Business Review.* November-December, 1966, 123–136.

Tarquin, A. J., and Blank, L. T. *Engineering Economy: A Behavioral Approach.* New York: McGraw-Hill, 1976.

Taylor, G. A. *Managerial and Engineering Economy,* 2d ed. Princeton, N.J.: D. van Nostrand Co., 1975.

Terborgh, G. *Business Investment Management.* Washington, D.C.: Machinery and Allied Products Institute, 1967.

Thuesen, H. G., Fabrycky, W. J., and Thuesen, G. J. *Engineering Economy,* 5th ed. Englewood Cliffs, N.J.: Prentice-Hall, Inc., 1977.

Weingartner, H. M. *Mathematical Programming and the Analysis of Capital Budgeting Problems.* Englewood Cliffs, N.J.: Prentice-Hall, Inc., 1963.

Wellington, A. M. *The Economic Theory of Railway Location.* New York: John Wiley & Sons, 1887.

Winfrey, R. *Economic Analysis for Highways.* New York: Intext Educational Publishers, 1969.

# Index

# DEPRECIATION

**STRAIGHT LINE (SL)**

$$\text{SL Depreciation In Any Year} = \frac{1}{N}(P - F) = \frac{\text{Book Value* at beginning of year} - F}{\text{Remaining Useful Life at beginning of year}}$$

**SUM-OF-YEARS DIGITS (SOYD)**

$$\text{SOYD Depreciation In Any Year} = \frac{\text{Remaining Useful Life at beginning of year}}{\frac{N}{2}(N + 1)}(P - F)$$

**DOUBLE DECLINING BALANCE (DDB)**

$$\text{DDB Depreciation In Any Year} = \frac{2}{N}(\text{Book Value* at beginning of year})$$

**SINKING FUND (SF)**

$$\text{SF Depreciation In Any Year} = (P - F)(A/F,i\%,N) + \left\{ \begin{array}{l} i\% \text{ interest on amount in} \\ \text{sinking fund at beginning} \\ \text{of year} \end{array} \right.$$

*Book Value = Cost minus depreciation charges to date.